LACAME 2008

LACAME 2008

Proceedings of the 11th Latin American Conference on the Applications of the Mössbauer Effect (LACAME 2008) held in La Plata, Argentina, 9–14 November 2008

Edited by

J. DESIMONI

Universidad Nacional de La Plata, Argentina

C. P. RAMOS

Centro Atómico Constituyentes, Argentina

B. ARCONDO

Universidad de Buenos Aires, Argentina

F. D. SACCONE

Universidad de Buenos Aires, Argentina

and

R. C. MERCADER

Universidad Nacional de La Plata, Argentina

Previously published in *Hyperfine Interactions*
Volume 195, Nos. 1–3 (2010)

Springer

Editors
J. Desimoni
Departamento de Física
Facultad de Ciencias Exactas
Universidad Nacional de La Plata
IFLP-CONICET
La Plata, Argentina

F. D. Saccone
Laboratorio de Sólidos Amorfos
Facultad de Ingeniería
Universidad de Buenos Aires
INTECIN-UBA-CONICET
Buenos Aires, Argentina

C. P. Ramos
Departamento de Física
Centro Atómico Constituyentes
Comisión Nacional de Energía Atómica
Buenos Aires, Argentina

R. C. Mercader
Departamento de Física
Facultad de Ciencias Exactas
Universidad Nacional de La Plata
IFLP-CONICET
La Plata, Argentina

B. Arcondo
Laboratorio de Sólidos Amorfos
Facultad de Ingeniería
Universidad de Buenos Aires
INTECIN-UBA-CONICET
Buenos Aires, Argentina

ISBN 978-3-642-10763-4
Springer Heidelberg Dordrecht London New York

Library of Congress Control Number: 2009942151

Printed on acid-free paper

Springer is part of Springer Science+Business Media (www.springer.com)

Table of Contents

Hyperfine Interact (2010) 195:1–2
DOI 10.1007/s10751-009-0092-8

Eleventh Latin American conference on the applications of the Mössbauer effect, LACAME 2008, La Plata, Argentina, 9–14 November 2008

Published online: 24 October 2009

Chairpersons

J. Desimoni *(La Plata)*
A. F. Cabrera *(La Plata)*
S. J. Stewart *(La Plata)*

Local Organizing Committee

R. C. Mercader *(La Plata)*
F. H. Sánchez *(La Plata)*
C. E. Rodríguez Torres *(La Plata)*
C. Saragovi *(Buenos Aires)*
C. P. Ramos *(Buenos Aires)*
B. Arcondo *(Buenos Aires)*
G. Punte *(La Plata)*
F. D. Saccone *(Buenos Aires)*

Latin American Committee

E. Baggio-Saitovich *(Brazil)*
N. R. Furet-Bridón *(Cuba)*
F. González-Jiménez *(Venezuela)*
J. A. Jaén*(Panama)*
R. C. Mercader *(Argentina)*
N. Nava *(Mexico)*
V. A. Peña *(Peru)*
G. A. Pérez *(Colombia)*

Scientific Committee

C. A. Barrero-Meneses *(Colombia)*
A. Bustamante *(Peru)*
N. R. Furet- Bridón *(Cuba)*

J. A. Jaén *(Panama)*
E. Passamani *(Brazil)*
T. Pérez *(Argentina)*
C. Pizarro *(Chile)*
F. H. Sánchez *(Argentina)*
C. Saragovi *(Argentina)*

SPONSORS

Universidad Nacional de La Plata (UNLP)
Consejo Nacional de Investigaciones Científicas y Técnicas (CONICET)—Argentina
Departamento de Física, Facultad de Ciencias Exactas UNLP
Instituto de Física La Plata-CONICET-UNLP
Facultad de Ciencias Exactas UNLP
Comisión Nacional de Energía Atómica (CNEA)—Argentina
Fundación para el Desarrollo Tecnológico—Argentina
Centro Latinoamericano de Física—Rio de Janeiro
Springer Science + Business Media B.V.—The Netherlands
Universidad de Buenos Aires—Argentina
Asociación Cooperadora de Física del Departamento de Física de la CNEA
Ministerio de Educación de la Nación—Argentina
Cámara de Diputados de la Provincia de Buenos Aires—Argentina
Municipalidad de La Plata—Argentina
Agencia Nacional de Promoción Científica y Tecnológica—Argentina
Comisión de Investigaciones Científicas de la Provincia de Buenos Aires—Argentina
Fundación Universidad de Salamanca—Spain

Hyperfine Interact (2010) 195:3–4
DOI 10.1007/s10751-009-0093-7

Preface

**J. Desimoni · C. P. Ramos · B. Arcondo ·
F. D. Saccone · R. C. Mercader**

Published online: 30 September 2009

The Mössbauer effect is the resonant emission and absorption of gamma rays. Through the hyperfine interactions of the probe nucleus with its solid state environment, it gives results highly sensitive to the local chemical and physical properties of the investigated systems. This information, at times irreplaceable, is of great importance to condensed matter and materials science. Because ^{57}Fe is the best Mössbauer isotope and forms part of $\approx 2\%$ of natural iron, which is the fourth most abundant element in the Earth's crust, the effect has been applied to a wide range of scientific subjects, like solid state physics, catalysis, magnetism, chemistry, mineralogy, geology, corrosion and environmental studies, new biological materials and medical drugs. Its relatively simple and not too expensive experimental setup has been of great importance to the scientific development in Latin America, whose Mössbauer community exhibits a growing number of laboratories across the region and shows the highest relative increase rate in publications and development in the world.

Although soon after the discovery of the Mössbauer effect several laboratories across Latin America were set up, Mössbauer spectroscopy (MS) did not catch up scientific momentum until the series of Latin American Conferences on the Applications of the Mössbauer Effect, LACAMEs, which have been held without interruption each two years since 1988. The venues have been Rio de Janeiro, Havana, Buenos Aires, Santiago de Chile, Cusco, Cartagena de Indias, Caracas, Panama City, and Mexico D.F. A new round has started with the LACAME 2006 in Rio de Janeiro. Consequently, MS in Latin America has consolidated into a well-known community that in particular, in Argentina alone, has produced in the last ten years more than 45 Ph.D. theses based mainly on results obtained through MS.

J. Desimoni · R. C. Mercader (✉)
La Plata, Argentina
e-mail: mercader@fisica.unlp.edu.ar

C. P. Ramos · B. Arcondo · F. D. Saccone
Buenos Aires, Argentina

This cycle of biennial conferences aims at encouraging the knowledge and scientific exchange between the research groups and applied science relevant to the needs of the Latin American countries. At the same time, it allows that young researchers, students and the societies in general, can meet with foreign visitors that come to the conferences and bring along their expertise and last developments in the different topics covered by the scientific programs. This kind of scientific achievements would not be possible to the broad majority of the young researchers because of the vast geography of the region, the high costs of transportation and the scant budgets that are common to almost all the laboratories in Latin America.

LACAME 2008 has been no exception and has proved a valuable means to improve and increase the cooperation between laboratories across the region. A total of 87 participants from 14 countries attended the conference, which was developed in a warmhearted and pleasant atmosphere that cooperated toward the mentioned goals and encouraged the participants to keep on strengthening their links and developing the series of Latin American meetings.

J. Desimoni and R. C. Mercader
La Plata

C. P. Ramos, B. Arcondo and F. D. Saccone
Buenos Aires

Hyperfine Interact (2010) 195:5–13
DOI 10.1007/s10751-009-0094-6

Fe/MCM-41 sylilated catalyst: structural changes determination during the Fischer–Tropsch reaction

J. F. Bengoa · N. A. Fellenz · M. V. Cagnoli · L. A. Cano · N. G. Gallegos · A. M. Alvarez · S. G. Marchetti

Published online: 16 October 2009

Abstract Two Fe/MCM-41 systems, one of them sylilated, were obtained to be used as catalysts in Fischer–Tropsch reaction. They have more than 90% of the iron species located inside the support channels, leading to a narrow crystal size distribution accessible to reactive gases. The samples were characterized by X-ray diffraction, atomic absorption spectroscopy, N_2 adsorption, Mössbauer spectroscopy and Fourier transformer infrared spectroscopy. Mössbauer spectroscopy allowed us to demonstrate that the catalytic active species were the same in both catalysts. The only difference between them was the surface hydrophobicity, which decreases the "water gas shift reaction" in the sylilated catalyst. Besides, this solid is more active for hydrocarbon production, with a lower methane yield.

Keywords Fischer–Tropsch synthesis · Fe/MCM-41 · Sylilation · Mössbauer spectroscopy · Hydrophobicity

1 Introduction

Fischer–Tropsch synthesis (FTS) is a process used to produce a broad distribution of "clean" hydrocarbons from syngas ($CO + H_2$) using principally Fe catalysts [1]. Although this process has been used industrially for decades, the main problem to be resolved is to reach a good selectivity towards desirable products. Two main approaches have been applied to this purpose: the control of the operative variables of the process (pressure, contact time, reactor selection or temperature) or the structural properties modification of the catalysts. The last option is the more promising because the reaction is kinetically controlled.

J. F. Bengoa · N. A. Fellenz · M. V. Cagnoli · L. A. Cano · N. G. Gallegos · A. M. Alvarez · S. G. Marchetti (✉)
CINDECA. Fac. Cs. Exactas, Fac. de Ingeniería,
UNLP–CICPBA–CCT–CONICET, 47 No. 257, La Plata, 1900, Argentina
e-mail: march@quimica.unlp.edu.ar

It is accepted that FTS is a structure-sensitive reaction for iron crystal sizes lower than 10 nm [2]. A strategy to increase the selectivity is to obtain a catalyst with a narrow size distribution of active phase crystals whose average sizes are within a determined range. If a mesoporous solid, like as MCM-41 [3], is used as iron support, and if it would be possible to introduce the iron species mainly inside its channels, small sizes of crystals would be obtained. They would have a high selectivity towards the production of light hydrocarbons (HC), since the growth chain would be limited [2]. Nevertheless, these crystals have a smaller activity than the larger ones. It can be speculated that the low activity of small crystallites might be due to "intrinsic" effects as it can be assumed that chain growth steps in Fischer–Tropsch synthesis only occur on sites which contain a certain number of metal atoms in a certain configuration. The density of these sites might be much lower in small crystallites, therefore rendering them less active [2]. For this reason, the aim of the present work is to modify the surface of Fe/MCM-41 catalyst, avoiding water adsorption, which is a product of the reaction that leads to the poisoning of active sites. This surface modification would compensate the decreasing of the activity due to the crystal size. With this purpose a sylilation treatment was carried out to eliminate the surface hydroxyls, increasing the surface hydrophobicity.

2 Experimental

2.1 Catalysts synthesis

The solid chosen as support was MCM-41. This solid is a system consisting of a two-dimensional hexagonal array of uni-dimensional channels whose walls are constituted by amorphous silica (SiO_2). It was prepared according to the methodology proposed by Ryoo and Kim [3]. Two catalysts of approximately 8% *w/w* of Fe, were obtained by incipient wetness impregnation method with an $Fe(NO_3)_3{\cdot}9H_2O$ aqueous solution using MCM-41 as support. The samples were dried and calcined in a dry N_2 stream. One of them was called p-Fe/MCM-41 and the other one was sylilated. For the latter treatment, the solid was outgassed and then was mixed with a solution of 1% (*v/v*) of hexametil disilazane in toluene. This process was carried out in Ar atmosphere in a glove box. After drying, this solid was called psil-Fe/MCM-41.

2.2 Catalysts characterization

The samples were characterized by X-ray diffraction (XRD), atomic absorption spectroscopy (AAS), N_2 adsorption (BET), Mössbauer spectroscopy (MS) at 298 and 25 K and Fourier transformer infrared spectroscopy (FT-IR).

Nitrogen adsorption isotherms were recorded at the temperature of liquid nitrogen (77 K) using a Micromeritics ASAP 2020 apparatus. Samples were degassed at 373 K for 12 h prior to the determination of the adsorption isotherm. Specific areas were calculated by applying the Brunaer–Emmett–Teller (BET) method to portions of the isotherms within the $0.05 < P/P_0 < 0.30$ relative pressure range. The total pore volume was calculated from the amount of vapour adsorbed at a relative pressure close to unity assuming that the pores are filled with the condensate in the liquid state.

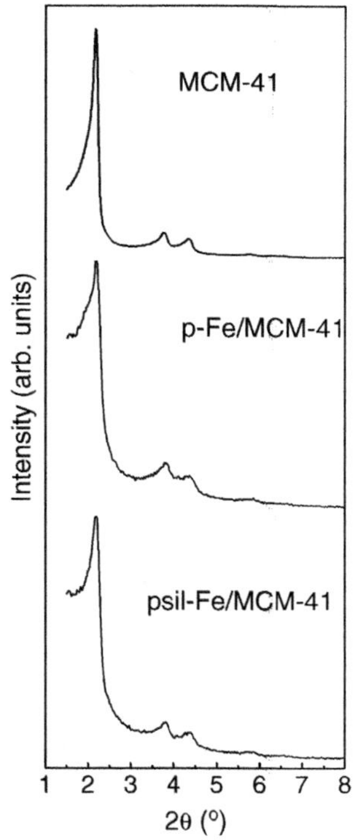

Fig. 1 XRD patterns of the support and both precursors

The Mössbauer spectra were obtained in transmission geometry with a 512-channel constant acceleration spectrometer. Velocity calibration was performed against a 12 μm-thick α-Fe foil. All isomer shifts (δ) mentioned in this paper are referred to this standard. The temperature between 25 and 298 K were varied using a Displex DE-202 Closed Cycle Cryogenic System. The Mössbauer spectra were evaluated using a fitting commercial program named Recoil [4]. The spectra were folded to minimize geometric effects. Lorentzian lines were considered with equal widths for each spectrum component. To fit the spectra with relaxing effects the dynamic shape analysis was used. The line shape used is the one derived by Blume and Tjon [5] which uses a stochastic model for the fluctuation of the magnetic hyperfine field. The model assumes that the hyperfine field fluctuates randomly between two directions along the same axis.

Activity and selectivity measurements in FTS were carried out in a stainless steel fixed bed reactor during 6 days run time, at 703 K, 1 bar of total pressure, with H_2/CO ratio of 2:1 and 20 cm^3/min of total flow. The catalysts were activated under continuous flow of the previous gas mixture, using a thermal program that raised the temperature from 298 to 703 K at 5 K/min.

3 Results and discussion

The structural properties of ordered mesoporous solid MCM-41, were verified by XRD (Fig. 1) and by textural measurements (Table 1).

Table 1 Textural properties and Fe contents

	Sg (m^2/g)	Dp (nm)	Vp (cm^3/g)	Fe (% *w*/*w*)
MCM-41	912	2.7	0.88	–
p-Fe/MCM-41	691	2.9	0.64	8.9
psil-Fe/MCM-41	494	2.4	0.59	8.6

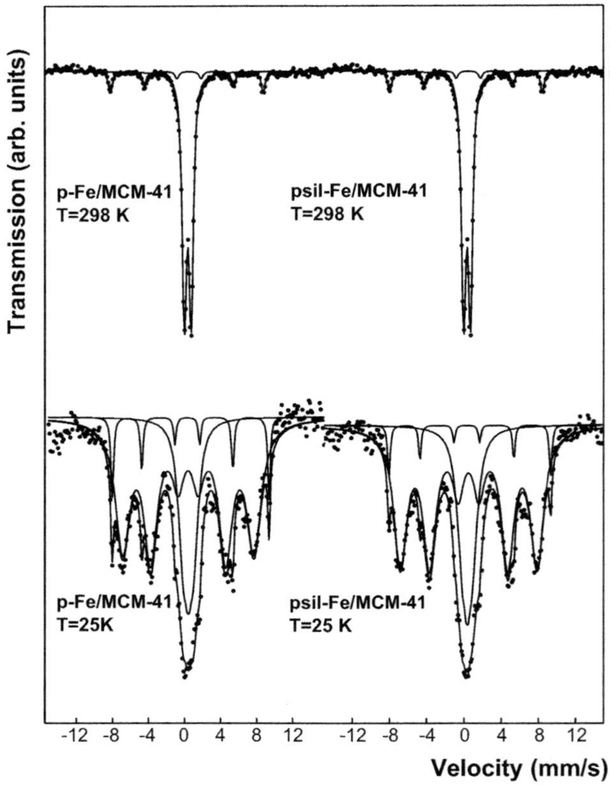

Fig. 2 Mössbauer spectra of the precursors at 298 and 25 K

The impregnation, calcination and sylilation treatments did not modify the support structure as it was verified trough XRD measurements (Fig. 1).

With respect to the textural properties, a diminution of the specific surface (Sg) and pore volume (Vp) after iron impregnation can be observed. This effect is attributed to a partial pore filling of the support with Fe oxide species (Table 1). After the sylilation treatment these results were more pronounced and they were attributed to the covering of the internal pore wall with trimethylsilyl groups, in agreement with the observations by [6, 7].

The Fe contents obtained by AAS were equal in both precursors within the experimental error (Table 1).

The Mössbauer spectra of p-Fe/MCM-41 and psil-Fe/MCM-41 (Fig. 2) are practically identical at both temperatures. Thus, at 298 K it can be seen a very intense central doublet and a small sextet. At 25 K both spectra displayed a broad central line and three peaks at both sides, two of them broad and the other narrow.

Table 2 Mössbauer parameters of the precursors at 25 K

Species	Parameters	p-Fe/MCM-41	psil-Fe/MCM-41
$\alpha-Fe_2O_3$ magnetically blocked	H (kOe)	540 ± 2	540 ± 2
	δ (mm/s)	0.48 ± 0.02	0.48 ± 0.03
	2ε (mm/s)	0.36 ± 0.05	0.31 ± 0.06
Relaxing α-Fe_2O_3 I	H (kOe)	462[a]	462[a]
	δ (mm/s)	0.40 ± 0.03	0.51 ± 0.03
	2ε (mm/s)	0.00[a]	0.00[a]
Relaxing α-Fe_2O_3 II	H (kOe)	415[a]	415[a]
	δ (mm/s)	0.42 ± 0.04	0.39 ± 0.03
	2ε (mm/s)	0.00[a]	0.00[a]

H hyperfine magnetic field, δ isomer shift (all the isomer shifts are referred to α-Fe at 298 K), 2ε quadrupole shift, Δ quadrupole splitting
[a] Parameters held fixed in fitting

The spectra at 25 K were fitted with one non relaxing sextet and two relaxing ones (Table 2). The first sextet has parameters of α-Fe_2O_3 [8] which has undergone Morin transition, with diameters larger than $\approx$20 nm [8] (called α-Fe_2O_3 magnetically blocked). The second sextet has not completed the magnetic blocked state, yet (called relaxing α-Fe_2O_3 I). Finally, the third sextet, is in relaxation process with a fluctuation speed much higher than the previous one (called relaxing α-Fe_2O_3 II). The original doublet at room temperature has disappeared so we can speculate that a further decrease in the measurement temperature, to 4 K, should block magnetically all samples. Therefore, there should be two fractions of α-Fe_2O_3 with very different sizes: a small percentage, between 5% and 10% of crystals larger than 20 nm and the great majority as extremely small crystals that even at 25 K does not get magnetically blocked, located inside MCM-41channels. The crystal size value of the large fraction is in agreement with that obtained by XRD. The XRD line profile analysis, using a Voight function, to fit the integral width of the peaks was conducted. The size of the crystals—after subtracting the instrumental broadening and the maximum strain—was calculated [9]. The values (very similar to those estimated by Mössbauer) were 30 nm for p-Fe/MCM-41 and 20 nm for psil-Fe/MCM-41. Therefore, this fraction close to 10% of the total α-Fe_2O_3 must be located outside the channels of the support in both precursors.

The OH substitution from silanol groups by trimethylsilyl groups can be monitored by FT-IR [7, 10]. To this end IR spectra of both precursors were obtained (Fig. 3). In psil-Fe/MCM-41 spectrum, the appearance of two new bands can be distinguished: in 758 cm^{-1} assignable to the stretching of Si–C bond and in 849 cm^{-1} assignable to rocking of CH_3 groups [7]. According to Lin et al. [10] these results indicate that trimethylsilyl groups of the sylilant agent are anchored on the mesoporous solid surface and our sylilation process was successful.

To characterize the activated samples, called c-Fe/MCM-41 and csil-Fe/MCM-41, Mössbauer spectra at 298 and 25 K in a controlled atmosphere (H_2/CO) were obtained (Fig. 4). In the Mössbauer spectra of both catalysts at 298 K, an asymmetrical doublet with broad lines can be observed. It would correspond to paramagnetic and/or superparamagnetic (sp) species. In order to realize the assignment of the species, the catalysts spectra at 25 K were obtained, resulting in a curved background, characteristic of the sp relaxation. It is important to remark that, due to the extreme

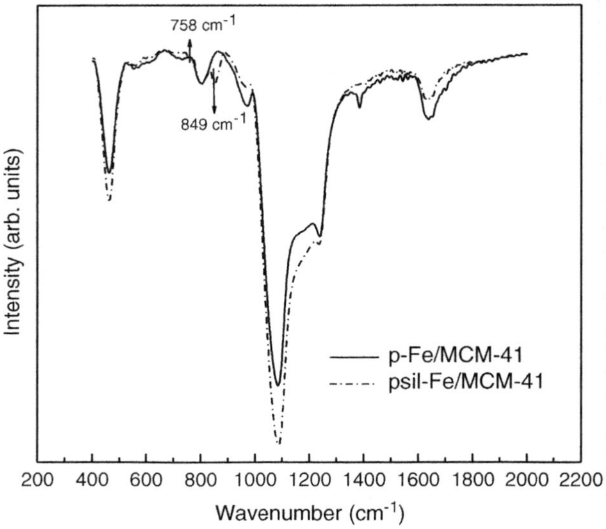

Fig. 3 FT-IR spectra of the precursors

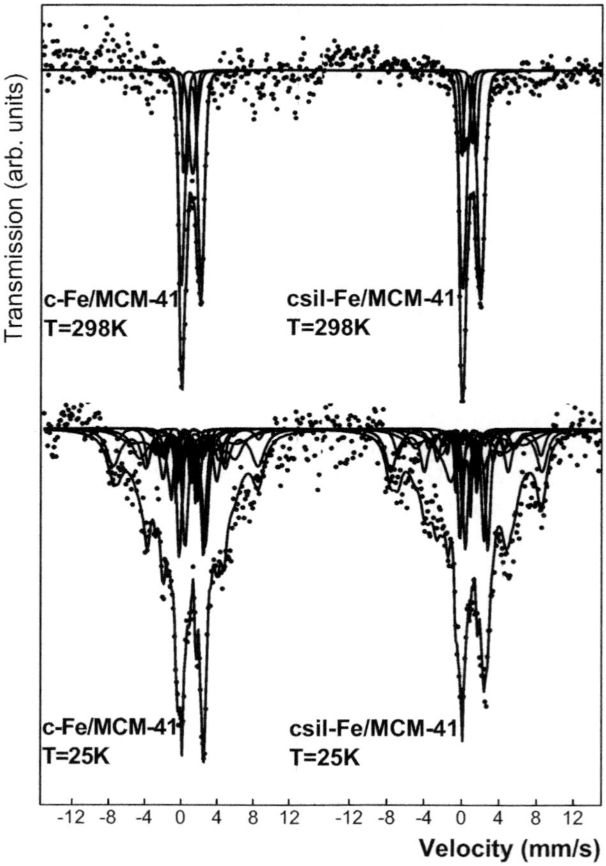

Fig. 4 Mössbauer spectra of the activated catalysts at 298 and 25 K

Table 3 Mössbauer parameters of the activated catalysts at 25 K

Species	Parameters	c-Fe/MCM-41	csil-Fe/MCM-41
Fe_3O_4 I site	H (kOe)	518 ± 42	512 ± 30
	δ (mm/s)	0.37[a]	0.37[a]
	2ε (mm/s)	−0.02[a]	−0.02[a]
Fe_3O_4 II site	H (kOe)	526 ± 14	523 ± 12
	δ (mm/s)	0.49[a]	0.49[a]
	2ε (mm/s)	0.00[a]	0.00[a]
Fe_3O_4 III site	H (kOe)	497 ± 18	498 ± 40
	δ (mm/s)	0.83[a]	0.83[a]
	2ε (mm/s)	−0.27[a]	−0.27[a]
Fe_3O_4 IV site	H (kOe)	482[a]	475 ± 17
	δ (mm/s)	1.03[a]	1.03[a]
	2ε (mm/s)	−0.41[a]	−0.41[a]
Fe_3O_4 V site	H (kOe)	370 ± 21	339 ± 41
	δ (mm/s)	0.96[a]	0.96[a]
	2ε (mm/s)	0.89[a]	0.89[a]
χ−Fe_5C_2 I site	H (kOe)	209 ± 8	210 ± 35
	δ (mm/s)	0.30[a]	0.30[a]
	2ε (mm/s)	−0.10[a]	−0.10[a]
χ−Fe_5C_2 II site	H (kOe)	264 ± 11	259 ± 16
	δ (mm/s)	0.38[a]	0.38[a]
	2ε (mm/s)	0.30[a]	0.30[a]
χ−Fe_5C_2 III site	H (kOe)	132 ± 15	132 ± 7
	δ (mm/s)	0.30[a]	0.30[a]
	2ε (mm/s)	−0.10[a]	−0.10[a]
Octahedral Fe^{2+}	δ (mm/s)	1.56 ± 0.07	1.6 ± 0.1
	Δ (mm/s)	2.2 ± 0.1	2.4 ± 0.1
Tetrahedral Fe^{2+}	δ (mm/s)	1.13 ± 0.07	1.1 ± 0.1
	Δ (mm/s)	2.7 ± 0.1	2.6 ± 0.1

H hyperfine magnetic field, δ isomer shift (all the isomer shifts are referred to α-Fe at 298 K), 2ε quadrupole shift, Δ quadrupole splitting
[a]Parameters held fixed in fitting

complexity of the spectra, produced by the low iron loading with various crystalline sites in a relaxation regime, a real fit process was not carried out. Initially the hyperfine parameters values were fixed and the areas and the relaxation times were allowed to vary freely. When these converged to a minimum, they were fixed and just then, the values of δ and quadrupole splitting (Δ) of the doublets and the hyperfine magnetic fields (H) of the sextets were allowed to vary freely. Therefore, it is possible that the minimum obtained it is not unequivocal, being able to exist other possible minima with very small differences. Therefore, to choose this set of values, the solid story and its physical and chemical properties are also taken into account. The spectra at 25 K were fitted with five sextets corresponding to the five Fe_3O_4 sites [11], three sextets assigned to the three χ-Fe_2C_5 carbide sites [12] and two doublets assignable to Fe^{2+} ions located in octahedral and tetrahedral sites within the SiO_2 wall [13] (Table 3). These species are present in the spectra at 298 K in sp regime: a doublet of sp carbide [12], a singlet of Fe_3O_4 sp [13] and two Fe^{2+} doublets assignable to ions diffused within the SiO_2 wall and/or Fe_3O_4 sp. The percentages of the species determined by the fitting at 25 K are equal for both catalysts within the experimental error. For c-Fe/MCM-41, the total Fe^{2+} is: 25 ± 4%, Fe_3O_4: 62 ± 21% and χ-Fe_2C_5: 13 ± 7%, while in csil-Fe/MCM-41 the total Fe^{2+} is: 22 ± 4%, Fe_3O_4: 68 ± 26% and χ-Fe_2C_5: 10 ± 9%. Considering that the activated fresh catalysts are structurally

Table 4 Activity and selectivity results

	c-Fe/MCM-41	csil-Fe/MCM-41
HC production (molecules/g s)	2.1×10^{17}	2.3×10^{17}
$R = \dfrac{C_2^= + C_3^=}{C_2 + C_3}$	0.85	0.67
% C_1 [a]	61	53
$\%C_4^{+}$ [a]	3	5
X_{CO} (%)	11	12
% CO_2 [b]	71	53

[a] $S_i = \dfrac{Mols\,(i) * Number.of.C\,(i)}{\sum_i Mols\,(i) * Number.of.C\,(i)}$

[b] % CO converted to CO_2

identical, the only difference between them is the hydrophobicity degree generated by the sylilation treatment.

The results of the activity tests after 100 h of reaction showed a total hydrocarbon production (HC) per gram for csil-Fe/MCM-41, 10% higher than the catalyst without sylilation (Table 4). It should be noted that csil-Fe/MCM-41 reached the activity value of the pseudo-stationary state for $t = 0$, while c-Fe/MCM-41 required about 2 h to obtain it. Both experimental results can be justified considering that the surface hydrophobicity in csil-Fe/MCM-41 inhibit the water adsorption on the catalyst active site, favouring the arrival of the reagents. The total CO conversion X_{CO}, analyzed at the same space velocity and mass of catalyst, is very similar for both solids. Therefore, it is possible to consider iso-conversion conditions, neglecting the influence of this variable on the selectivity results. In these conditions, the olefins/paraffins ratio (R) is slightly higher for c-Fe/MCM-41 than csil-Fe/MCM-41 (Table 4). Besides, csil-Fe/MCM-41 produces approximately 13% less CH_4 (C_1) and 25% less CO_2 than c-Fe/MCM-41. In the FTS, the CO_2 can come from the water gas shift reaction ($CO + H_2O \rightarrow CO_2 + H_2$) or from the CO disproportionation (Boudouard reaction) ($2CO \rightarrow C + CO_2$). This last reaction would not have to be altered by a hydrophobicity difference; therefore, the higher water rejection in csil-Fe/MCM-41 must be responsible for the smaller CO_2 production in this catalyst. In addition, it is observed that HC higher than C_4 (C_4^+) are produced in minimum percentage by both catalysts. These results can be attributed to the effect of structural properties (very small crystallite size [2]), and the operative variables used (high reaction temperature). Both of them contribute to increase the light HC production.

4 Conclusions

Two Fe/MCM-41 systems, with more than 90% of Fe species located inside the channels of the support were obtained, which produced a narrow small crystals distribution, accessible to reactive gases.

In the activated catalysts, the extremely small crystal sizes and their location inside the support channels would modify the activation energy and/or would generate diffusion impediments. To overcome these obstacles it is necessary to use a temperature higher than that traditionally used in the FTS. These conditions generate a high selectivity towards light HC in both catalysts.

On the other hand, the activated solids only differ in the surface hydrophobicity, produced by the sylilation treatment, which would decrease the "water gas shift reaction" in csil-Fe/MCM-41. Consequently, the undesirable CO_2 production is diminished. Besides, this solid is more active for HC production, with 13% less of CH_4 generation. All these results improve the catalyst performance in FTS.

Acknowledgements The authors acknowledge support of this work by the Consejo Nacional de Investigaciones Científicas y Técnicas, Comisión de Investigaciones Científicas de la Provincia de Buenos Aires, Universidad Nacional de La Plata and Agencia Nacional de Promoción Científica y Tecnológica (PICT No. 14-11267).

References

1. Fischer–Tropsch Technology: Stud. in Surf. Sci. and Catal. In: Steynberg and Dry (ed.). Elsevier (2004)
2. Mabaso, E.I., van Steen, E.I., Clasys, M.: D.G.M.K. Tagungsbericht. **4**, 93 (2006)
3. Ryoo, R., Kim, J.M.: J. Chem. Soc. Chem. Commun. 711 (1995)
4. Lagarec, K., Rancourt, D.G.: Mossbauer Spectral Analysis Software, Ver. 1.0. Department of Physics, University of Otawa (1998)
5. Blume, M., Tjon, J.A.: Phys. Rev. **165**, 446 (1968)
6. Sever, R.R., Alcala, R., Dumesic, J., Root, T.W.: Microporous Mesoporous Mater. **66**, 53 (2003)
7. Bu, J., Rhee, H.K.: Catal. Letters **66**, 245 (2000)
8. Vandenberghe, R.E., De Grave, E., Landuydt, C., Bowen, L.H.: Hyperfine Interact. **53**, 175 (1990)
9. Delhez, R., de Keijser, Th.H., Mittemeijer, E.J., Fresenius, Z.: Anal. Chem. **312**, 1 (1982)
10. Lin, K., Wang, L., Meng, F., Sun, Z., Yang, Q., Cui, Y., Jiang, D., Xiao, F.S.: J. Catal. **235**, 423 (2005)
11. Berry, F.J., Skinner, S., Thomas, M.F.: J. Phys., Condens. Matter **10**, 215 (1998)
12. Lin, S.C.h., Phillips, J.: J. Appl. Phys. **58**(5), 1943 (1985)
13. Clausen, B.S., Topsøe, H., Mørup, S.: Appl. Catal. **48**, 327 (1989)

Hyperfine Interact (2010) 195:15–19
DOI 10.1007/s10751-009-0095-5

Reduction of hematite with ethanol to produce magnetic nanoparticles of Fe_3O_4, $Fe_{1-x}O$ or Fe^0 coated with carbon

Juliana C. Tristão · José D. Ardisson · Maria Terezinha C. Sansiviero · Rochel M. Lago

Published online: 2 October 2009

Abstract The production of magnetic nanoparticles of Fe_3O_4 or Fe^0 coated with carbon and carbon nanotubes was investigated by the reduction of hematite with ethanol in a Temperature Programmed Reaction up to 950°C. XRD and Mössbauer measurements showed after reaction at 350°C the partial reduction of hematite to magnetite. At 600°C the hematite is completely reduced to magnetite (59%), wüstite (39%) and metallic iron (7%). At higher temperatures, carbide and metallic iron are the only phases present. TG weight losses suggested the formation of 3–56 wt.% carbon deposits after reaction with ethanol. It was observed by SEM images a high concentration of nanometric carbon filaments on the material surface. BET analyses showed a slight increase in the surface area after reaction. These materials have potential application as catalyst support and removal of spilled oil contaminants.

Keywords Ethanol · Iron oxide · Magnetic particles · Carbon nanostructures

1 Introduction

The combination of magnetic nanomaterials and carbon nanotubes (CNTs) is a promising research with a wide range of applications such as catalyst support, adsorption process for technological and environmental application, magnetic storage and drug delivery. Their small size provides new physic-chemical properties as superparamagnetism and high surface area. Different coats and magnetic phases have

J. C. Tristão · M. T. C. Sansiviero · R. M. Lago (✉)
Departamento de Química—ICEx, Universidade Federal de Minas Gerais,
Belo Horizonte, MG 31270-901, Brazil
e-mail: rochel@ufmg.br

J. D. Ardisson
Laboratório de Física Aplicada, Centro de Desenvolvimento de Tecnologia Nuclear, CDTN,
Belo Horizonte, MG 30123-970, Brazil

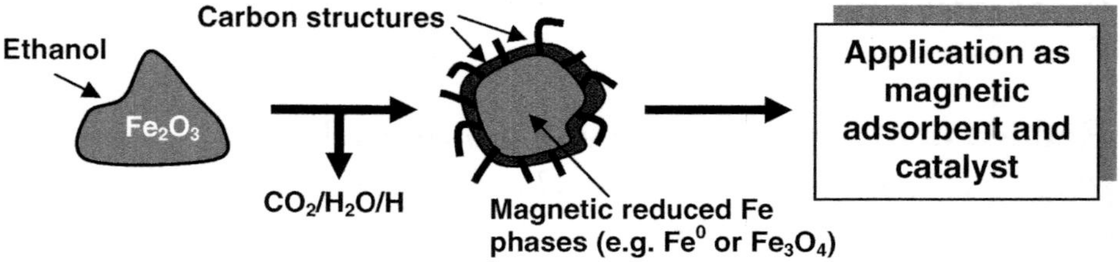

Fig. 1 Formation of magnetic nanoparticles with carbon structures coated iron phases by reaction with ethanol

been reported, e.g. C/SiO_2@Fe [1], Al_2O_3/γ-Fe_2O_3 [2], FeNi@C [3], Fe@Au [4] and polyacrylamide/Fe_3O_4 [5].

Mössbauer spectroscopy has been used for studies of carbon-coated ferromagnetic nanoparticles, such as nanocrystals produced by carbon arc discharge [6, 7] and a randomly network of carbon nanotubes produced by the pyrolysis of polymers and ferrocene [8].

In this work, it was investigated the reduction of hematite with ethanol using the Temperature Programmed Reaction (TPRe) technique to produce magnetic nanoparticles with the catalytic growth of carbon structures on the surface (Fig. 1). The proposed method is technically simple and very promising to produce magnetic nanoparticles coated with carbon as well as CNTs.

2 Experimental

The hematite was prepared by heating of 5 g of $Fe(NO_3)_3 \cdot 9H_2O$ in air at 5°C min^{-1} up to 400°C for 3 h. The TPRe (Temperature Programmed Reaction) studies were carried out with ethanol at ca. 6% in N_2 (30 mL min^{-1}) and hematite placed in a quartz tube of 7 mm diameter with a heating rate of 5°C min^{-1} up to 950°C. To guarantee a homogeneous temperature throughout the sample and uniform reaction only 200 mg hematite was used. The samples before and after TPRe were characterized by XRD (Rigaku Geigerflex using Cu Kα radiation scanning from 10 to 80° (2θ)), transmission Mössbauer spectroscopy (spectrometer CMTE MA250 with a 57Co/Rh source at room temperature using α-Fe as a reference), thermal analysis (Shimadzu TGA-60, constant heating rate of 10°C min^{-1} under air flow of 100 mL min^{-1}), SEM (Jeol JSM 840A), Raman spectroscopy (IN Via Renishaw micro-Raman with the excitation wavelength of 514.5 nm) and BET surface area (nitrogen adsorption with a 22 cycles N_2 adsorption/desorption in an Autosorb 1 Quantachrome).

3 Results and discussion

The reduction of hematite with ethanol was interrupted at different temperatures, i.e. 350, 600, 700, 800, 900 and 950°C. Mössbauer spectra of the samples are shown in Fig. 2 and the analyses of the hyperfine parameters are given in Table 1. Mössbauer and XRD showed for pure hematite (Ht) the presence of a sextet relative to the

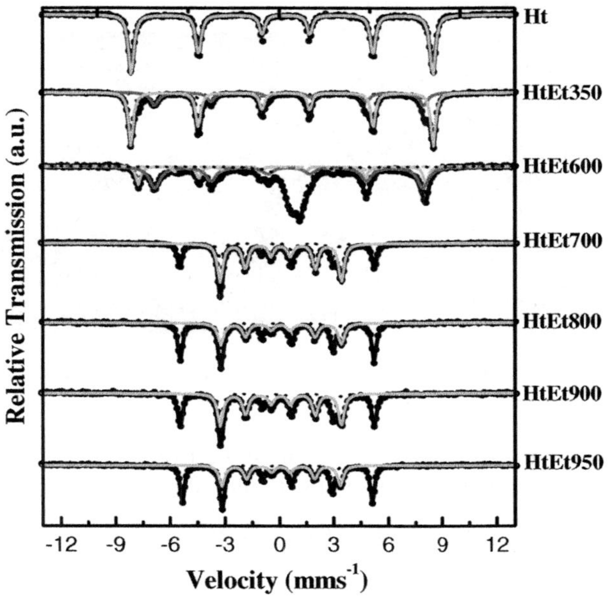

Fig. 2 Mössbauer spectra from reaction of ethanol and hematite (HtEt) interrupted at different temperatures: 600, 700, 800, 900 and 950°C

Table 1 Mössbauer hyperfine parameters for the materials after reaction of ethanol and hematite interrupted at different temperatures: 350, 600, 700, 800, 900 and 950°C

Sample	Phase	δ/mm s^{-1} (0.05)	ε, Δ/mm s^{-1} (0.05)	B_{hf}/T (0.05)	AR/% (1)
Ht	α-Fe_2O_3	0.36	−0.21	51.4	100
HtEt350	α-Fe_2O_3	0.36	−0.21	51.4	60
	{Fe_3O_4}	0.63	0.02	45.92	24
	[Fe_3O_4]	0.29	−0.05	49.05	16
HtEt600	{Fe_3O_4}	0.63	0.01	45.78	43
	[Fe_3O_4]	0.27	−0.06	49.09	16
	$Fe_{1-x}O$	0.97	0.54	–	34
	Fe	0.00	0.00	33.00	7
HtEt700	Fe	0.00	0.00	33.00	35
	Fe_3C	0.17	0.03	20.60	65
HtEt800	Fe	0.00	0.00	33.00	56
	Fe_3C	0.18	0.03	20.63	44
HtEt900	Fe	0.00	0.00	33.00	46
	Fe_3C	0.17	0.03	20.60	54
HtEt950	Fe	0.00	0.00	33.00	57
	Fe_3C	0.17	0.03	20.60	43

δ isomer shift with respect to α-Fe; ε quadrupole shift; Δ quadrupole splitting; B_{hf} hyperfine field; RA relative sub-spectral area; *[]* tetrahedric site; *{}* octahedric site

phase α-Fe_2O_3 and crystallite size of 21 nm, determined by the Scherrer equation. After TPRe at 350°C, XRD and Mössbauer measurements showed the reduction of 40% of hematite to magnetite phase (two news sextets relative to octahedric and tetrahedric sites of magnetite) with particle size of 26 nm. At 600°C the hematite is completely reduced to magnetite (59%), wüstite (39%) that appears as a central doublet and small amounts of metallic iron (7%). At higher temperatures it is observed only two phases present as two sextets identified as carbide and metallic

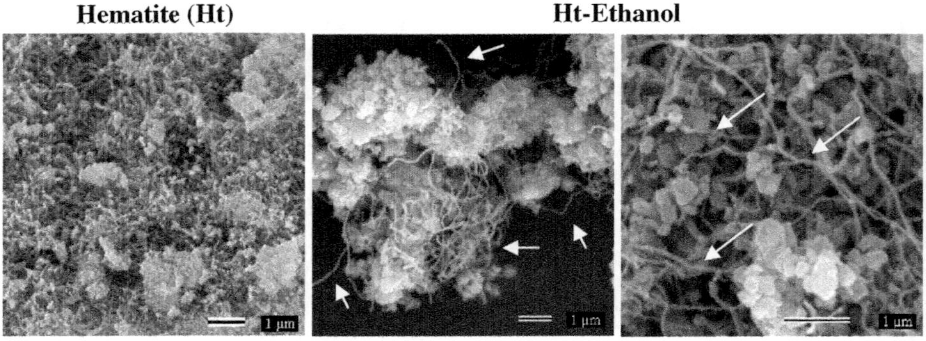

Fig. 3 SEM of the hematite before (*Ht*) and after TPRe at 800°C with ethanol

iron with relative area of metallic iron of 57% and iron carbide with 43%. The reduction of hematite with ethanol likely follows the steps below.

$$3Fe_2O_3 + CH_3CH_2OH \rightarrow 2Fe_3O_4 \quad + \text{ products of ethanol oxidation}$$
$$Fe_3O_4 + CH_3CH_2OH \rightarrow 3Fe_{1-x}O \quad + \text{ products of ethanol oxidation}$$
$$Fe_3O_4/Fe_{1-x}O + CH_3CH_2OH \rightarrow 4Fe^0 \quad + \text{ products of ethanol oxidation}$$
$$3Fe^0 + C \rightarrow Fe_3C$$

TG profiles (in air) of products after TPRe with ethanol indicate weight losses of ca. 24, 39, 25 and 30% between 450 and 700°C for the samples at 700, 800, 900 and 950°C, respectively. These weight losses are related to the oxidation of carbon deposits. These results suggest carbon contents of ca. 44–56 wt.% at temperatures higher than 700°C. SEM images (Fig. 3) showed agglomerate particles of ca. 0.2–1.0 μm with a high concentration of nanometric carbon filaments with diameter around 100 nm on the material surface after reaction with ethanol. Raman spectroscopy is a typical technique in the characterization of carbon materials such as amorphous carbon, graphite and carbon nanotubes. The obtained spectra for samples after TPRe above 700°C showed two peaks related to carbon forms: (1) at 1,583 cm^{-1}, assigned to Tangential Mode band (TM-band) related C–C modes of graphitic layers and (2) at 1,353 cm^{-1}, the disordered band (D-band). After TPRe with ethanol it can be observed an increase on the BET surface area from 17 $m^2\ g^{-1}$ for pure hematite to 45 $m^2\ g^{-1}$ at 700°C. This increase on the surface area occurs due to an increase of mesopores lower than 10 nm in the materials. At higher temperatures the surface area suffers a slight decrease (34–24 $m^2\ g^{-1}$).

4 Conclusions

Carbon coated magnetic nanoparticles of Fe_3O_4 and/or Fe^0 can be produced by reaction of hematite and ethanol in a TPRe system. The results show that under the reaction conditions ethanol converts α-Fe_2O_3 to Fe_3O_4, $Fe_{1-x}O$ and Fe^0 with the formation of carbon structures as well as CNTs. These materials have potential application as catalyst support and removal of spilled oil contaminants. We are currently investigating activation process to produce high surface area magnetic activated carbon.

Acknowledgements The authors are grateful to PRPq/UFMG, FAPEMIG, CNPq.

References

1. Veintemillas-Verdaguer, S., Leconte, Y., Costo, R., Bomati-Miguel, O., Bouchet-Fabre, B., Morales, M.P., Bonville, P., Perez-Rial, S., Rodriguez, I., Herlin-Boime, N.: Continuous production of inorganic magnetic nanocomposites for biomedical applications by laser pyrolysis. J. Magn. Magn. Mater. **311**, 120–124 (2007)
2. Shi, G.M., Zhang, Z.D., Yang, H.C.: Al_2O_3/Fe_2O_3 composite-coated polyhedral Fe nanoparticles prepared by arc discharge. J. Alloys Compd. **384**, 296–299 (2004)
3. Wei, X.W., Zhu, G.X., Xia, C.J., Ye, Y.: A solution phase fabrication of magnetic nanoparticles encapsulated in carbon. Nanotechnology **17**, 4307–4311 (2006)
4. Wang, Z.F., Mao, P.F., He, N.Y.: Synthesis and characteristics of carbon encapsulated magnetic nanoparticles produced by a hydrothermal reaction. Carbon **44**, 3277–3284 (2006)
5. Hong, M.K., Park, B.J., Choi, H.J.: Preparation and physical characterization of polyacrylamide coated magnetite particles. Phys. Status Solidi A-Appl. Mat. Sci. **204**, 4182–4185 (2007)
6. Zhang, H.: The Mössbauer spectra of graphite-encapsulated iron and iron compound nanocrystals prepared in carbon arc method. J. Phys. Chem. Solids **60**, 1845–1847 (1999)
7. Rechenberg, H.R., Coaquira, J.A.H., Marquina, C., Garcia-Landa, B., Ibarra, M.R., Benito, A.M., Maser, W., Munoz, E., Martinez, M.T.: Mössbauer and magnetic characterisation of carbon-coated small iron particles. J. Magn. Magn. Mater. **226**, 1930–1932 (2001)
8. Morozan, A., Figiel, H., Zukrowski, J., Japa, E., Tokarz, W., Dumitru, A., Stamatin, L.: Structure and magnetic properties of nanoparticles trapped in a carbon matrix along with the catalytic growth of carbon nanotubes. Mater. Sci. Eng. C. Biomin. Supra. Syst. **27**, 1167–1170 (2007)

Hyperfine Interact (2010) 195:21–25
DOI 10.1007/s10751-009-0096-4

Magnetic nanoparticles based on iron coated carbon produced from the reaction of Fe_2O_3 with CH_4: a Mössbauer study

Juliana C. Tristão · Aline A. Silva · José D. Ardisson · Rochel M. Lago

Published online: 3 October 2009

Abstract In this work, it was investigated the production of magnetic nanoparticles encapsulated with carbon by the reaction of hematite and methane by Temperature Programmed Reaction up to 950°C. XRD and Mössbauer analyses showed that the materials prepared at 600°C and 700°C are mainly composed of magnetite and small amounts of hematite α-Fe_2O_3 with particle size of 30–40 nm. At higher temperatures, the spectra also display two central doublets corresponding to wüstite phase ($Fe_{1-x}O$). The materials were also characterized by magnetization measurements, BET surface area, thermal analysis (TG) and SEM. These materials can be prepared by a simple and low cost process and show great potential to be used as adsorbents and catalyst support.

Keywords Methane · Iron oxide · Magnetic particles · Carbon nanostructures

1 Introduction

Magnetic nanoparticles have been used for many different applications as catalyst support and magnetic adsorbent. Their small size provides new physical and chemical properties as superparamagnetism and high surface area. Some examples of application are the use of coated magnetic nanoparticles in magnetic recovery catalyst in hydrogenation reactions [1], in pre-concentration of phenolic compounds from environmental water samples [2] and even as contrast agent for MR imaging [3].

J. C. Tristão · A. A. Silva · R. M. Lago (✉)
Departamento de Química—ICEx, Universidade Federal de Minas Gerais,
Belo Horizonte, MG, Brazil, 31270-901
e-mail: rochel@ufmg.br

J. D. Ardisson
Laboratório de Física Aplicada, Centro de Desenvolvimento de Tecnologia Nuclear, CDTN,
Belo Horizonte, MG, Brazil, 30123-970

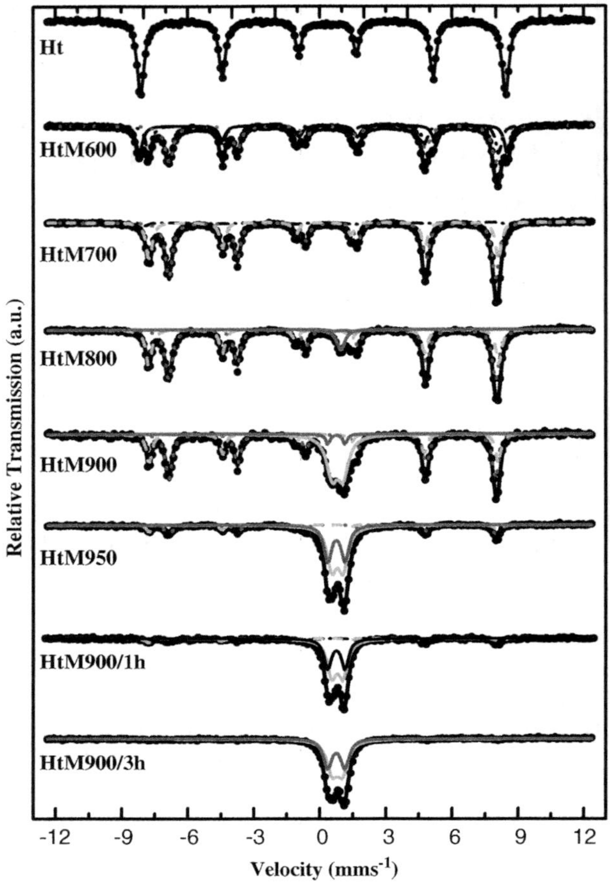

Fig. 1 Mössbauer spectra from reaction of methane and hematite interrupted at different temperatures and times: 600°C, 700°C, 800°C, 900°C, 950°C and 900°C 1 and 3 h

Different encapsulation and magnetic phases are reported such as C/SiO_2@Fe [4], Al_2O_3/γ-Fe_2O_3 [5], FeNi@C [6], Fe@Au [7] and polyacrylamide/Fe_3O_4 [8].

In this work, the production of magnetic nanoparticles was investigated using hematite and methane. This process involves the reduction of α-Fe_2O_3 with CH_4 via a TPRe (Temperature Programmed Reaction up to 950°C) to form a magnetic nucleus of Fe_3O_4 with the catalytic growth of a carbon layer on the surface [9].

2 Experimental

The hematite was prepared by heating of 5 g of $Fe(NO_3)_3.9H_2O$ in air at 5°C min^{-1} up to 400°C over 3 h. The TPRe (Temperature Programmed Reaction) studies were carried out with methane at ca. 5% in N_2 (30 mL min^{-1}) and hematite placed in a quartz tube of 7 mm diameter with a heating rate of 5°C min^{-1} up to 950°C. To guarantee a homogeneous temperature throughout the sample and uniform reaction only 100 mg hematite was used. The samples before and after TPRe were characterized by XRD (Rigaku Geigerflex using Co Kα radiation scanning from 10 to 80° (2θ)), transmission Mössbauer spectroscopy (spectrometer CMTE MA250 with

Table 1 Mössbauer hyperfine parameters for the materials after reaction of methane and hematite interrupted at different temperatures and times: 600°C, 700°C, 800°C, 900°C, 950°C and 900°C 1 h and 3 h

Sample	Phase	δ/mm s^{-1} (0.05)	ε, Δ/mm s^{-1} (0.05)	B_{hf}/T (0.05)	AR/% (1)
Ht	α-Fe_2O_3	0.36	−0.20	51.3	100
HtM600	α-Fe_2O_3	0.36	−0.20	51.3	27
	{Fe_3O_4}	0.66	−0.01	46.1	47
	[Fe_3O_4]	0.27	−0.01	49.1	26
HtM700	α-Fe_2O_3	0.36	−0.20	51.3	7
	{Fe_3O_4}	0.66	−0.01	46.1	60
	[Fe_3O_4]	0.27	−0.01	49.1	33
HtM800	{Fe_3O_4}	0.66	−0.01	46.1	58
	[Fe_3O_4]	0.27	−0.01	49.1	33
	$Fe_{1-x}O$	0.87	0.65	–	9
HtM900	{Fe_3O_4}	0.66	−0.01	46.1	43
	[Fe_3O_4]	0.27	−0.01	49.1	23
	$Fe_{1-x}O$	0.87	0.65	–	34
HtM950	{Fe_3O_4}	0.66	−0.01	46.1	15
	[Fe_3O_4]	0.27	−0.01	49.1	8
	$Fe_{1-x}O$	0.87	0.65	–	46
	$Fe_{1-x}O$	1.04	0.60	–	31
HtM900/1 h	{Fe_3O_4}	0.64	−0.01	45.9	7
	[Fe_3O_4]	0.25	−0.05	49.1	7
	$Fe_{1-x}O$	0.87	0.65	–	52
	$Fe_{1-x}O$	1.04	0.60	–	34
HtM900/3 h	$Fe_{1-x}O$	0.87	0.65	–	70
	$Fe_{1-x}O$	1.04	0.60	–	30

δ isomer shift with respect to α-Fe, ε quadrupole shift, Δ quadrupole splitting, B_{hf} hyperfine field, *RA* relative sub-spectral area, [] tetrahedric site, {} octahedric site

a 57Co/Rh source at room temperature using α-Fe as a reference), magnetization measurements (portable magnetometer with a fixed magnetic field of 0.3 T), BET surface area (nitrogen adsorption with a 22 cycles N_2 adsorption/desorption in an Autosorb 1 Quantachrome), thermal analysis (Shimadzu TGA-60, constant heating rate of 10°C min^{-1} under air flow of 100 mL min^{-1}) and SEM (Jeol JSM 840A).

3 Results and discussion

It was studied the reaction of hematite and methane at different temperatures, i.e. 600°C, 700°C, 800°C, 900°C, 950°C and 900°C for 1 and 3 h. XRD analyses showed for pure hematite (Ht) the presence of the phase α-Fe_2O_3 (PDF 1–1053), with crystallite size of 22 nm, determined by the Scherrer equation. After TPRe at 600°C and 700°C the materials prepared are mainly composed of magnetite (Fe_3O_4, PDF 19–629) and small amounts of hematite α-Fe_2O_3 with particle size of 30–40 nm. At higher temperatures, wüstite phase ($Fe_{1-x}O$, PDF 6–615) was formed.

Mössbauer spectra in Fig. 1 and Table 1 show a sextet related for the pure hematite. After TPRe at 600°C, this sextet decreases the relative area (from 100% to 27%) and splits into two news sextets assigned to octahedric and tetrahedric sites of the magnetite phase, with 47% and 26% of area, respectively. At 700°C of reaction

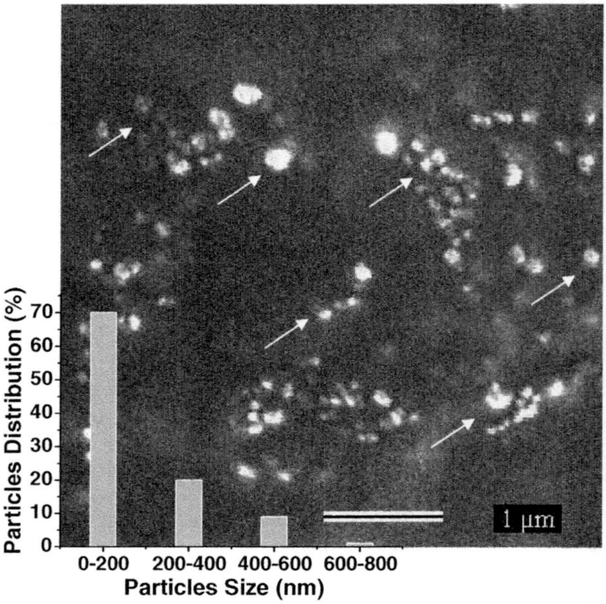

Fig. 2 SEM image and particle size distribution of HtM700 dispersed in water

with methane, the reduction to magnetite is almost complete (93% of relative area). For higher temperatures it can be observed one or two duplets related to the presence of Fe^{2+} and Fe^{3+} from wüstite phase. It is observed that higher temperature and longer times of reaction increase the relative area of wüstite phase, from 9% at 800°C to 100% at 900°C/3 h. The reactions taking place of hematite with methane are show below.

$$CH_4 + 12Fe_2O_3 \rightarrow 8Fe_3O_4 + CO_2 + 2H_2O$$
$$CH_4 + 4Fe_3O_4 \rightarrow 12Fe_{1-x}O + CO_2 + 2H_2O$$

Magnetization measurements showed that all materials produced were magnetic proportional to the content of magnetite phase. The highest magnetizations were found for the samples at 700°C and 800°C with 91 and 88 J T^{-1} kg^{-1}, respectively. BET area showed a drastically reduction of surface area after TPRe at higher temperature and longer times, that suggest the materials agglomeration or the carbon deposition over the porous structures. The surface area changed from 17 m^2 g^{-1} for pure hematite (Ht) to 6.0, 3.0, 0.4, 0.4 and 0.2 m^2 g^{-1} at 700, 900, 950, 900°C/1 h and 900°C/3 h, respectively. TG weight losses suggested carbon contents of 3–4 wt.% after reaction with methane. Raman spectroscopy is a typical technique in the characterization of carbon materials such as amorphous carbon, graphite and carbon nanotubes. At temperatures above of 700°C, the obtained spectra showed two weak signals at 1,300 and 1,395 cm^{-1} related to amorphous carbon. The magnetic materials were dispersed in water and the suspension analyzed by SEM. The SEM images (Fig. 2) showed the presence of a large fraction of particles with nanometric dimension of 100–200 nm that remain highly dispersed in different solvents favoring the catalytic process. Organic molecules, i.e. methylene blue dye and chlorobenzene, are strongly adsorbed due to the hydrophobic carbon surface.

4 Conclusions

Hematite can be reduced by methane to produce magnetic materials encapsulated with a layer of carbon. XRD and Mössbauer spectroscopy showed that under the reaction conditions methane converts α-Fe_2O_3 to Fe_3O_4 and $Fe_{1-x}O$. After dispersion in solvent a nanometric magnetic particles fraction can be separated and used as adsorbent and catalyst support which can be removed from the medium by simple magnetic separation process.

Acknowledgements The authors would like to thank the support from PRPq/UFMG, FAPEMIG, CNPq.

References

1. Jacinto, M.J., Kiyohara, P.K., Masunaga, S.H., Jardim, R.F., Rossi, L.M.: Recoverable rhodium nanoparticles: synthesis, characterization and catalytic performance in hydrogenation reactions. Appl. Catal., A Gen. **338**, 52–57 (2008)
2. Zhao, X.L., Shi, Y.L., Wang, T., Cai, Y.Q., Jiang, G.B.: Preparation of silica–magnetite nanoparticle mixed hemimicelle sorbents for extraction of several typical phenolic compounds from environmental water samples. J. Chromatogr. A **1188**, 140–147 (2008)
3. Sosnovik, D.E., Nahrendorf, M., Weissleder, R.: Magnetic nanoparticles for MR imaging: agents, techniques and cardiovascular applications. Basic Res. Cardiol. **103**, 122–130 (2008)
4. Veintemillas-Verdaguer, S., Leconte, Y., Costo, R., Bomati-Miguel, O., Bouchet-Fabre, B., Morales, M.P., Bonville, P., Perez-Rial, S., Rodriguez, I., Herlin-Boime, N.: Continuous production of inorganic magnetic nanocomposites for biomedical applications by laser pyrolysis. J. Magn. Magn. Mater. **311**, 120–124 (2007)
5. Shi, G.M., Zhang, Z.D., Yang, H.C.: Al_2O_3/Fe_2O_3 composite-coated polyhedral Fe nanoparticles prepared by arc discharge. J. Alloys Compd. **384**, 296–299 (2004)
6. Wei, X.W., Zhu, G.X., Xia, C.J., Ye, Y.: A solution phase fabrication of magnetic nanoparticles encapsulated in carbon. Nanotech. **17**, 4307–4311 (2006)
7. Wang, Z.F., Mao, P.F., He, N.Y.: Synthesis and characteristics of carbon encapsulated magnetic nanoparticles produced by a hydrothermal reaction. Carbon **44**, 3277–3284 (2006)
8. Hong, M.K., Park, B.J., Choi, H.J.: Preparation and physical characterization of polyacrylamide coated magnetite particles. Phys. Status Solidi A, Appl. Mater. Sci. **204**, 4182–4185 (2007)
9. Tristão, J.C., Ardisson, J.D., Sansiviero, M.T.C., Lago, R.M.: Reduction of hematite with ethanol to produce magnetic nanoparticles of Fe_3O_4, $Fe_{1-x}O$ or Fe^0 coated with carbon. Hyperfine Interact. (2009). doi:10.1007/s10751-009-0095-5

Hyperfine Interact (2010) 195:27–34
DOI 10.1007/s10751-009-0097-3

Niobian iron oxides as heterogeneous Fenton catalysts for environmental remediation

Diana Q. L. Oliveira · Luiz C. A. Oliveira · Enver Murad · José D. Fabris · Adilson C. Silva · Lucas Morais de Menezes

Published online: 26 September 2009

Abstract Heterogeneous Fenton or Fenton-like reagents consist of a mixture of an iron-containing solid matrix and a liquid medium with H_2O_2. The Fenton system is based on the reaction between Fe^{2+} and H_2O_2 to produce highly reactive intermediate hydroxyl radicals ($^{\bullet}OH$), which are able to oxidize organic contaminants, whereas the Fenton-like reaction is based on the reaction between Fe^{3+} and H_2O_2. These heterogeneous systems offer several advantages over their homogeneous counterparts, such as no sludge formation, operation at near-neutral pH and the possibility of recycling the iron promoter. Some doping transition cations in the iron oxide structure are believed to enhance the catalytic efficiency for the oxidation of organic substrates in water. In this work, goethites synthesized in presence of niobium served as precursors for the preparation of magnetites (niobian magnetites) via chemical reduction with hydrogen at 400°C. These materials were used as Fenton-like catalysts. Both groups of (Nb, Fe)-oxide samples were characterized by ^{57}Fe Mössbauer spectroscopy at 298 K. The results show that increasing niobium contents raise the catalytic potential for decomposition of methylene blue, which was, in this work, used as a model molecule for organic substrates in water.

Keywords Magnetite · Niobium · Fenton reaction

D. Q. L. Oliveira (✉) · E. Murad · J. D. Fabris · L. M. de Menezes
Department of Chemistry, Federal University of Minas Gerais, 31270-901 Belo Horizonte, Minas Gerais, Brazil
e-mail: dianaquintao@ufmg.br

L. C. A. Oliveira · A. C. Silva
Department of Chemistry, Federal University of Lavras, 37200-000 Lavras, Minas Gerais, Brazil

1 Introduction

Spinel iron oxides are of technological importance because of their structural, electronic, magnetic and catalytic properties [1]. Other metal cations can isomorphously replace iron in magnetite (Fe_3O_4), thereby changing selected physico-chemical properties of the material, depending on the nature and amount of the metal and on the structural site on which the metal is incorporated [1].

A novel and promising catalytic application of iron oxides is the chemical decomposition of organic contaminants in wastewaters, using H_2O_2 in a heterogeneous Fenton system [2–6]. In this heterogeneous Fenton system the iron oxide activates H_2O_2 to generate radicals, especially $^{\bullet}OH$, which can completely oxidize organics present in the aqueous medium [7]. Magnetite has been observed to be particularly active for the Fenton oxidation of organics by H_2O_2. This activity was assigned to the presence of Fe^{2+} species in the magnetite structure, which can activate H_2O_2 by a Haber–Weiss mechanism [8]. Recent work has shown that the activity of magnetite is strongly influenced by the presence of different metals in the spinel structure. Thus the introduction of cobalt and manganese in the magnetite structure remarkably increased the reactivity towards the Fenton chemistry, whereas nickel showed an inhibitory effect [9–11].

In this work, niobium associated with magnetite ("niobian magnetite") has been used to produce an active heterogeneous spinel system. Nb^{5+} shows interesting features for this system such as an ionic radius (64 pm in octahedral coordination) [12] that is comparable to that of high-spin octahedral Fe^{3+} (65 pm) and, based alone on size, might be structurally incorporated in magnetite, and a high reactivity towards H_2O_2 activation. To our knowledge, no systematic studies on iron-rich spinels prepared in a niobium-containing medium or their catalytic properties have been reported so far.

2 Experimental

2.1 Synthesis and characterization

The goethite samples were prepared by precipitation of $Fe(NO_3)_3 \cdot 6H_2O$ (1 mol L^{-1}) with potassium hydroxide (5 mol L^{-1}) and the addition of 2, 7 and 17 mol% niobium that is present in $NH_4[NbO(C_2O_4)_2(H_2O)](H_2O)_n$ supplied by CBMM (Araxá, MG, Brazil). The precipitates were washed with water until the pH became adjusted to 7; they were then transferred to a 2 L-beaker with distilled water and aged for 72 h at 60°C. The so produced goethites were finally reduced with hydrogen for 30 min at 400°C to produce magnetites.

The total iron was determined by $K_2Cr_2O_7$ titration and energy dispersive spectrometry (EDS) analyses were performed using a Jeol JXA-8900RL microscope. Mössbauer spectra were taken at room temperature (298 K) on a CMTE MA250 spectrometer with a constant acceleration drive and a $^{57}Co/Rh$ source at room temperature. The experimental data were fitted using Lorentzian functions with a least-squares procedure based on the NORMOS program, and calibration was effected and isomer shifts are given relative to α-Fe. Powder X-ray diffraction (XRD) was carried out on a Rigaku Geigerflex 3064 diffractometer equipped with

Table 1 Compositions of the niobian magnetites from chemical analyses and EDS data

Sample	Total Fe/%	Nb/%
(a) Pure Mt	70(3)	0
(b) Mt-Nb2	67(1)	2.1(2)
(c) Mt-Nb7	61(1)	7(1)
(d) Mt-Nb17	59(1)	17(2)

a Cu tube and a graphite diffracted-beam monochromator. To improve accuracy, additional scans of the samples were taken with Si as internal standard.

2.2 Reactions

The hydrogen peroxide decomposition was carried out by mixing 10 mL H_2O_2 2.9 mol L^{-1} and 10 mg catalyst and measuring the formed gaseous O_2 in a volumetric glass system. The oxidation of the methylene blue dye (50 mg L^{-1}) with H_2O_2 (0.3 mol L^{-1}) at pH 6.0 (the pH of the sole H_2O_2 solution) was carried out with a total volume of 10 mL and 10 mg of the oxide catalyst. The reactions were monitored by UV–vis measurements. All reactions were carried out under magnetic stirring in a recirculating controlled-temperature bath kept at 25 ± 1°C.

In an attempt to identify the intermediate products, the decomposition of methylene blue was also monitored with the positive ion mode ESI-MS in an Agilent MS-ion trap mass spectrometer (1100 Series). The reaction samples were analyzed by introducing aliquots into the ESI source with a syringe pump at a flow rate of 5 L min^{-1}. The spectra were obtained as an average of five scans of 0.2 s each. Typical ESI conditions were as follows: heated capillary temperature 325°C; sheath gas (N_2) at a flow rate of 20 units (ca. 4 L min^{-1}); spray voltage 4 kV; capillary voltage 25 V; tube lens offset voltage 25 V.

3 Results and discussion

3.1 Characterization of materials

Not unexpectedly, the chemical compositions of the niobian magnetites (Table 1) show a concurrent increase of niobium and decrease of iron.

Room-temperature Mössbauer spectra of the unsubstituted magnetite and the niobian series are shown in Fig. 1. While spectra of the pure magnetite and the samples prepared in presence of 2 and 7 mol% niobium show two characteristic sextets that can be assigned to Fe^{3+} on the A sites and Fe^{2+}/Fe^{3+} on the B sites of magnetite [13], the sample prepared in the presence of 17 mol% niobium shows only a minor indication of magnetite.

Analysis of the Mössbauer spectra (Table 2) shows the tetrahedral (A site) of the unsubstituted magnetite to have a hyperfine field (B_{hf}) of 49.0 T and the B (octahedral) site to have a hyperfine field of 45.8 T with isomer shifts (δ) relative to α-Fe of 0.28 and 0.67 mm s^{-1}, respectively. While these parameters resemble those of pure, stoichiometric magnetite [13], the area ratio of the B-site resonance to that of the A-site resonance (1.39) indicates the unsubstituted magnetite to have a composition given by $Fe_{2.94}O_4$, and thus to have undergone noticeable oxidation.

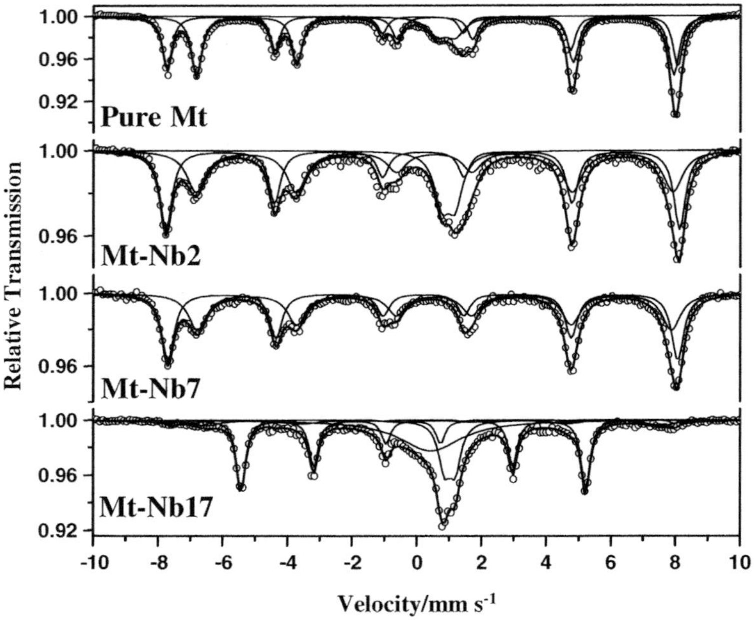

Fig. 1 Room-temperature Mössbauer spectra of the niobian magnetites

Table 2 Room-temperature (~298 K) Mössbauer parameters for pure magnetite and magnetites prepared in the presence of niobium

Sample	^{57}Fe site	δ/mm s^{-1}	2ε, Δ/mm s^{-1}	B_{hf}/T	Γ/mm s^{-1}	RA/%
Pure Mt	A	0.2831(8)	−0.010(1)	48.960(6)	0.349(2)	36.6(2)
	B	0.6680(7)	0.010(1)	45.786(5)	0.388(2)	50.8(2)
	Fe^{2+}	1.000(4)	0.571(7)	–	0.70[a]	12.6(1)
Mt-Nb2	A	0.295(2)	−0.013(3)	49.27(1)	0.433(5)	40.6(4)
	B	0.649(3)	0.012(6)	45.75(2)	0.72(1)	43.5(1)
	Fe^{2+}	1.076(3)	0.408(6)	–	0.55[a]	15.9(1)
Mt-Nb7	A	0.304(3)	−0.020(5)	48.90(2)	0.458(9)	50(1)
	B	0.642(5)	0.020(9)	45.46(4)	0.69(2)	50(1)
Mt-Nb17	A	0.30[a]	−0.02[a]	48.2(1)	0.42(6)	3.0(4)
	B	0.63[a]	0[a]	43.7(1)	0.71[a]	7.2(3)
	Fe^{2+}	1.124(3)	0.326(4)	–	0.42(1)	16.6(5)
	Fe^{0}	−0.0020(8)	0[a]	33.034(6)	0.323(3)	41.0(3)
	Doublet	0.59(2)	0[a]	–	2.29(4)	32.2(6)

δ isomer shift relative to α-Fe, 2ε quadrupole shift, Δ quadrupole splitting, B_{hf} magnetic hyperfine field, RA: relative sub-spectral area

[a]Parameter held fixed during fitting

The degree of oxidation rises with increasing niobium concentration during synthesis from the above value to $Fe_{2.92}O_4$ at 2% niobium in the synthesis solution and to $Fe_{2.91}O_4$ at 7% niobium. The line widths of the magnetites prepared in the presence of niobium, particularly those of the B sites, are furthermore significantly higher

Fig. 2 X-ray diffraction diagrams of the niobian magnetites

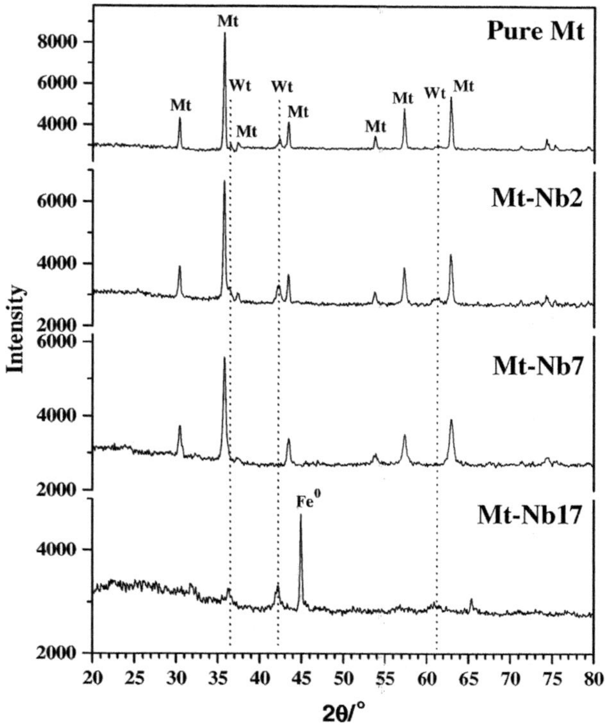

than those of the pure magnetite. At a niobium concentration of 17%, magnetite formation is largely inhibited, but the Mössbauer spectrum indicates the presence of magnetically ordered metallic iron and paramagnetic Fe^{2+}. The magnetic hyperfine fields of the A site show no and the B-fields show little systematic variation as a function of the niobium concentration during synthesis. Because Nb^{5+} is a diamagnetic ion with a krypton configuration, this does not provide conclusive evidence for the concept that niobium has substituted iron in the magnetite structure. The lines of the magnetite components become broader as the concentration of niobium in the synthesis solution increases, indicating decreasing particle sizes. Mean coherence lengths calculated from broadening of the 311 diffraction line indicates a significant decrease from 62 to 45, 24 and 8 nm from the pure magnetite to the 2%, 7%, and 17% Nb samples, respectively.

Powder XRD (Fig. 2) also shows the samples to contain various proportions of magnetite (the sample prepared in the presence of 17% niobium again showing the lowest proportion of magnetite), and the samples prepared in the presence of 2% and 17% niobium show the presence of some wüstite ($Fe_{1-x}O$). Diffraction diagrams (not shown) were also taken under addition of Si as an internal standard for 2θ. The resulting patterns do not show a systematic variation of the position of the strongest magnetite peak (311) as a function of the niobium concentration, so that XRD also does not provide conclusive evidence for the incorporation of niobium in the magnetite structure.

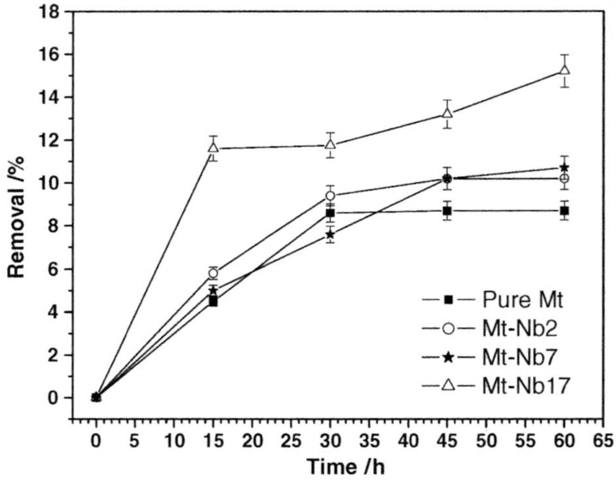

Fig. 3 Methylene blue discoloration in the presence of niobian magnetites

3.2 H_2O_2 decomposition and dye oxidation

The catalytic activity of the niobian magnetites was studied using two reactions: (1) the decomposition of H_2O_2 to O_2 ($H_2O_2 \rightarrow H_2O + 0.5\ O_2$) and (2) the oxidation of the model contaminant (methylene blue dye) with H_2O_2 in an aqueous medium. The reaction kinetics was investigated by UV/vis discoloration measurements to form non-colored intermediates. The discoloration of the solution of the dye is shown in Fig. 3.

The control experiment (only methylene blue and H_2O_2 but no catalyst) showed no significant discoloration even after a reaction time of 60 min. On the other hand, in the presence of the niobian magnetite catalyst and H_2O_2 as oxidant, discoloration can be observed. For the system niobian magnetite/H_2O_2, the discoloration is moderate in the presence of the pure magnetite and the magnetites prepared at low niobium contents. However, the sample prepared at high niobium content (Mt-Nb17) produces a catalytic discoloration of more than 15% after 60 min. This behavior can have several causes: sample mineralogy (e.g. the presence of metallic iron in the Mt-Nb17 sample), a catalytic enhancement due to the presence of Nb, and particle size changes. While some or all of these effects can act together to bring about the higher catalytic activity of the Mt-Nb17 sample, the drastic particle size reduction of magnetite in this sample indicates particle size to play a key role. Although H_2O_2 decomposition kinetics is a complex reaction, the linear behavior of the decomposition plots between 0 and 25 min suggests that the process can be approximated to pseudo-zero order kinetics [8].

Discoloration measured by UV–vis spectroscopy does not provide any information about reaction intermediates of the methylene blue oxidation. Such identification could be achieved by using mass spectroscopy with electrospray ionization. The identification of intermediates was studied obtaining spectra after 60 min of reaction in the presence of H_2O_2 (Fig. 4). The spectrum of methylene blue solution (without catalysts) indicated only one peak at $m/z = 284$ due to its cationic structure. After 1 h of reaction, intermediates were observed in the spectra, showing that the mineralization was not total. An intense peak at $m/z = 270$ suggested the beginning

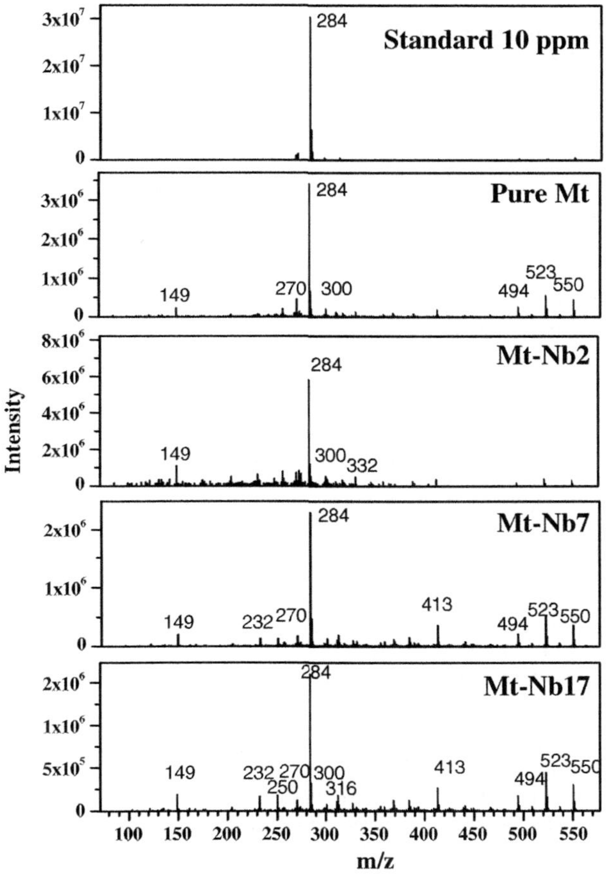

Fig. 4 Mass spectra of the reaction products of methylene blue oxidation with electrospray ionization as a function of the niobium concentration

of dye structure degradation. A signal at $m/z = 300$ and 316 is probably due to successive hydroxylation of the aromatic ring [14]. The peak at $m/z = 149$ is indicative of ring rupture and subsequent total mineralization [15].

4 Conclusions

The presence of niobium associated with magnetite has a significant effect on the crystallinity and catalytic activity of the oxide. In this study, the oxidation of organic compounds with H_2O_2 has been shown to probably take place via radicals, as suggested by ESI-MS data. An oxidation mechanism was indicated to occur by attack of the free radical *OH over the molecule, giving rise to hydroxylation products as principal by-products, although compounds resulting from ring cleavage are also detected. The presence of niobium increases the activity of the oxide for the H_2O_2 decomposition, and the surface niobian species seem to act directly in the catalytic properties of the niobian magnetites.

Acknowledgements Work supported by CAPES, CNPq and FAPEMIG (Brazil). The authors are indebted to the Brazilian Company of Metallurgy and Mining (CBMM) for providing the ammonium niobium oxalate salt from its industrial line. CNPq/Prosul (grant # 490132/2006-5) also supported the international mission by JDF to attend the Latin American Conference on the Applications of the Mössbauer Effect, LACAME 2008, in La Plata, Argentina.

References

1. Cornell, R.M., Schwertmann, U.: The Iron Oxides, 2nd edn. Wiley-VCH, Weinheim (2003)
2. Moura, F.C.C., Araújo, M.H., Costa, C.C., Fabris, J.D., Ardisson, J.D., Macedo, W.A.A., Lago, R.M.: Efficient use of Fe metal as an electron transfer agent in a heterogeneous Fenton system based on Fe/Fe_3O_4 composites. Chemosphere **60**, 1118–1123 (2005)
3. Lee, S., Oh, J., Park, Y.: Degradation of phenol with Fenton-like treatment by using heterogeneous catalyst (modified iron oxide) and hydrogen peroxide. Bull. Korean Chem. Soc. **27**, 489–494 (2006)
4. Baldrian, P., Merhautova, V., Gabriel, J., Nerud, F., Stopka, P., Hruby, M., Benes, M.J.: Decolorization of synthetic dyes by hydrogen peroxide with heterogeneous catalysis by mixed iron oxides. Appl. Catal. B **66**, 258–264 (2006)
5. Wu, R.C., Qu, J.H.: Removal of azo dye from water by magnetite adsorption–Fenton oxidation. Water Environ. Res. **76**, 2637–2642 (2004)
6. Lago, R.M., Moura, F.C.C., Ardisson, J.D., Macedo, W.A., Araujo, M.H., Silva, C.N.: Formation of highly reactive species at the interface Fe^0-iron oxides particles by mechanical alloying and thermal treatment: potential application in environmental remediation processes. Chem. Lett. **34**, 1172–1174 (2005)
7. Moura, F.C.C., Araujo, M.H., Dalmazio, I., Alves, T.M.A., Santos, L.S., Eberlin, M.N., Augusti, R., Lago, R.M.: Investigation of reaction mechanisms by electrospray ionization mass spectrometry: characterization of intermediates in the degradation of phenol by a novel iron/magnetite/hydrogen peroxide heterogeneous oxidation system. Rapid Commun. Mass Spectrom. **20**, 1859–1863 (2006)
8. Moura, F.C.C., Oliveira, G.C., Araujo, M.H., Ardisson, J.D., Macedo, W.A.A., Lago, R.M.: Highly reactive species formed by interface reaction between Fe^0-iron oxides particles: an efficient electron transfer system for environmental applications. Appl. Catal. A Gen. **307**, 195–204 (2006)
9. Costa, R.C.C., Lelis, M.F.F., Oliveira, L.C.A., Fabris, J.D., Ardisson, J.D., Rios, R.V.A., Silva, C.N., Lago, R.M.: Remarkable effect of Co and Mn on the activity of $Fe_{3-x}M_xO_4$ promoted oxidation of organic contaminants in aqueous medium with H_2O_2. Catal. Commun. **4**, 525–529 (2003)
10. Costa, R.C.C., Lelis, M.F.F., Oliveira, L.C.A., Fabris, J.D., Ardisson, J.D., Rios, R.V.A., Silva, C.N., Lago, R.M.: Novel active heterogeneous Fenton system based on $Fe_{3-x}M_xO_4$ (Fe, Co, Mn, Ni): the role of M^{2+} species on the reactivity towards H_2O_2 reactions. J. Hazard. Mater. **129**, 171–178 (2006)
11. Nejad, M.A., Jonsson, M.: Reactivity of hydrogen peroxide towards Fe_3O_4, Fe_2CoO_4 and Fe_2NiO_4. J. Nucl. Mater. **334**, 28–34 (2004)
12. Kaye, G.W., Laby, T.H.: Tables of Physical and Chemical Constants and some Mathematical Functions, p. 386. Longmans, Green and Co., London (1975)
13. Murad, E., Cashion, J.: Mössbauer Spectroscopy of Environmental Materials and their Industrial Utilization. Kluwer, Norwell (2004)
14. Oliveira, L.C.A., Ramalho, T.C., Gonçalves, M., Cereda, F., Carvalho, K.T., Nazzarro, M.S., Sapag, K.: Pure niobia as catalyst for the oxidation of organic contaminants: mechanism study via ESI-MS and theoretical calculations. Chem. Phys. Lett. **446**, 133–137 (2007)
15. Oliveira, L.C.A., Gonçalves, M., Guerreiro, M.C., Ramalho, T.C., Fabris, J.D., Pereira, M.C., Sapag, K.: A new catalyst material based on niobia/iron oxide composite on the oxidation of organic contaminants in water *via* heterogeneous Fenton mechanisms. Appl. Catal. A Gen. **316**, 117–124 (2007)

Hyperfine Interact (2010) 195:35–41
DOI 10.1007/s10751-009-0098-2

Magnetic minerals from volcanic Ultisols as heterogeneous Fenton catalysts

S. Aravena · C. Pizarro · M. A. Rubio · L. C. D. Cavalcante · V. K. Garg · M. C. Pereira · J. D. Fabris

Published online: 30 September 2009

Abstract This study was devoted to the evaluation of the effectiveness of Fenton catalysts, based on magnetically-concentrated portions of iron oxide-rich sand fractions from two magnetic Ultisols, derived from volcanic materials of southern Chile. The samples were labeled according to the municipality where the sample sites are geographically located, namely Metrenco and Collipulli, and were characterized with Mössbauer spectroscopy at 298 K and saturation magnetization (σ) measurements. Mössbauer data revealed a complex magnetic hyperfine structure for these magnetic portions from both soil-sand materials, suggesting relatively complex mineral assemblages. The monitored rate of H_2O_2 decomposition *via* heterogeneous Fenton reaction revealed that materials from the Collipulli soil are more efficient Fenton catalyst than are those from the Metrenco soil. The reasons for these differences are from now on being explored on basis of a more detailed chemical investigation of these samples.

Keywords Iron oxides · Environment · Magnetite · Catalysis · Soil

S. Aravena (✉) · C. Pizarro · M. A. Rubio
Facultad de Química y Biología, USACH, Av. L. B. O'Higgins 3363, Santiago 7254758, Chile
e-mail: susana.aq@gmail.com

L. C. D. Cavalcante
Centro de Ciências da Natureza, Universidade Federal do Piauí, 64049-550 Teresina, PI, Brazil

V. K. Garg
Instituto de Física, Universidade de Brasília, 70919-970 Brasília, DF, Brazil

L. C. D. Cavalcante · M. C. Pereira · J. D. Fabris
Departamento de Química, ICEx-UFMG, Pampulha, 31270-901 Belo Horizonte, MG, Brazil

1 Introduction

A heterogeneous Fenton reagent typically consists of a mixture of a Fe^{2+}-containing solid matrix in a liquid medium with H_2O_2. It is highly active in promoting oxidation of organic substrates, being used, for instance, to remove organic pollutants in water. The heterogeneous reaction presents some practical advantages over the homogeneous system, as no sludge is formed, by operating at near-neutral pH, and enables recycling the iron promoter [1, 2]. Some iron-bearing geomaterials are potentially very interesting alternatives as Fenton catalyst, from the technological and economical viewpoints. This is mainly true if the reaction is directed to environmental remediation of contaminated water bodies with organic residues, as from industrial effluents or with agriculture chemical residues, including those based on phenols and formaldehydes, or from most commonly bio-active substances from commercial pesticides [3, 4].

Ultisols are mainly characterized by having high clay and iron oxide contents. It has been considered an increasing interest to know about their potential applicability, specifically of their clay fraction, as inexpensive heterogeneous catalysts, for different reactions of environmental and industrial uses, as water gas shift, Fenton and Fenton-like reactions [5, 6]. Ultisols are particularly rich in magnetic iron oxides, which are mainly found in their sand fractions. This feature suggests that Ultisol materials are presumably good technological alternative as catalysts for decomposition of organic substrate in water *via* Fenton or Fenton-like processes.

This study was devoted to the evaluation of the effectiveness of Fenton catalysts, based on the magnetic portion of an iron oxide-rich sand fraction from two Ultisols derived from volcanic materials in southern Chile.

2 Materials and methods

The two soil samples were collected at a depth between 15 and 30 cm, one from each of two Ultisol profiles located in southern Chile (geographical coordinates of the sampling sites: Collipulli at 36°58′ S 72°09′ W and Metrenco at 38°50′ S 72°37′ W). The collected soil material was sieved in the field, so to obtain a fraction with mean particle diameter ≤2 mm (whole soil), and packed under roughly the corresponding field moisture atmosphere [7, 8].

The chemical compositions of samples were determined by digesting 0.100 g of the dry sample with HF and *aqua regia* (at 10:1.5 ratio), in a PARR bomb digester, set at 110°C, during 12 h. The resulting digestion mixture was treated with ultra pure graded (Merck) boric acid, diluted to 100 mL with water, and stored in a plastic flask at 5°C, prior to elemental chemical dosages of Fe, Si, Al, Mg, Mn and Ti, from readings on a flame atomic absorption spectrometer (Perkin Elmer, model 2380). Data presented in this work are expressed as averaged values of two sample-replicas.

The specific soil magnetization was measured with a portable soil magnetometer [9]. The equipment allows direct digital readings of the magnetic moment of soil samples, expressed J T^{-1}. From the sample mass, the specific saturation magnetization in J T^{-1} kg^{-1} is deduced. An amount of about 0.3 g of each sample was placed into a plastic container and the saturation magnetization was obtained as the averaged value of about 20 readings.

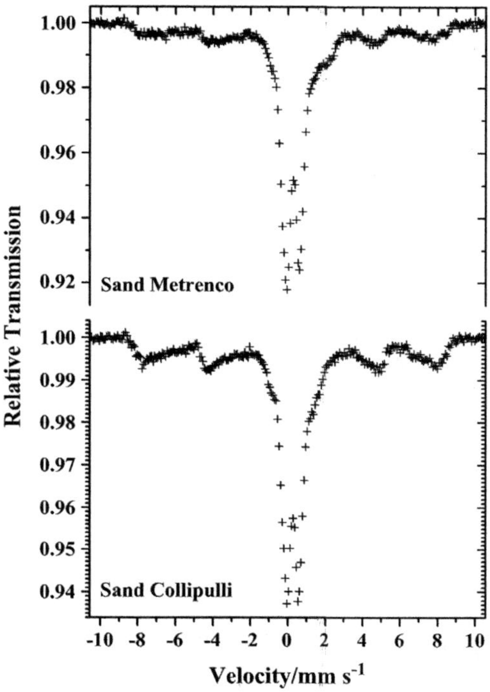

Fig. 1 298 K-^{57}Fe-Mössbauer spectra for the whole sand fractions from Metrenco and Collipuli soil materials

The Fenton activity was evaluated by measuring the formation of gaseous O_2 from the decomposition of 3.5 mol L^{-1} hydrogen peroxide, in a volumetric gas system [4] at 20 ± 1°C and pH 5.7, using 15 mg of the solid sample as catalyst. The volumetric gas system consisted of a glass ball as reactor containing 50 mL of the solid sample, 2 mL 30% *w*/*v* hydrogen peroxide and 5 mL of water. The reactor system was connected to an inverted burette immersed in a 2 L cylinder containing water. The oxygen formation was followed each minute by measuring the hydrogen peroxide volume decrease. An analytical blank without solid sample was measured in similar conditions.

The Mössbauer spectra were collected at 298 K in constant acceleration transmission mode with a ∼30 mCi ^{57}Co/Rh source. Data were stored in a 512-channel MCS memory unit, with Doppler velocities ranging between approximately ±10 mm s^{-1}. The experimental data were fitted using Lorentzian functions, with least-squares fitting procedure based on the NORMOS™ computer program.

3 Results and discussion

Magnetic portions of the sand fraction of Metrenco and Collipulli soil materials were found to be $\sigma = 31.3(1)$ J T^{-1} kg^{-1} and $\sigma = 34.0(3)$ J T^{-1} kg^{-1}, respectively. Comparing these saturation magnetization values with those for the pure magnetic phase, it would be possible to directly infer about its quantitative proportion in each soil sample. However, to identify, characterize and quantify the magnetic mineral(s) in such a complex natural mixture is a real experimental challenge and often requires

Table 1 Chemical composition (data expressed in mass % of the corresponding element, on an oxide basis) for the whole soil and sand (including its magnetic extract) and clay fractions from the Collipulli and Metrenco soil samples

Sample	Fraction	Fe_2O_3	Si_2O	Al_2O_3	TiO_2	MnO_2
Metrenco	Soil	12.5	42.9	10.4	1.8	nd
	Sand	16.9	51.9	19.0	2.4	nd
	Clay	11.9	56.9	10.0	1.7	nd
	Magnetic portion from sand	57.0	nd	nd	3.1	1.6
Collipulli	Soil	17.8	39.8	23.2	2.3	0.4
	Sand	14.6	43.2	23.1	2.0	0.3
	Clay	15.8	32.5	30.2	1.2	0.2
	Magnetic portion from sand	56.7	nd	nd	3.6	1.5

nd not determined

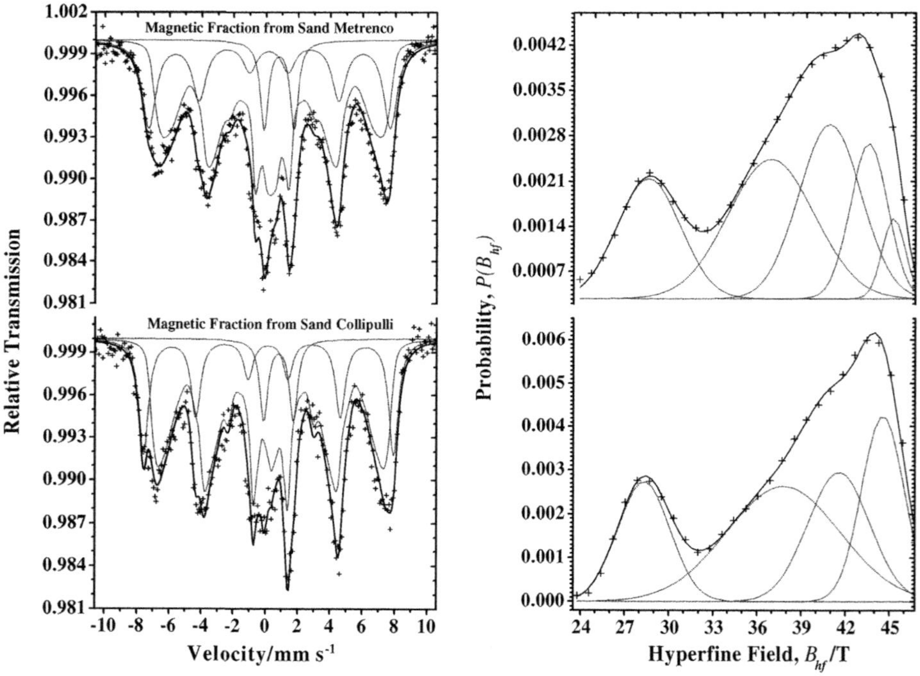

Fig. 2 298 K-^{57}Fe-Mössbauer spectra and corresponding profiles of hyperfine distributions for the magnetic portions from sand fractions of Metrenco and Collipuli soil materials

information from several experimental techniques. As it would be expected for most soil samples, room temperature Mössbauer analyses revealed rather complex patterns for both sand fractions, Metrenco and Collipulli, with an intense central doublet and low-intensity and broad resonance lines (Fig. 1). From results of chemical composition for the soil, sand and clay fractions, silicon appears in the highest proportion (Table 1). For their corresponding magnetic fractions, the amount of iron is, as expected, significantly higher than in the whole sand soil fractions. Mössbauer spectra for the corresponding magnetic portions reveal that resonance signals due

Table 2 298 K-Mössbauer fitted parameters for the magnetic portions from sand fractions of Metrenco and Collipuli soil materials

Sample	Assignment	δ/mm s^{-1}	ε, Δ/mm s^{-1}	Γ/mm s^{-1}	B_{hf}/T	AR/%
Metrenco	Mt-A	0.29 (2)	0.03 (3)	1.03 (6)	46.4 (1)	26 (2)
	Mt-B-Dist	0.48 (1)	0[a]	0.31[a]	27.8 (2)	13 (1)
		0.49 (1)	0[a]	0.31[a]	36 (6)	21 (1)
		0.49 (1)	0[a]	0.31[a]	41 (1)	19 (2)
		0.50 (1)	0[a]	0.31[a]	43.4 (6)	10 (1)
		0.50 (1)	0[a]	0.31[a]	45.2 (4)	3 (1)
	Paramagnetic-Fe^{2+}	0.91 (1)	1.84 (3)	0.60[a]		8 (1)
Collipulli	Mt-A	0.28 (1)	0.02 (2)	0.71 (4)	48.0 (1)	24 (2)
	Mt-B-Dist	0.46 (2)	0[a]	0.31[a]	28.3 (1)	12 (1)
		0.45 (2)	0[a]	0.31[a]	38 (4)	26 (3)
		0.45 (2)	0[a]	0.31[a]	41 (1)	16 (1)
		0.44 (2)	0[a]	0.31[a]	44.6 (3)	15 (1)
	Paramagnetic-Fe^{2+}	0.92 (2)	1.82 (4)	0.60[a]		7 (1)

A magnetic hyperfine spectrum fitted with Lorentzian-shaped lines, *B-Dist* model-independent hyperfine field distribution (for this case, B_{hf} = hyperfine field at maximum probability), δ isomer shift relative to αFe, ε quadrupole shift, Δ quadrupole splitting, Γ line width, B_{hf} magnetic hyperfine field, *AR* relative subspectral area
[a]Fixed parameter during the fitting procedure

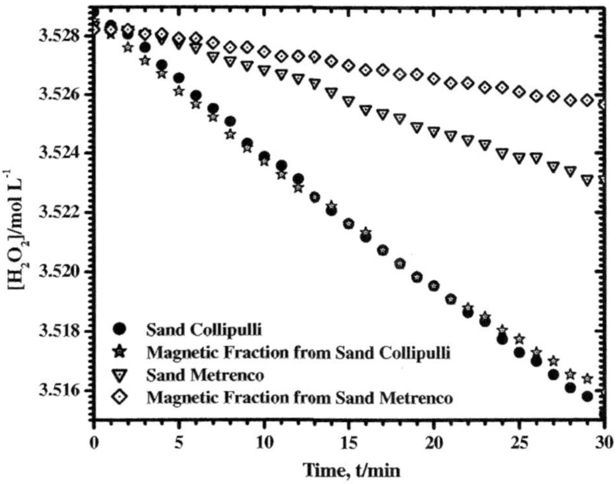

Fig. 3 Kinetic curves for hydrogen peroxide decomposition by using different Fenton catalysts from the two Ultisol materials

to the magnetic hyperfine structure are clearer more intense then those obtained for the whole sand fractions (Fig. 2). A ferrous doublet also appears in spectra for samples from both sand fractions. Corresponding fitted parameters are presented in Table 2. Non-stoichiometry magnetite has been previously identified by Mössbauer analysis, supporting optical microscope image observations that reveal typical spinel grains assignable to magnetite, in this magnetic portion from the sand fraction of this Collipuli soil material [10]. Microprobe analysis also revealed a relatively wide range of chemical composition of the spinel phases: Mg–Ti–Mn–magnetite formulae were allocated from some averaged data obtained from optically similar probing zones on individual grains [10].

Figure 3 shows results obtained for the decomposition measurements of hydrogen peroxide for the Metrenco and Collipulli soil materials, carried out in a volumetric gas system. The hydrogen peroxide decomposition rate for the sand and magnetic fractions from Collipulli soils are found to be approximately 8.0 and 6.0 times that of the blank, respectively, meaning that a significant decomposition rate of H_2O_2 occurred, by using those fractions as heterogeneous Fenton catalysts. For all samples it was taken in account that the free oxygen volume is due only to the oxidation of the hydrogen peroxide decomposition.

The decomposition rates either with whole sand fractions or its corresponding magnetic portion for the Collipulli are significantly higher than those for the Metrenco soil samples. The reasons for these kinetic results will from now be analyzed on basis of the elemental chemical composition for both soils, as there is no clear difference on their iron oxides mineralogy, basing on the Mössbauer patterns only for the two magnetic portions. In spite of the occurrence of Ti and Mn in these soil materials (Table 1), the catalytic activity in the magnetic portion of sand materials is being accounted to the iron-rich spinels only. From the allocated formulae based on the chemical microprobe analysis [10], Ti and Mn presumably contribute very few to the catalytic activities, as they represent only 2.9% and 0.8%, respectively, of all Fe content in the magnetic portion from the sand fraction of the Collipulli sample. Chemical Analysis for these magnetic portions reveals that the Ti and Mn contents are about 6.0% and 2.7%, respectively, in both samples (Collipulli and Metrenco) relatively to the corresponding iron contents (Table 1).

4 Conclusions

The sand fraction and its magnetically concentrated extract from the Collipulli soil is by far more efficient than those from Metrenco, as Fenton catalysts, as it could be evaluated, from the decomposition of H_2O_2. The saturation magnetization and Mössbaer patterns for both soil materials are nearly comparable. Reasons for differences in the catalytic behavior on the decomposition of H_2O_2 are now being explored on basis of a more detailed chemical investigation of these samples.

Acknowledgements Work supported by FONDECYT (Chile; grant 1040272), DICYT-USACH (Chile), MIDEPLAN (Chile; grant Presidente de la República), Prosul-CNPq (Brazil; grant 490132/2006-5), FAPEMIG (Brazil), Fulbright (USA)/CAPES (Brazil), and the ERSP program (U.S. DOE/BER, grant DE-FG02-7ER64374, NSF grant EAR 01-26308).

References

1. Segura, C., Zaror, C., Mansilla, H.D., Mondaca, M.A.: Imidacloprid oxidation by photo-Fenton reaction. J. Hazard. Mater. **150**(3), 679–686 (2008)
2. Bigda, R.J.: Consider Fenton's chemistry for wastewater treatment. Chem. Eng. Prog. **91**(12), 62–66 (1995)
3. Kabita, D., Subrata, M., Sekhar, B., Basab, Ch.: Chemical oxidation of methylene blue using a Fenton-like reaction. J. Hazard. Mater. **B84**, 57–71 (2001)
4. Pereira, M.C., Tavares, C.M., Fabris, J.D., Lago, R.M., Murad, E., Criscuolo, P.S.: Characterization of a tropical soil and a waste from kaolin mining and their suitability as heterogeneous catalysts for Fenton and Fenton-like reactions. Clay Miner. **42**, 299–306 (2007)

5. Moya, S.A., Flores, A., Escudey, M., Venegas, R., Castello, G., Latorre, R.: Mössbauer spectroscopy for the characterization of soils used as catalysts. Hyperfine Interact. **67**, 581–586 (1991)
6. Pizarro, C., Escudey, M., Moya, S.A., Fabris, J.D.: Iron oxides from volcanic soils as potential catalysts in the water gas shift reaction. In: American Institute of Physics, AIP, vol. 765, pp. 56–59. Melville, NY (2005)
7. Pizarro, C., Escudey, M., Fabris, J.D.: Influence of organic matter on the iron oxide mineralogy of volcanic soils. Hyperfine Interact. **148/149**, 53–59 (2003)
8. Borquez, M.: Extracción de óxidos de hierro no cristalinos en suelos volcánicos a través de métodos químicos disolutivos. Chemical Engineering Thesis, p. 73, Universidad de Santiago de Chile (2005)
9. Coey, J.M.D., Cugat, O., McCauley, J., Fabris, J.D.: A portable soil magnetometer. Rev. Fis. Apl. Instrum. **7**(1), 25–29 (1992)
10. Pizarro, C., Escudey, M., Fabris, J.D., Almeida, A.B.: Iron-rich spinel phases from the sand fraction of three Chilean soils developing on volcanic materials. Commun. Soil Sci. Plant Anal. **32**(17–18), 2741–2754 (2001)

Hyperfine Interact (2010) 195:43–48
DOI 10.1007/s10751-009-0099-1

Hematite reaction with tar to produce carbon/iron composites for the reduction of Cr(VI) contaminant

Fabiano Magalhães · Márcio César Pereira · José Domingos Fabris · Sue Ellen Costa Bottrel · Alejandro Amaya · Nestor Mogliazza · Rochel Montero Lago

Published online: 1 October 2009

Abstract In this work, highly reactive carbon–iron composites were prepared using a waste, i.e. tar, as carbon precursor and a simple iron oxide, i.e. hematite. Tar was impregnated on Fe_2O_3 with different tar/hematite weight ratios of 1:1; 2:1 and 4:1, and thermally treated under N_2 atmosphere (400°C, 600°C and 800°C). Mössbauer, XRD and magnetization measurements suggested that treatment at 400°C and 600°C produces Fe_3O_4 but treatment at 800°C produced mainly Fe°. Raman and TG analyses of the different composites suggested the formation of carbon contents of 18, 24 and 32 wt.% as amorphous and graphitic highly dispersed on the Fe surface. The composites obtained at 800°C showed high efficiency to reduce Cr(VI) as CrO_4^{2-} in aqueous medium with much better results compared to finely ground commercial Fe°.

Keywords Iron oxides · Carbon · Mössbauer spectroscopy · Cr(VI) contamination

1 Introduction

Iron is a naturally occurring material and is in abundance in the earth's crust. This metal has a very interesting chemical because of its magnetic properties, electrical,

F. Magalhães
Departamento de Ciências Exatas, Universidade Federal de Alfenas, Minas Gerais 37.130-000, Alfenas, Brazil
e-mail: fabiano@unifal-mg.edu.br

M. C. Pereira · J. D. Fabris · S. E. Costa Bottrel · R. M. Lago (✉)
Departamento de Química, ICEx, Universidade Federal de Minas Gerais, Belo Horizonte, 31270-901, Minas Gerais, Brazil
e-mail: rochel@ufmg.br

A. Amaya · N. Mogliazza
Facultad de Química, Universidad de la República Oriental del Uruguay, cc1157, Montevideo, Uruguay

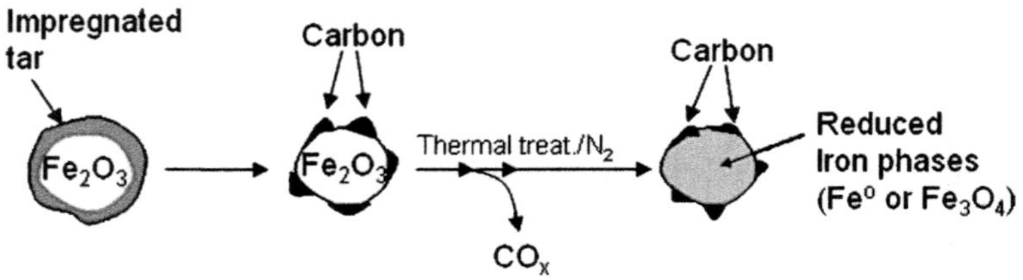

Fig. 1 Schematic representation of the reduced Fe/carbon composites prepared from tar and Fe_2O_3

physical, chemical and morphological. Is an important reducing agent due to its low toxicity, cost and availability. For these reasons, Fe° has been extensively used in several environmental remediation processes, like permeable reactive barriers [1], reduction of different contaminants such as organochloro [2], nitroaromatics [3], dyes [4], Cr(VI), Pb(II), Ag(I) and Cd(II) [5, 6].

In this work, highly active reduction systems based on composites of carbon and different Fe phases, have been investigated for environmental remediation. These composites can be prepared from any available and low cost iron oxide, e.g. Fe_2O_3, from several origins (synthetic, geological or from industrial mining waste), and a carbon source. In the present case, a pinus tar, a well known waste from the coal industry was used (Fig. 1). Upon thermal treatment, tar mixed with Fe_2O_3 decomposes on the iron oxide surface to produce elemental carbon, which readily reacts with Fe^{3+} to produce variably reduced chemical species, according to the simplified reaction equations:

$$0.5\,C + 3Fe_2O_3 \rightarrow 2Fe_3O_4 + 0.5\,CO_2 \quad (1)$$

$$0.5\,C + Fe_3O_4 \rightarrow 3FeO + 0.5\,CO_2 \quad (2)$$

$$0.5\,C + FeO \rightarrow Fe + 0.5\,CO_2 \quad (3)$$

2 Experimental

The hematite was obtained by thermal treatment of $Fe(NO_3)_3{\cdot}9H_2O$ (Aldrich) until 400°C for 3 h. The pinus tar pre-treated at 100°C (to remove volatile compounds) under vacuum was dissolved in acetone and impregnated on the Fe_2O_3. The composites were prepared with different tar/hematite weight ratios of 1:1; 2:1 and 4:1. The tar/hematite composites were separated into three parts treated at three different temperatures 400°C, 600°C and 800°C under N_2 flow (50 mL min^{-1}). After treatment the composites were stored in N_2 atmosphere. Mössbauer spectra were collected in a constant acceleration transmission mode with a 50 mCi Co57/Rh source. A spectrometer equipped with a transducer (CMTE model MA250) controlled by a linear function driving unit (CMTE model MR351) was used to obtain the spectra at a temperature of 298 K. Data were store in a 512 channel-MCS memory unit, with the Doppler velocity ranging between about ±10 mm s^{-1}. Mössbauer isomer shifts are quoted relative to α-Fe. The surface area was determined by the B.E.T. method using a 22 point N_2 adsorption/desorption in an Autosorb 1 Quantachrome. Magnetization measurements were carried out in a portable magnetometer with

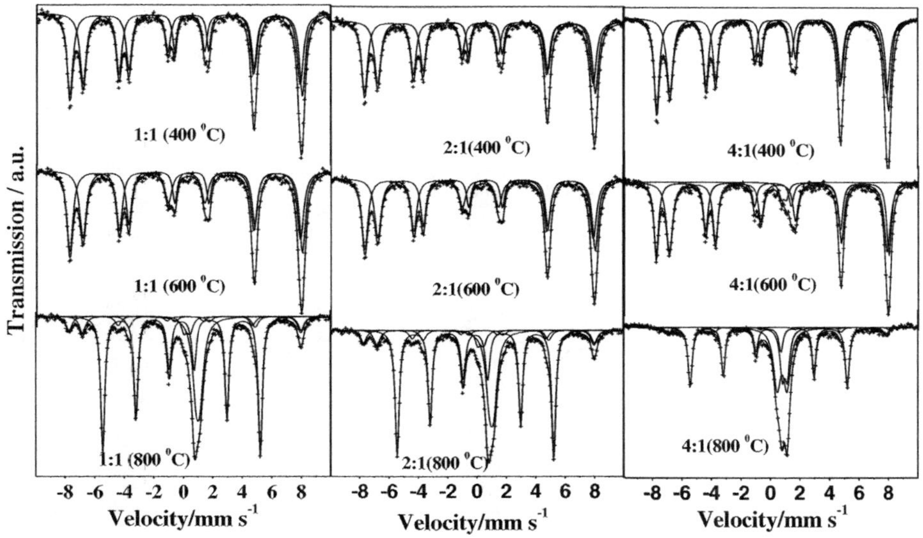

Fig. 2 Mössbauer spectra of the tar/iron oxide 1:1, 2:1, and 4:1 composites treated at different temperatures

magnetic field of 0.3 T calibrated with Ni metal. The reduction of Cr(VI) was performed using 5 mL solution 50 ppm of Cr(VI) with 60 mg of composite at 28 ± 2°C. During the reaction, a magnet was used to retain the solid while aliquots of 0.1 mL were collected and Cr(VI) was complexed with 1.5-diphenylcarbazide (5 g L^{-1} in acetone) in 25 mL pH 1 (H_2SO_4). The concentration of the red complex formed was determined by spectrophotometric analyses at $\lambda = 545$ nm in an UV–Vis Analyser 800 M spectrophotometer. Cr(VI) reductions were also carried out with reactions with commercial metallic iron (Synth) and were used to compare with the composites activity.

3 Results and discussion

Figure 2 shows the Mössbauer spectra (MS) obtained to all treated composites. MS for the composites treated at 400°C is composed of a signal related to magnetite suggesting that Fe_2O_3 is reduced by the carbon formed from tar (Eq. 1). At 600°C Fe_3O_4 with small amounts of wüstite (FeO) are produced. On the other hand, treatment of composites at 800°C produced large amounts of metallic iron with small amounts of magnetite and wüstite (Fig. 2). MS also suggested that at 800°C besides α-Fe, small amounts of γ-Fe were observed. Detailed XRD investigation confirmed all Mössbauer results.

Figure 3 shows a general idea of phase compositions for the different composites treated at 400°C, 600°C and 800°C.

Saturation magnetization was obtained for the composites treated at 400°C, 600°C and 800°C. As Fe_2O_3 is reduced to Fe_3O_4 (400°C and 600°C) a magnetization of 60–70 J K^{-1} T^{-1} was obtained. When Fe° is produced, the magnetization tends to

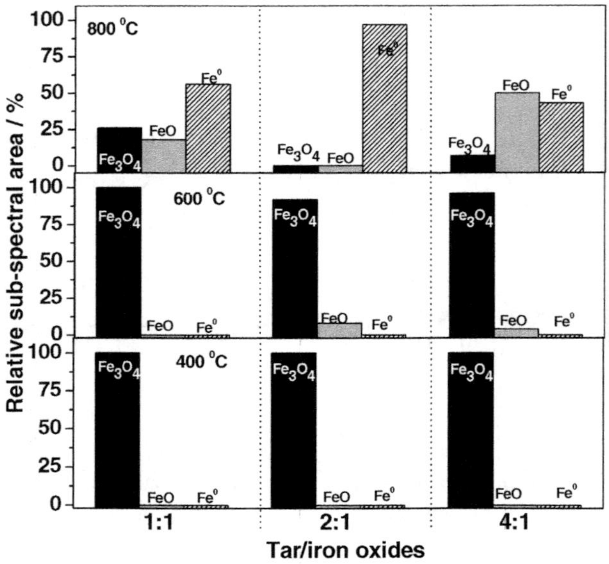

Fig. 3 Phase compositions obtained from the MS data for the different composites treated at 400°C, 600°C and 800°C

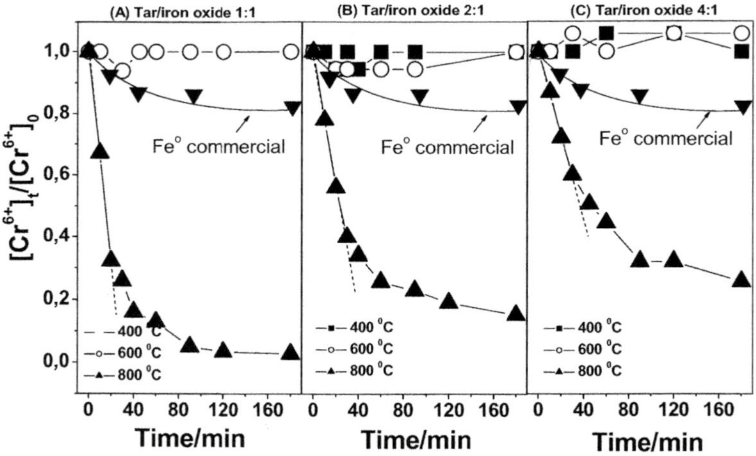

Fig. 4 Cr(VI) reduction by tar/iron oxide composites **a** 1:1, **b** 2:1 and **c** 4:1 treated at 400°C, 600°C and 800°C

increase, reaching values of 139 J K^{-1} T^{-1} since pure standard Fe° shows higher saturation magnetization of 180 J K^{-1} T^{-1} compared to Fe_3O_4 (100 J K^{-1} T^{-1}) [7].

BET surface areas were obtained from the N_2 adsorption isotherms on the treated tar/iron oxide composites. The original Fe_2O_3 showed low surface area of 4 m^2 g^{-1}. On the other hand, as the composites are thermally treated the surface area increases to 25, 36 and 90 m^2 g^{-1} (1:1 tar/iron oxide), 6, 10 and 86 m^2 g^{-1} (2:1 tar/iron oxide) and 2, 4 and 46 m^2 g^{-1} (4:1 tar/iron oxide) to composites treated on 400°C, 600°C and 800°C. This increase in surface area is likely related to the formation of carbon on

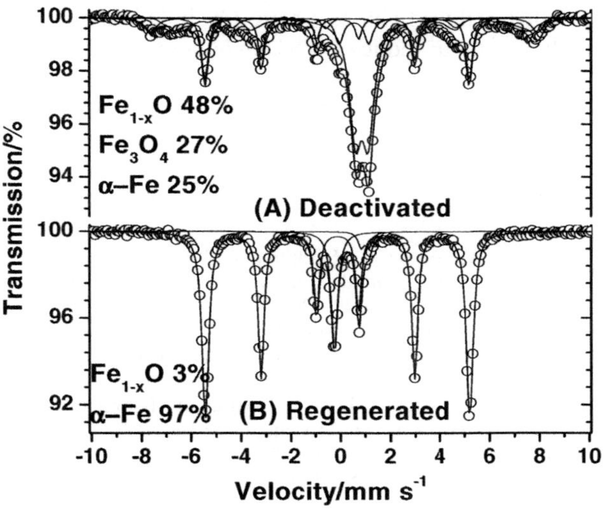

Fig. 5 Mösbauer spectroscopy of the 4:1 tar/iron oxides composite treated at 800°C after deactivation (**a**) and regenerated at 800°C on N_2 flow (**b**)

the Fe_2O_3 surface. Upon thermal treatment, part of this carbon reacts with the iron oxide developing some porosity and producing an increase in the surface area. In fact, DTF pore distribution suggests that after treatment at 800°C meso and macropores are created.

The thermally treated tar/iron oxide composites were used to reduce the Cr(VI), as CrO_4^{2-}, solution at an initial pH of 5. Figure 4 shows that composites treated at 400°C and 600°C are nearly inactive for the Cr(VI) reduction. Reactions with commercial Fe° powder showed slightly higher efficiency of 20% Cr(VI) conversion after 3 h. On the other hand, for the composites treated at 800°C a remarkable higher efficiency was observed with almost complete chromium reduction. After the reactions, all composites were exposed to a phosphate solution in order to desorb any Cr(VI) (CrO_4^{2-}) from the composite surface [8].

The deactivation of the composites was investigated by exposing the same composite sample consecutively to different $HCrO_4^-$ solutions (50 ppm) until no Cr(VI) was observed. These studies showed that after five consecutive reactions with Cr(VI) solutions, the composite strongly deactivates, likely due to an oxidation of Fe° on the composite surface. To investigate the surface modifications of the composite the MS was obtained after the 5th reaction (Fig. 5). The MS shows that the Fe° content in the composite decreases from 43% to 25% whereas the phase magnetite increased from 7% to 27%. Wüstite present in the composite at 50% did not change significantly after reaction. The deactivate composite was then regenerated at 800°C under N_2. After this thermal treatment, the activity of the composite was completely restored. The MS obtained for the regenerated composite (Fig. 5) shows a strong increase in the Fe° content to 96% which can explain the activity for Cr(VI) reduction.

4 Conclusion

The results obtained from Mössbauer spectroscopy and saturation magnetization showed that tar decomposes on the hematite surface and reacts with Fe_2O_3 to produce reduced phases, i.e. Fe_3O_4, FeO and Fe°. These phases, mainly Fe°, showed high activity to reduce Cr(VI) in aqueous medium.

Acknowledgements The authors are grateful to UFMG, FAPEMIG, CNPq and CAPES.

References

1. Wilopo, W., Sasaki, K., Hirajima, T., Yamanaka, T.: Immobilization of arsenic and manganese in contaminated groundwater by permeable reactive barriers using zero valent iron and sheep manure. Mater. Trans. **49**, 2265–2274 (2008)
2. Dries, J., Bastiaens, L., Springael, D., Agathos, S.N., Diels, L.: Competition for sorption and degradation of chlorinated ethenes in batch zero-valent iron systems. Environ. Sci. Technol. **38**, 2879–2884 (2004)
3. Sanchez, I., Stuber, F., Font, J., Fortuny, A., Fabregat, A., Bengoa, C.: Elimination of phenol and aromatic compounds by zero valent iron and EDTA at low temperature and atmospheric pressure. Chemosphere **68**, 338–344 (2007)
4. Tratnyek, P.G., Mam, S.: Reduction of azo dyes with zerovalent iron. Water Res. **34**, 1837–1745 (2000)
5. Yang, J.E., Kim, J.S., Ok, Y.S., Yoo, K.R.: Mechanistic evidence and efficiency of the Cr(VI) reduction in water by different sources of zerovalent irons. Water Sci. Technol. **55**, 197–202 (2007)
6. Li, X., Zhang, W.: Sequestration of metal cations with zerovalent iron nanoparticles: a study with high resolution X-ray photoelectron spectroscopy (HR-XPS). J. Phys. Chem., C **111**, 6939–6946 (2007)
7. Schwertmann, U., Cornell, R.M.: Electronic, electrical and magnetic properties. In: Schwertmann, U., Cornell, R.M. (eds.) The Iron Oxides: Structure, Properties, Reactions, Occurrence and Uses, pp. 103–125. VCH, New York (1996)
8. Zayed, A.M., Terry, N.: Chromium in the environment: factors affecting biological remediation. Plant Soil **249**, 139–156 (2003)

Hyperfine Interact (2010) 195:49–54
DOI 10.1007/s10751-009-0100-z

Supported iron based redox systems for hydrogen production and storage from ethanol

M. G. Rosmaninho · L. R. Souza · G. M. Gomes · R. F. Zica · J. S. Nascimento · M. C. Pereira · J. D. Fabris · J. D. Ardison · F. C. C. Moura · R. M. Lago · M. H. Araujo

Published online: 30 September 2009

Abstract Hydrogen production using ethanol and Fe_2O_3/support in a redox cycle was investigated. The composites were prepared by impregnation of Al_2O_3 and SiO_2 with $Fe(NO_3)_3$, with different proportions of iron, i.e. 10, 30 and 50 wt.%, calcinated at 450°C and characterized by Mössbauer spectroscopy, XRD, SEM, BET and TG. The redox cycle to produce and/or store hydrogen is a two step process (1) initially the ethanol is used to reduce the iron oxide to Fe°; (2) and when H_2 is needed, Fe° reacts with H_2O to produce CO-free hydrogen, and the iron oxide is reduced again to Fe° making this system cyclic. After the reactions it was interesting to observe that ethanol can directly reduce the iron oxide to produce metallic iron, with carbon deposition and iron–carbon as side product. Preliminary results indicate that it is possible to perform multiple redox cycles with the supported iron oxide without deactivation.

Keywords Hydrogen production · Hydrogen storage · Ethanol · Supported iron oxide

M. G. Rosmaninho (✉) · L. R. Souza · J. S. Nascimento · M. C. Pereira · J. D. Fabris · R. M. Lago · M. H. Araujo
Departamento de Química; ICEx, Universidade Federal de Minas Gerais (UFMG), Belo Horizonte, MG, Brazil
e-mail: mgrosmaninho@ufmg.br

G. M. Gomes · R. F. Zica
Departamento de Engenharia Química; Escola de Engenharia, Universidade Federal de Minas Gerais, Belo Horizonte, MG, Brazil

J. D. Ardison
Centro Nacional de Desenvolvimento da Tecnologia Nuclear (CDTN), Belo Horizonte, MG, Brazil

F. C. C. Moura
Departamento de Química; ICEB, Universidade Federal de Ouro Preto (UFOP), Ouro Preto, MG, Brazil

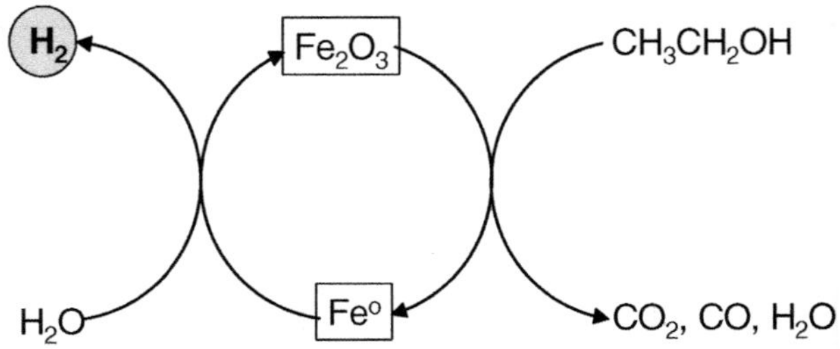

Fig. 1 Iron-based redox system to hydrogen production from ethanol

1 Introduction

Hydrogen, the "fuel of the future", will probably replace fossil energy sources in few years, with higher efficiency and several environmental advantages [1–3]. Today, the large-scale use of H_2 faces two main challenges: its production and storage [2, 4–7].

Currently, for the production of H_2, the steam methane reforming is the largest and generally the most economical way, however, producing hydrogen from fossil fuels makes H_2 an indirect source of CO_2. Therefore, the real opportunity for the future is the non-fossil fuel approaches for the production of H_2. For hydrogen storage, the use of solid state materials is considered to be one of the most efficient and secure and in accordance with the International Energy Agency (IEA) an ideal solid should store at least 5 wt.% and 50 kg H_2/m^3 of H_2 and the supply and storage should work in a cyclic process [8–10]. To the best of our knowledge, no materials show all these properties. Otsuka et al. have proposed the use of a reversibly redox system to store and supply hydrogen to fuel cell based on the reduction of iron oxide by hydrogen. The metallic iron produced could be easily stored and, when necessary, the iron could be re-oxidized by water, producing pure hydrogen [11–15].

In this work, we present an innovative method for H_2 production using ethanol and Fe°/Fe_2O_3 redox system. This is a two step process (Fig. 1): (1) initially the ethanol is used to reduce the iron oxide to Fe°; (2) when H_2 is needed, Fe° reacts with H_2O to produce CO-free hydrogen, and the iron oxide is reduced again to Fe° making this system reversible. In order to produce a more reactive system, Fe_2O_3 was dispersed on Al_2O_3 and SiO_2, which have a high surface area. The samples were submitted to Temperature Programmed Reaction (TPRe) cycles, using a mixture of $EtOH/N_2$ for reduction and H_2O/N_2 for re-oxidation, from 30°C to 900°C.

2 Experimental

2.1 Preparation of Fe_2O_3/Al_2O_3 and Fe_2O_3/SiO_2

The composites were prepared by impregnation of Al_2O_3 or SiO_2 (Aldrich) with $Fe(NO_3)_3$ in water, slowly dried at 60°C and calcinated in a tubular furnace (Lindberb Blue M) at 10°C min^{-1} until 450°C and held in this temperature for 4 h. The amount of iron was calculated to produce materials with 10, 30 and 50 wt.% of supported hematite.

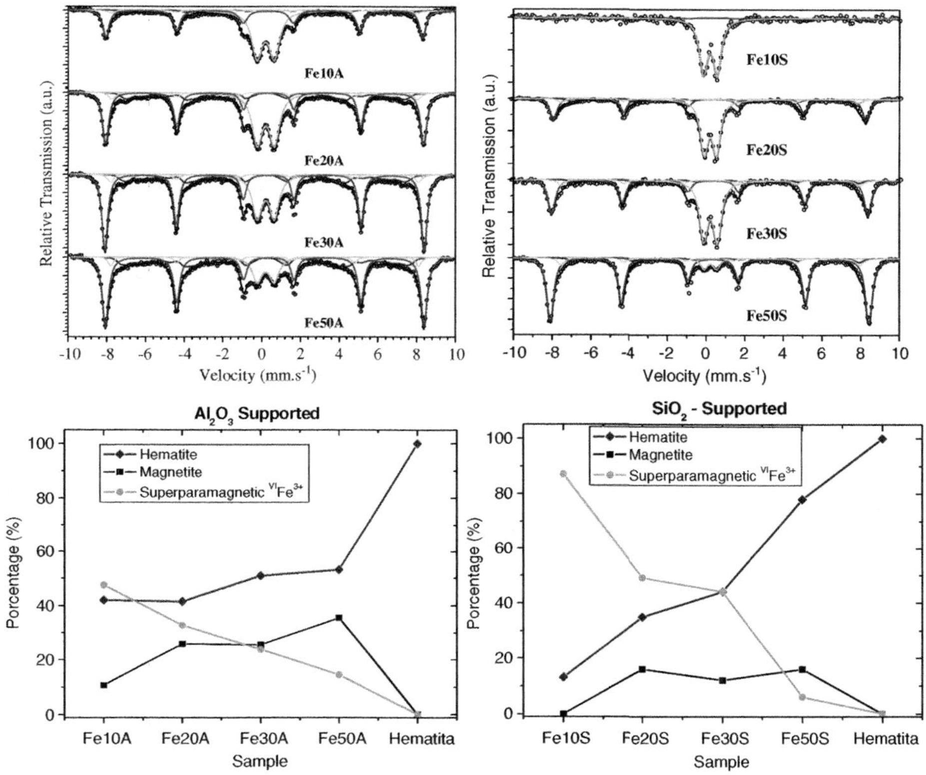

Fig. 2 Mössbauer spectra and relative area of the iron phases for the composites with alumina (*left*) and silica (*right*) before TPRe

2.2 Temperature programmed reactions

The reactions with H_2 were carried out with a flow of 5% H_2/N_2 (30 mL min^{-1}). The ethanol used for the reduction reaction was kept at 0°C. When water was used it was kept at 25°C. In both reactions N_2 (30 mL min^{-1}) was used as carrier gas.

The reactions between ethanol and the iron oxide composites (30 mg) were carried out in a tubular furnace between 30°C and 900°C, with a heating rate of 5°C min^{-1} or 10°C min^{-1}. The gas products were analyzed by a Shimadzu 17A gas chromatograph equipped with a flame ionization detector (FID) or by a ChemBET3000 TPR of the Quantachorme Instruments with a thermal conductivity detector (TCD).

2.3 Characterization

Mössbauer spectroscopy experiments were carried out in a spectrometer CMTE model MA250 with a $^{57}Co/Rh$ source at room temperature using α-Fe as a reference. The powder XRD data were obtained in a Philips 1830 equipment using Cu Kα radiation scanning from 2° to 75° at a scan rate of 4° min^{-1}. Scanning electron microscopy (SEM) analysis was done using a Jeol JSM840A. The surface area was determined by the B.E.T. method using a 22 cycles N_2 adsorption/desorption in an

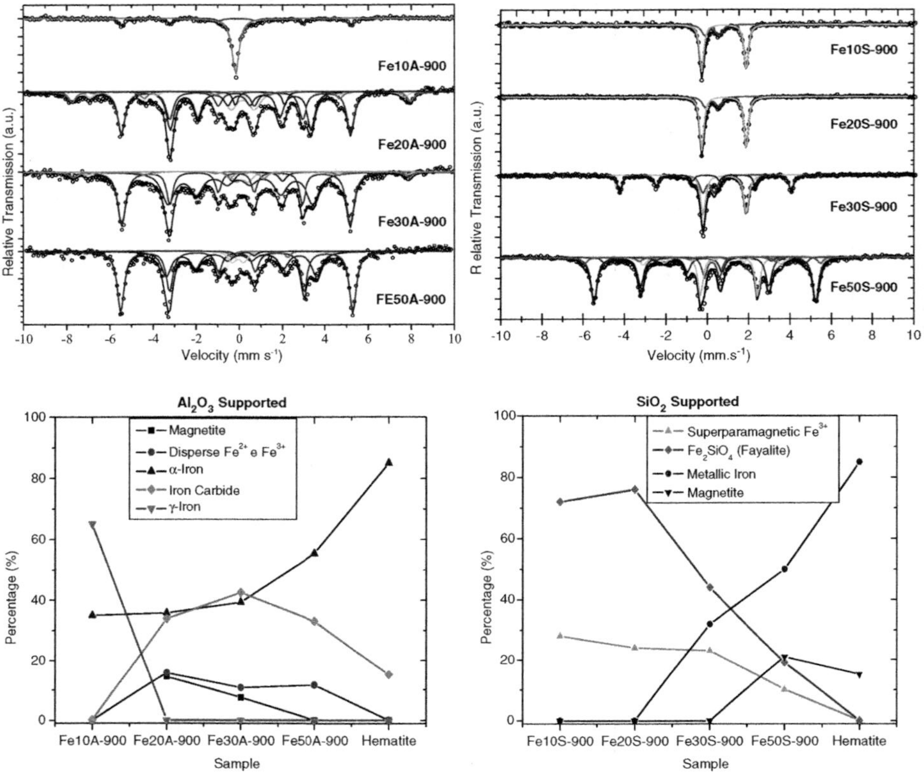

Fig. 3 Mössbauer spectra and relative area of the iron phases for the composites with alumina (*left*) and silica (*right*) after TPRe

Autosorb 1 Quantachrome instrument. TG analyses were carried out in a Shimadzu TGA60 simultaneous TG/DTA.

3 Results and discussion

XRD and Mössbauer Spectroscopy (Fig. 2), for the composites before the reactions, show mainly hematite and the superparamagnetic $^{VI}Fe^{3+}$ phase. The amount of this superparamagnetic phase, attributed to nanometric hematite particles, decreases with the increase of the total amount of iron on the composites, and it is greater on the SiO_2 support, which has higher surface area.

α-Iron and iron carbide are the main species observed in the Mössbauer spectra for the Al_2O_3 composites, after the first reduction with ethanol at 900°C (Fig. 3). Small amount of disperse superparamagnetic Fe^{2+} and Fe^{3+} could also be observed, as well γ-iron (probably stabilized by carbon) on the sample with 10% of initial iron oxide content. For the samples supported on SiO_2 a fayalite (Fe_2SiO_4) phase was observed after the reaction. It was also observed that the amount of metallic iron phase increase with the increase of the total percentage of iron on the sample, while the oxidized phases (including fayalite and iron carbide) decrease with the increase

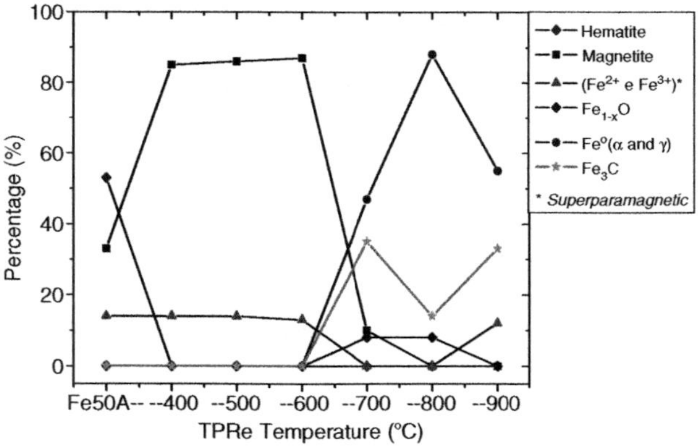

Fig. 4 Changes on the iron phases after TPRe of the 50 wt.% Fe_2O_3/Al_2O_3 composites

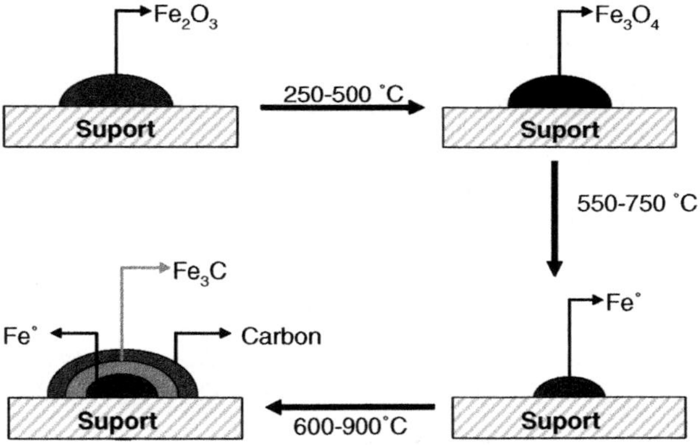

Fig. 5 A proposal for the changes on the iron phases during the TPRe with ethanol

of iron percentage. XDR shows carbon deposition over the composites which was confirmed by SEM and TG analyses. The others side-products observed during the TPRe with ethanol were ethene and small amount of ethylic ether (produced by intra and inter dehydration of ethanol, respectively).

The changes on the iron phases after TPRe of the 50 wt.% Fe_2O_3/Al_2O_3 composites are shown in Fig. 4. Fe° is produced at temperatures over 600°C with the best yield at 800°C. In Fig. 5 a proposal for the changes on the iron phases during the TPRe with ethanol is presented.

After one redox cycle with the composites (reduction with ethanol and oxidation with water) Fe_2O_3/support and/or Fe_3O_4/support were observed. To better understand this process several redox cycles will have to be carried out for these composites, considering that with pure hematite five redox cycles were successfully obtained with only small deactivation.

4 Conclusions

It was interesting to observe that ethanol can directly reduce the iron oxide to produce metallic iron, with carbon deposition and iron–carbon as side products.

When SiO_2 was used as support a high concentration of fayalite was observed, especially for the composites with less than 50% of iron oxide. This silicate phase could be stable during the oxidation step of the cycle, deactivating the composite after some cycles.

Preliminary results indicate that it is possible to perform multiple redox cycles with the supported iron oxide without deactivation. More detailed studies have been carried out to determinate the life-time and the efficiency of these composites for hydrogen production and/or storage.

Acknowledgements The authors are grateful to FAPEMIG and CNPq for the financial support.

References

1. Armor, J.N.: The multiple roles for catalysis in the production of H_2. Appl. Catal. A Gen. **176**, 159–176 (1999)
2. Cook, B.: Introduction to fuel cells and hydrogen technology. Eng. Sci. Educ. J. **11**, 205–216 (2002)
3. Hacker, V., Faleschini, G., Fuchs, H., Fankhauser, R., Simader, G., Ghaemi, M., Spreitz, B., Friedrich, K.: Usage of biomass gas for fuel cells by the SIR process. J. Power Sources **71**, 226–230 (1998)
4. Ogden, J.M., Steinbugler, M.M., Kreutz, T.G.: A comparison of hydrogen, methanol and gasoline as fuels for fuel cell vehicles: implications for vehicle design and infrastructure development. J. Power Sources **79**, 143–168 (1999)
5. Prigent, M.: On board hydrogen generation for fuel cell powered electric cars—a review of various available techniques. Rev. Inst. Fr. Pet. **52**, 349–360 (1997)
6. Thomas, C.E., Kuhn, I.F., James, B.D., Lomax, F.D., Baum, G.N.: Affordable hydrogen supply pathways for fuel cell vehicles. Int. J. Hydrogen Energy **23**, 507–516 (1998)
7. Wendt, W., Götz, M., Linardi, M.: Tecnologia de Células a Combustível. Quim. Nova **23**, 538–546 (2000).
8. Hacker, V.: A novel process for stationary hydrogen production: the reformer sponge iron cycle (RESC). J. Power Sources **118**, 311–314 (2003)
9. Grochala, W., Edwards, P.P.: Thermal decomposition of the non-interstitial hydrides for the storage and production of hydrogen. Chem. Rev. **104**, 1283–1315 (2004)
10. Schlapbach, L., Zuettel, A.: Hydrogen-storage materials for mobile applications. Nature (London, United Kingdom) **414**, 353–358 (2001)
11. Otsuka, K., Kaburagi, T., Yamada, C., Takenaka, S.: Chemical storage of hydrogen by modified iron oxides. J. Power Sources **122**, 111–121 (2003)
12. Otsuka, K., Yamada, C., Kaburagi, T., Takenaka, S.: Hydrogen storage and production by redox of iron oxide for polymer electrolyte fuel cell vehicles. Int. J. Hydrogen Energy **28**, 335–342 (2003)
13. Hui, W., Takenaka, S., Otsuka, K.: Hydrogen storage properties of modified fumed-Fe-dust generated from a revolving furnace at a steel industry. Int. J. Hydrogen Energy **31**, 1732–1746 (2006)
14. Takenaka, S., Nomura, K., Hanaizumi, N., Otsuka, K.: Storage and formation of pure hydrogen mediated by the redox of modified iron oxides. Appl. Catal. A Gen. **282**, 333–341 (2005)
15. Wang, H., Wang, G., Wang, X., Bai, J.: Hydrogen production by redox of cation-modified iron oxide. J. Phys. Chem. C **112**, 5679–5688 (2008)

Hyperfine Interact (2010) 195:55–62
DOI 10.1007/s10751-009-0101-y

Mineralogical characterization of Greda clays and monitoring of their phase transformations on thermal treatment

E. Chavez Panduro · J. Bravo Cabrejos

Published online: 12 November 2009

Abstract The mineralogical characterization of two clay samples from the Central Andean Region of Peru, denominated White Greda and Red Greda, is reported. These clays contain the clay minerals mica and illite respectively. Both clays were treated thermally in an oxidising atmosphere under controlled conditions up to 1,100°C with the purpose of obtaining information about structural changes that may be useful for pottery manufacture. X-ray fluorescence was used for the elemental characterization of the samples and X-ray diffractometry was used to determine the collapse and formation of the mineral phases present in the samples caused by thermal treatment. At temperatures above 1,000°C it is observed the formation of spinel in the case of White Greda and of hematite, corundum and cristobalite in the case of Red Greda. Room temperature transmission Mössbauer spectroscopy allowed the monitoring of the variation of the hyperfine parameters with the thermal treatment temperature; In the case of the evolution of the quadruple splitting of the paramagnetic Fe^{3+} sites with temperature, in both clays, the analyses reproduced results such as the "camel back" curve shape, found by other workers (Wagner and Wagner, Hyperfine Interact 154:35–82, 2004; Wagner and Kyek, Hyperfine Interact 154:5–33, 2004).

Keywords Greda · Mica · Illite · X-ray diffractometry · Transmission Mössbauer spectroscopy · Thermal treatment

E. C. Panduro (✉) · J. B. Cabrejos
Facultad de Ciencias Físicas, Universidad Nacional Mayor de San Marcos, Apartado Postal 14-0149, Lima 14, Perú
e-mail: 04130127@unmsm.edu.pe

J. B. Cabrejos
e-mail: jbravoc@unmsm.edu.pe

1 Introduction

The effects of firing a clay sample can be studied by monitoring the changes in the hyperfine parameters. Mössbauer spectroscopy (ME) is the only technique that provides direct information about the content and the transformation of the different phases of iron containing minerals during the thermal treatment. The thermal treatments applied to the clay samples cause a sequence of chemical and structural modifications, such as dehydration, oxidation, dihydroxilation, decomposition, formation of new phases and vitrification, all of which in general improve some of the physical properties such as mechanical resistance, durability, etc. X-ray diffractometry (XRD), as a characterization technique, provides evidence about the early reactions in the process showing changes in the basal spacing of the philosilicates when water is removed from the interlayer; the more radical changes such as the fracture of the crystalline structure due to the loss of structural hydroxyls, and the formation of new phases at high temperature can also be monitored by DRX. ME sees selectively the structural sites occupied by iron atoms, which frequently are hard to detect by XRD. Several papers [1, 3, 4] have shown that the changes observed in the Mössbauer spectra of clay minerals caused by thermal treatment from 100°C to 400°C are due to the loss of hydroxyl radicals in mineral clays, which produces a distortion in the crystalline electric field which manifests itself in the increase of the electric field gradient (QUA) at the structural sites occupied by Fe^{3+} cations in the clay minerals. At higher temperatures, between 700°C and 800°C, the laminar structure collapses eventually into amorphous phases in order to form new crystalline structures and QUA starts to decrease.

2 Experimental methods

2.1 Previous preparation and measurements

The pH was measured of both samples as found. The clay fractions, <2 μm, were obtained by sedimentation. The textural class of the samples as found was obtained by the hydrometer method. Previous to the XRD analysis organic matter was eliminated from the samples using a conventional chemical technique. The sample was mounted in the sample holder using the glace slade method; this consists in preparing a solution of water and clay, and then allowed to precipitate on the sample holder; then the sample is dried at room temperature. This procedure produces a preferential orientation of the (00ℓ) planes of the clay minerals along the vertical direction.

2.2 Applied analytical techniques

For the elemental characterization by X-ray fluorescence an energy dispersive system, manufactured by Amptek, was used. This uses a Si solid state detector cooled to −30°C using the Peltier effect, and an X-ray source that uses a Ag cathode operated at 25 kV. The sample was prepared in the form of a compressed tablet 5 mm in diameter and 3 mm in thickness. The spectra were taken in air and room temperature. The energy detection threshold is about 1.3 keV. The program QXAS, available from the IAEA web page, was used for the qualitative analysis and calculation of the areas under the spectral peaks of the characteristic X-rays observed in the spectra.

Table 1 Results of the elemental quantitative analysis

Sample	K (%)	Ca (%)	Ti (%)	Mn (%)	Fe (%)	Zn (%)	Rb (%)
San Joaquin	2.030	1.890	0.342	0.054	3.500	0.011	0.010
White Greda	3.624	0.015	0.297	0.016	1.478	0.008	0.016
Red Greda	1.206	0.727	0.407	0.054	5.262	0.011	0.023

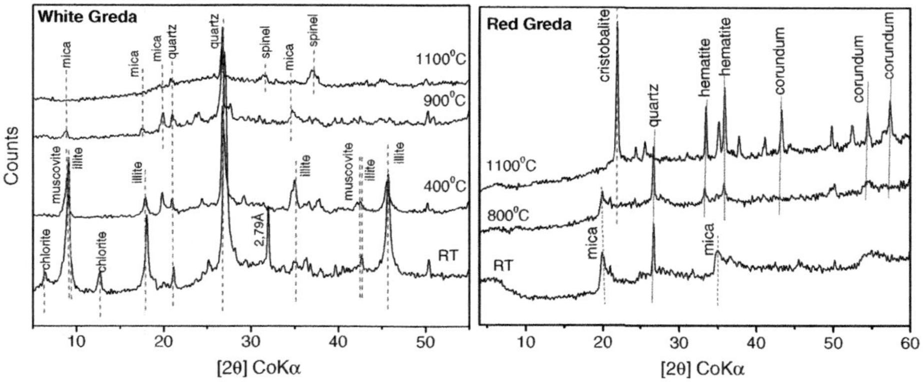

Fig. 1 Diffractograms of White Greda and Red Greda samples obtained at RT and after heating in an oxidizing atmosphere for 3 h, from 400°C to 1,100°C

For the structural analysis an X-ray diffractometer (RIGAKU, Miniflex model, Cu-Kα radiation, with a Ni filter and with no monochromator) was used. The operating conditions were: 30 kV, 15 mA, interval $4^\circ < 2\theta < 70^\circ$ with a 2θ step of 0.03° per step of 2 s. For the identification of the mineralogical components the data base JCPDS-ICDD (PCPDFWIN, the programa Origin 7.0 and the program Crystallographica were used.

For the characterization of the local structural order transmission Mössbauer spectroscopy was used. A radioactive ^{57}Co source in a rhodium matrix was used. The spectra were fitted using the program PCMOSII in its crystalline site version. The samples were placed in a circular sample holder with an internal diameter of 1.68 cm, which is placed between the source and the detector. For this sample holder 200 mg of sample in powder was required to obtain a total optical thickness of about 1.60 at 14.4 keV [7]. The spectra were taken at room temperature.

A tubular kiln with flow of oxygen was used for the thermal treatment. The range of temperatures used was from 100°C to 1,100°C. The temperature was increased at a rate of 1.7°C/min until the required maximum temperature was reached; then this was maintained during 3 h.

3 Results and discussion

3.1 Quantitative elemental analysis by X-ray fluorescence

X-ray fluorescence spectra were taken of the two Greda clays and of a reference soil sample, San Joaquin (SRM 2709, NIST), under the same irradiation and detection conditions. The accumulated counts in the spectra were normalized for the same time of irradiation taking into account the different counting rates and dead time effects. The resulting areas under the identified characteristic X-ray peaks, using the criteria that the areas are approximately proportional to the concentration of the identified elements in each of the clay samples and the reference sample, can be used to estimate the concentrations of the identified elements (Table 1).

3.2 Characterization by X-ray diffractometry

XRD was used for the characterization and identification of structural phases present in the White and Red Gredas with no thermal treatment. The White Greda sample was mounted using the glace slade method and the Red Greda was prepared by the powder method. The clay minerals found in the White Greda were muscovite, illite and chlorite, whereas mica was found in the Red Greda. Moreover, it is observed that due to the higher Fe content in the Red Greda, the diffractogram shows a background with a slight positive gradient due to the detection of characteristic X-rays of Fe produced by fluorescence. In Fig. 1 it is shown the diffractograms of the White Greda sample mounted using the glace slade method (sample at RT); using the latter method it was possible to identify the presence of chlorite. Moreover, there appears a reflection peak with an interplanar distance of 2.79 Å, enhanced by the preferential orientation, that has not been possible to identify.

3.3 Monitoring of the thermal treatment by X-ray diffractometry

After thermal treatment of White Greda in an oxidizing atmosphere up to 1,100°C, of the initial minerals present in the sample only quartz remains exhibiting a reduced intensity in its peaks. Due to dehydration of the clay structure, the interplanar distances (00ℓ) of muscovite and illite diminish up to 400°C; likewise, the line intensities were affected also. From this temperature on, the changes in the structures are due to dihydroxilation, by which water molecules are formed from the (OH) radicals in the samples and free O_2. At 900°C illite converts to mica (Fig. 1). Above 900°C the structure of mica collapses and there appears a new phase called spinel ($MgAl_2O_4$).

In the case of Red Greda no notable change was observed up to 800°C, at which the clay structure begins to collapse; mica remains present with a reduced intensity but still it has not collapsed as the (110) peak appears in the diffractogram. The recrystallization process appears more pronounced in the Red Greda above 1,100°C. Eventually, the Fe^{3+} cations will depart from the clay structure to form magnetic iron oxides which are observed as sextets in the Mössbauer spectra; at the end of the process these become the more prominent characteristic of the Mössbauer spectra in Red Greda. Moreover, it is also observed that as a product of recrystallization, silicon oxide changes to a new phase called cristobalite, and aluminium oxide (Al_2O_3), denominated corundum, is formed.

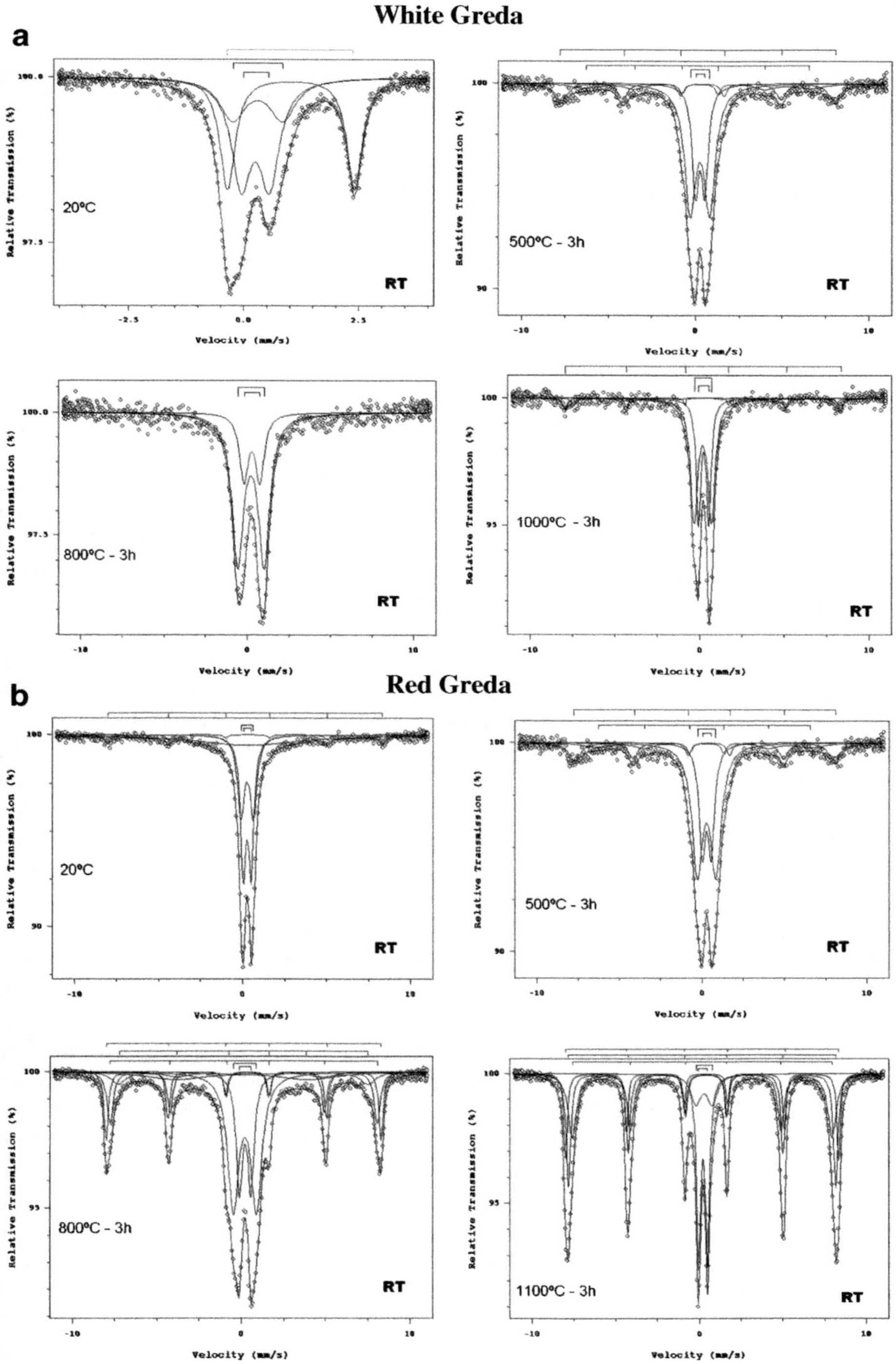

Fig. 2 **a** Mössbauer spectra at RT of White Greda samples thermally treated in an oxidizing atmosphere at different temperatures. **b** Mössbauer spectra at RT of Red Greda samples thermally treated in an oxidizing atmosphere at different temperatures

3.4 Characterization by Mössbauer spectroscopy

The results of the analyses by Mössbauer spectroscopy at RT of samples of White and Red Gredas, with no thermal treatment, are shown in Fig. 2a, b. In White Greda the spectrum shows two paramagnetic Fe^{3+} sites (with 60.0% of the total resonance absorption area) and a paramagnetic site that is assigned to illite (35.9%). In Red Greda the spectrum shows two paramagnetic Fe^{3+} sites (65.3%) and a paramagnetic Fe^{2+} site with a large quadrupolar splitting that can be associated with an amorphous phase in the clay (30.0%); in addition, it shows a magnetic site (4.7%) with a field of 50.82 T and it is assigned to hematite.

3.5 Monitoring of the thermal treatment by Mössbauer spectroscopy

The Mössbauer spectra of White Greda, with thermal treatment, is shown in Fig. 2a; in the temperature interval 100–400°C there is a decrease of the absorption area for the paramagnetic Fe^{2+} site due to oxidation of Fe. In the case of Fe^{3+} sites, their absorption areas increase up to 600°C (Fig. 3); from this temperature up to 800°C the hyperfine parameters and absorption areas behave almost constant; above this temperature a marked decrease in the absorption area of Fe^{3+} is observed (Fig. 3); this decrease is associated to the collapse of crystalline structures and the formation of new phases; at 900°C it is observed the appearance of a magnetic site associated hematite (Fig. 2a); then as the temperature temperature increases the absorption area associated with this magnetic site decreases from 20% to 5%; this is due to the new spinel phase (identified by XRD at 1,100°C in the White Greda). This behaviour is characteristic of hematite if the sample is free of iron oxides before it reaches the temperature of 800°C. For the case of Red Greda it is observed in Fig. 2b that as the temperature increases the absorption area of the paramagnetic site of Fe^{3+} increases; a similar behavior is observed for the parameter QUA (0.58–0.133 mm/s) for this Fe^{3+} site; which is due to the loss of water from the interlayer sheet and the dihydroxilation of the layers. These processes of collapse and formation of new structures undergo a transition state characterized by the presence of amorphous phases, which are associated to the appearance of undefined iron sites responsible for a relative absorption area less than about 20% in the case of Red Greda; then a reduction of QUA for Fe^{3+} is observed starting at 800°C (Fig. 2b), behavior associated to higher crystallinity. Starting at 300°C, as the temperature of thermal treatment increases, the relative area of the magnetic site increases from 4.7% to 71.74%.

In Fig. 4. it is observed that upon heating of the samples up to 100°C, the total absorption area diminishes by about 35% in the case of Red Greda, and in the case of White Greda it increases by about 36%; this behavior is related to a decrease of the f_A factor in the case of Red Greda and an increases in the case of White Greda. Figure 4 shows also the dependence of average line width of the Fe^{3+} doublets with the temperature of the thermal treatment; this dependence exhibits also a "camel back" curve shape and it is more pronounced in the case of the Red Greda.

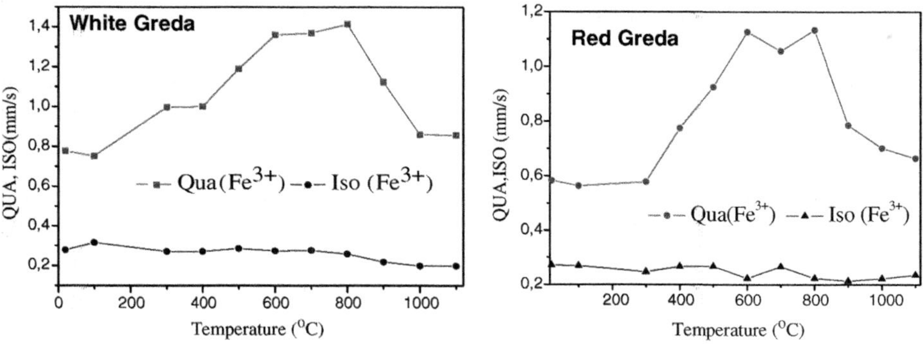

Fig. 3 Variation of hyperfine parameters with temperature for White Greda and Red Greda. *QUA* (Fe^{3+}) is the quadrupolar splitting, *ISO* (Fe^{3+}) is the isomeric shift

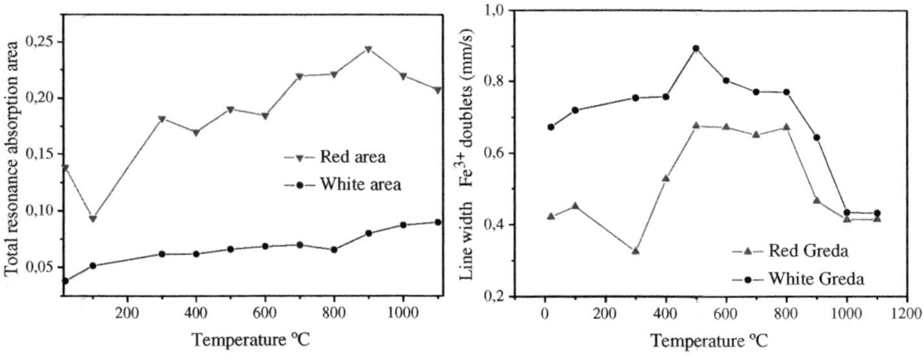

Fig. 4 *Left* variation of the total resonance absorption area as a function of the temperature of the thermal treatment for White Greda and Red Greda. *Right* variation of the line width of Fe^{3+} doublets with temperature for White Greda and Red Greda

4 Conclusions

The elemental analysis allows to deduce that the concentration of Fe in Red Greda is larger than in White Greda, but in a lower ratio than the respective total resonance absorption areas. This indicates that the effective f_A factor may be slightly larger in the White Greda than in the Red Greda. By XRD it is found that in the case of Red Greda, at high temperatures, hematite is formed and that quartz transforms into cristobalite. In the case of White Greda, at high temperature, it is observed the formation of spinel and hematite. The monitoring of the variation of the hyperfine parameter QUA for the Fe^{3+} sites with temperature in both clay samples reproduces a behavior obtained by other researchers that is known as the "camel back effect" [5, 6]. The evolution of the total resonance absorption areas with temperature shows an increase of the f_A factor of the clays after the thermal treatment and this takes place in a larger proportion in the case of Red Greda; this may indicate that the crystalline structures, after thermal treatment, become more rigid. The monitoring of the variation of the quadrupole splitting of the paramagnetic Fe^{3+} sites shows an increase up to about 800°C and then diminishes at higher temperatures. This behavior is related to structural changes that are produced in the samples during the first stage of the treatment which is followed by a stage of higher structural ordering associated to crystallization and vitrification.

Acknowledgements We acknowledge the support given to this project by the Research Initiation Program for undergraduate students in Physics which is administered by our Academic Vicerectorate at Universidad Nacional Mayor de San Marcos.

References

1. Yuanfu, H., Hongbo, H.: Mössbauer studies in Chinese archaeology. Hyperfine Interact. **150**, 33–50 (2004)
2. Carrol, D.: Clay minerals: a guide to their X-ray identification. The Geological Society of America Special Paper 126 (1970)
3. Wagner, F.E., Wagner, U.: Mössbauer spectra of clays and ceramics. Hyperfine Interact. **154**, 35–82 (2004)
4. Wagner, F.E., Kyek, A.: Application of Mössbauer spectroscopy to archaeology: introduction and experimental considerations. Hyperfine Interact. **154**, 5–33 (2004)
5. Moore, D.M., Reynolds, R.C.: X-ray Diffraction and the Identification and Analysis of Clay Materials, 2nd edn. Oxford University Press, Oxford (1997)
6. Brown, G.: The X-ray Identification and Crystal of Clay Mineral. Mineralogical Society, London (1961)
7. Bravo, J.A., Cerón, M.L., Fabián, J.: Optimization criteria in Mössbauer spectroscopy. Hyperfine Interact. **148/149**, 253–261 (2003)

Hyperfine Interact (2010) 195:63–68
DOI 10.1007/s10751-009-0102-x

Mössbauer spectroscopy study of a mineral sample from Oshno Hill, District of Chavín de Pariarca, Huanuco Region, Peru

A. Bustamante · D. Lovera · R. Quille ·
A. V. Arias · J. Quiñones

Published online: 30 September 2009

Abstract The analysis by X-ray diffraction of a mining sample collected from Oshno hill, which is located in the District of Chavín de Pariarca, Huamalies Province, Huanuco, Peru, indicates the presence of lepidocrocite (γ-FeOOH) and goethite (α-FeOOH). The room temperature Mössbauer spectrum (MS) doublet with broad lines displays hyperfine parameters corresponding to the presence of particles of iron hydroxides smaller than 100 Å in a superparamagnetic regime. The measurement of a MS at 4.2 K allowed confirming the presence of goethite and lepidocrocite (with average magnetic fields of 49.21 T and 44.59 T, respectively).

Keywords Goethite · Lepidocrocite · Mössbauer spectroscopy

1 Introduction

In the Marañón Basin, Peru, there is a Regional Geology named Complejo Marañón, consisting of metal minerals such as copper, silver and gold in schists (which are rocks that can easily be split in thin layers) that is a good candidate for mining exploitation. The minerals form part of a group of medium-grade metamorphic rocks mainly prominent by the preponderance of laminate minerals such as mica, chlorite, talc, hornblende, graphite, and others in contact with intrusions, as well as with nickel, cobalt, titanium, and platinum minerals, among others.

The metallurgy tests carried out in our laboratory indicate a good metallurgy recovery of the concentrate obtained of Chalcopyrite mineral, which means a good

A. Bustamante (✉) · R. Quille
Laboratorio de Cerámicos y Nanomateriales, Facultad de Ciencias Físicas,
Universidad Nacional Mayor de San Marcos, Lima, Perú
e-mail: abustamanted@unmsm.edu.pe

D. Lovera · A. V. Arias · J. Quiñones
Facultad de Ingeniería Geológica, Minera, Metalúrgica y Geográfica,
Universidad Nacional Mayor de San Marcos, Lima, Perú

Fig. 1 Oshno part of the hill where the sample is collected

economic possibility for the Marañón Basin [1]. The samples of Oshno as metallic powder grain are of interest to small and medium steel industry.

Although the usual mineralogical techniques allow identifying the sample minerals as well as the presence of iron oxides with a magnetization degree that indicates the presence of metals in areas where the samples are taken from, it is not possible to study with more detail the system because of the very small particles that make up the iron oxides. In this work, we apply Mössbauer spectroscopy to understand the size and nature of the particles of the iron oxides found in one mineral sample.

2 Experimental

The sample was collected at the Oshno hill, which is located in the Chavín de Pariarca District, Province of Huamalies, Huanuco Region, Peru. Figure 1 shows the place where the sample was collected. The characterization of minerals of this area is made by optical microscopy, Mössbauer spectroscopy at RT, and multi-elemental chemical analysis.

In the Laboratory of Extractive Metallurgy-UNMSM we proceeded to develop metallurgical processes and operations commissioned: crushing, grinding and sifting to take a size suitable for the chemical and physical characterization.

The X-ray powder diffraction patterns were recorded with a BRUKER D8 Focus using a graphite monochromator to select the Cu$K\alpha$-doublet, in the range 8° to 80° with steps of 0.02°. The chemical composition was determined by inductively coupled plasma optical emission spectroscopy (ICP-OES) using the method ICP-40B that consist in multi-acid (four acid) digestions that is a combination of HCl (hydrochloric acid), HNO_3 (nitric acid), HF (hydrofluoric acid) and $HClO_4$ (perchloric acid). Because hydrofluoric acid breaks down silicates, these digestions are often referred to as "near-total digestions". There can be a loss of volatiles during digestion (e.g. B, As, Pb, Ge, and Sb). Multi-acid digestion is a very effective dissolution

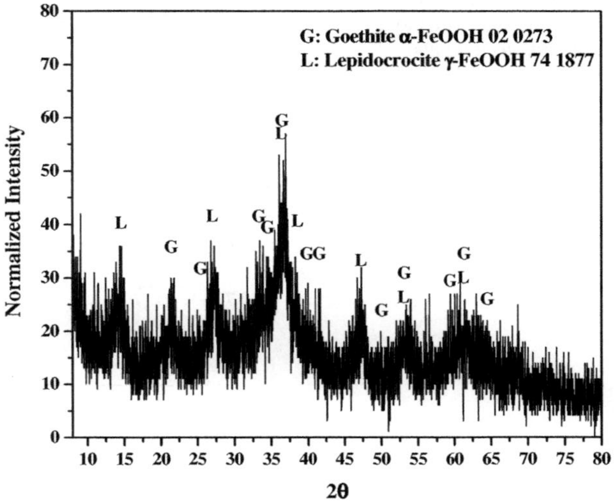

Fig. 2 X-ray diffraction of sample Oshno. *G* Goethite, *L* Lepidocrocite

Table 1 Chemical analysis of the sample Oshno

Major elements (ppm)	
Cu	968
Zn	3,996
As	367
Mn	1,118
Ba	282
Minor elements (ppm)	
Cr	92
V	66
W	34
Sr	197.8
Cd	23
Co	23
Ag	6.8
La	38.7
Sc	12.3
Sn	24
Ni	18
Fe	7.39

procedure for a large number of mineral species and is suitable for a wide range of elements.

For the sample presented in this paper, Mössbauer spectra were recorded in transmission geometry using a constant acceleration spectrometer and a 512 multichannel analyser. The source is of ^{57}Co in Rh matrix. Spectra were recorded at 300 K and, in this particular case, also in a liquid helium cryostat at 4.2 K with a sinusoidal velocity variation. Isomer shifts are given relative to the centroid of the spectrum of an absorber of α-Fe at room temperature. Spectra were analyzed by least-squares fits using lorentzian line shapes. For the spectrum at 4.2 K the fittings were performed with the NORMOS program [2] that allows analyzing the data with hyperfine field distributions.

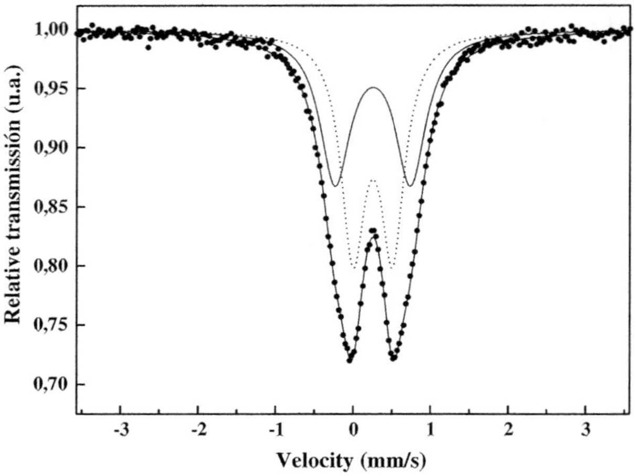

Fig. 3 Mössbauer spectrum at RT showing the Goethite (*continuous line*) and Lepidocrocite (*dotted line*) sub spectra

Table 2 Hyperfine parameters for fitted Mössbauer spectra

	δ(mm/s) ± 0.03(mm/s)	Δ or 2ε(mm/s) ± 0.03(mm/s)	B (T) ± 0.5(T)	Γ(mm/s) ± 0.02(mm/s)	A(%) ± 2%
RT					
Lepidocrocite	0.37	0.51		0.37	53
Goethite	0.37	0.98		0.48	47
4.2 K					
Lepidocrocite	0.36		44.6	0.35	
Goethite	0.36	−0.10	49.2	0.35	

δ is the isomer shift relative to α iron, Δ is the quadrupole splitting, 2ε is the quadrupole shift, Γ is the line width, A is the spectral fractional area as obtained from the fit, B is the hyperfine field

3 Results and discussion

3.1 X-ray difraction

The X-ray diffraction (XRD) pattern is shown in Fig. 2. The analysis indicates the presence of two mineralogical phases, namely, lepidocrocite (γ-FeOOH ASTM 74-1877) and goethite (α-FeOOH ASTM 02-0273).

3.2 Chemical analysis

The chemical analysis, shown in Table 1, indicates the presence of major elements Cu, Zn, As, Mn y Ba and Cr, V, W, Sr, Cd, Co, Ag, La, Sc, Sn, Fe y Ni as minor elements (Certificate No. 42876 of SGS company).

3.3 Mössbauer spectroscopy

The Mössbauer spectrum (MS) at room temperature of the sample from Oshno hill shows a broadened doublet (Fig. 3). Two doublets are required to obtain a

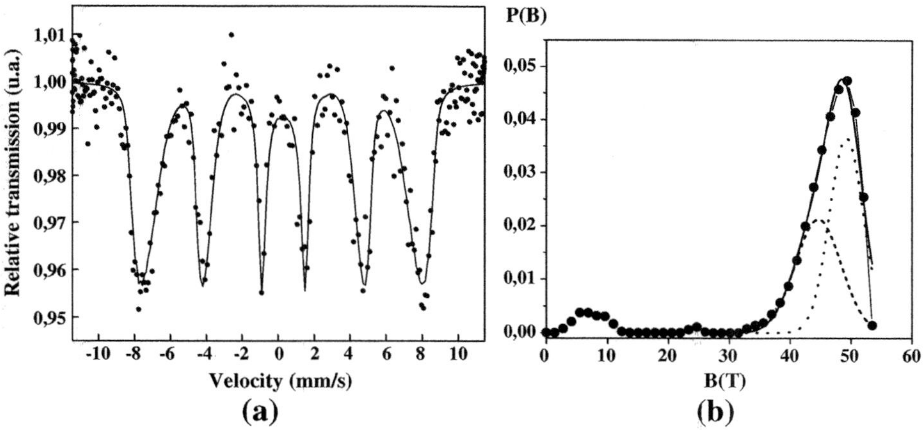

Fig. 4 **a** Mössbauer spectrum at 4.2 K and, **b** the corresponding hyperfine field distributions P(B) that was fitted with two Gaussian lines. The *dotted line* corresponds to goethite and *dashed* one to lepidocrocite

good fitting. The first doublet has hyperfine parameters $\delta = 0.37 \pm 0.03$ mm/s, $\Delta = 0.51 \pm 0.03$ mm/s and $\Gamma = 0.37 \pm 0.02$ mm/s and corresponds to the presence of lepidocrocite [3]. The second doublet has $\delta = 0.37 \pm 0.03$ mm/s, $\Delta = 0.98 \pm 0.03$ mm/s and $\Gamma = 0.48 \pm 0.02$ mm/s and a wide line. It corresponds to particles of oxides or hydroxides with average diameters smaller than ≈100 Å in a superparamagnetic regime [4] (see Table 2). Figure 4a shows the MS at 4.2 K in zero external magnetic field. It displays a broadened sextet which was numerically analyzed using a hyperfine field distribution computer program that yielded an average isomer shift of 0.36 mm/s and a quadrupole shift close to −0.10 mm/s. The mean value of the magnetic field was determined from the hyperfine field distribution P(B) by fitting the histogram with two lines of Gaussian shape in the range of 30 to 53 T. The data show an average field of 49.21 T, confirming the presence of goethite, as showed in Fig. 4b (dotted line) (see reference [3] page 366 that reports $\delta = 0.34$ mm/s, $2\varepsilon = -0.25$ mm/s and $B = 49.2$ T) and lepidocrocite with an average magnetic field of 44.59 T (dashed line) (see reference [3] page 436 reports that $\delta = 0.342$ mm/s, $2\varepsilon = 0.018$ mm/s and $B = 45.6$ T).

4 Conclusions

Using Mössbauer spectroscopy performed at liquid helium temperature it was possible to show that the iron material from Oshno hill is composed of superparamagnetic particles of Goethite and Lepidocrocite.

Acknowledgements A.B.D thanks to the UNMSM for supporting this research, through the SEGUSM system and Dra. Rosa B Scorzelli from CBPF for the measurement at liquid helium temperature.

References

1. Lovera, D., Bustamante, A., Gagliuffi, P., Quiñones, J., Puente, L., Gaudencio, L., Diego, J., Romero, A.: Rev. Inst. Investig. FIGMMG **8**, 44–50 (2005)
2. Brand, R.A.: Normos Programs. Duisburg University (1989)
3. Mössbauer Mineral Handbook: In: Stevens, J.G., Khasanov, A.M., Miller, J.W., Pollak, H., Li, Z. (eds.) Mössbauer Effect Data Center (2002)
4. Künding, W., Bömmel, H., Constabaris, G., Lindquist, R.H.: Phys. Rev. **142**, 327–333 (1966)

Hyperfine Interact (2010) 195:69–76
DOI 10.1007/s10751-009-0103-9

Hematite from a mining area in the east border of Quadrilátero Ferrífero, Minas Gerais, Brazil

Ana Rosa Passos Pereira · José Domingos Fabris · Francisco Javier Rios · Carlos Alberto Rosière · Paulo Rogério da Costa Couceiro · Fábio Furlan Ferreira · Lucas Morais de Menezes

Published online: 26 September 2009

Abstract Iron oxides are dominant minerals in many geo-domains of economical interest, as iron ore mines. Knowing the main mineral transformation pathways is a fundamental step to plan prospecting new mineral deposits. This study aimed at contributing to a better understanding of the chemical and mineralogical processes related to the genesis and transformations of iron oxides involving hematite in an iron-ore mine of the east border of Quadrilátero Ferrífero, Minas Gerais, Brazil. Two representative geo-samples were analyzed with synchrotron radiation X-ray diffraction (XRD), ^{57}Fe Mössbauer spectroscopy, X-ray fluorescence and saturation magnetization (σ) measurements. The iron content varied from 65 to 69 mass% Fe. From XRD data, hematite is indeed the major mineral for all samples but characteristic reflections of goethite and magnetite also appear. For the magnetic

A. R. P. Pereira (✉) · J. D. Fabris · L. M. de Menezes
Departamento de Química, Universidade Federal de Minas Gerais,
31270-901 Belo Horizonte, Minas Gerais, Brazil
e-mail: arpp@qui.dout.ufmg.br

A. R. P. Pereira · F. J. Rios
Centro de Desenvolvimento da Tecnologia Nuclear, 31270-901 Belo Horizonte,
Minas Gerais, Brazil

C. A. Rosière
Instituto de Geociências, Universidade Federal de Minas Gerais,
31270-901 Belo Horizonte, Minas Gerais, Brazil

P. R. da Costa Couceiro
Departamento de Química, Universidade Federal do Amazonas, Manaus, Brazil

F. F. Ferreira
Laboratório Nacional de Luz Síncrotron, Campinas, São Paulo, Brazil

sample, $\sigma = 6.9$ J T^{-1} kg^{-1}. 298 K- and 110 K-Mössbauer data allow characterizing hematite in these iron-rich geo-materials.

Keywords Iron oxides · Synchrotron radiation · Saturation magnetization · Mössbauer spectroscopy

1 Introduction

Iron oxides may be a very important scientific and technological issue in economically-based agriculture systems or in industrial mining activities [1], particularly to countries for which ore commodities represent an important item in its economy. Actually, about 60% of all industrial iron in Brazil is exploited from Quadrilátero Ferrífero, Minas Gerais state, Brazil. Knowing the mineral transformations pathways in such natural systems is the first fundament for planning exploitation, prospecting new mining areas, predicting ore deposit reserves and directing geological researches with technological interest [2]. Setting up optimal parameters in metallurgical industrial processes as those for ore agglomeration and iron metal production depends upon the mineralogical characteristics of the ore, including size and crystal arrangement. The study of existing fluid inclusions in crystalline formations is also of particular interest as it can tell us about various physical–chemical conditions governing the crystallization or re-crystallization of minerals, including any metamorphic events, occurring subsequently to the mineral genesis [3–6]. The nature of fluids forming or somehow involved in subsequent genetic events of deposits may also evidence chemical and physical conditions governing the transformation mechanisms of primary minerals to form hematite or specularite, a black-shinning plated form of metamorphic hematite. For instance, the change of magnetite to hematite is not yet completely understood. There are two general accepted chemical mechanisms: (1) one involving the formation of an intermediate iron oxide, namely maghemite [7], and (2) by other route, magnetite is directly oxidized to hematite [8–11]. In both cases, some factors are presumably associated to the differentiation process of magnetite oxidation, i.e. the hosted-rock, isomorphic substitution, particle size and temperature. The main purpose of this still ongoing work is to contribute to a better understanding of the main chemical and mineralogical pathways related to the iron oxides formation and transformations involving hematite in an iron oxide-rich geo-domain of the east border of Quadrilátero Ferrífero, Minas Gerais state, Brazil.

2 Experimental

Lumps of hematite-rich materials from a itabirite lithology (Figs. 1 and 2) were collected from two sampling sites in a mining area for iron (geographical coordinates of the sampling site 20°07′51.71″ S 43°25′00.09″ W), located in the east border of Quadrilátero Ferrífero, Minas Gerais state, Brazil. The non-magnetic (FZ-01) and magnetic (FZ-11) materials were identified as a first sampling criterion, simply by their ability to be stuck to a hand magnet. Their chemical composition was determined by conventional chemical analytical methods, as redox volumetry, for total

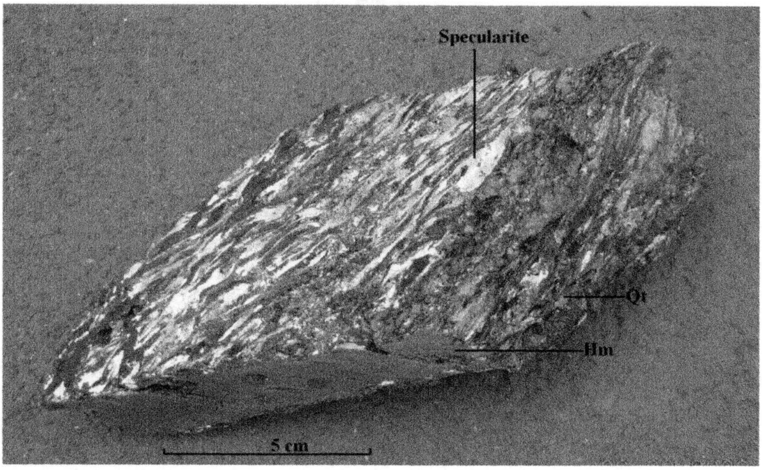

Fig. 1 Sample FZ-01: specularite and hematite (*Hm* hematite, *Qt* quartz)

Fig. 2 Sample FZ-11: hematite and magnetite (*Hm* hematite, *Mt* magnetite)

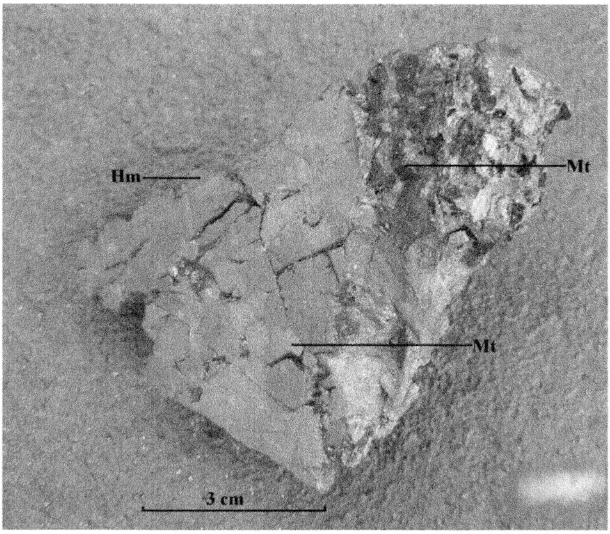

iron, and X-ray fluorescence. The mineralogy was first assessed with synchrotron radiation X-ray diffractometry (XRD), at 298 and ~12 K. The ^{57}Fe Mössbauer analysis was performed with the samples at 110 and 298 K; the saturation magnetization (σ) was measured with a portable soil magnetometer, under a fixed magnetic field of ~0.3 T [12].

3 Results and discussion

Synchrotron X-ray diffraction data revealed hematite to be by far the main mineralogical constituent in both samples, followed by a cubic structure of a spinel phase, which is putatively magnetite [13], along with weaker reflections of goethite

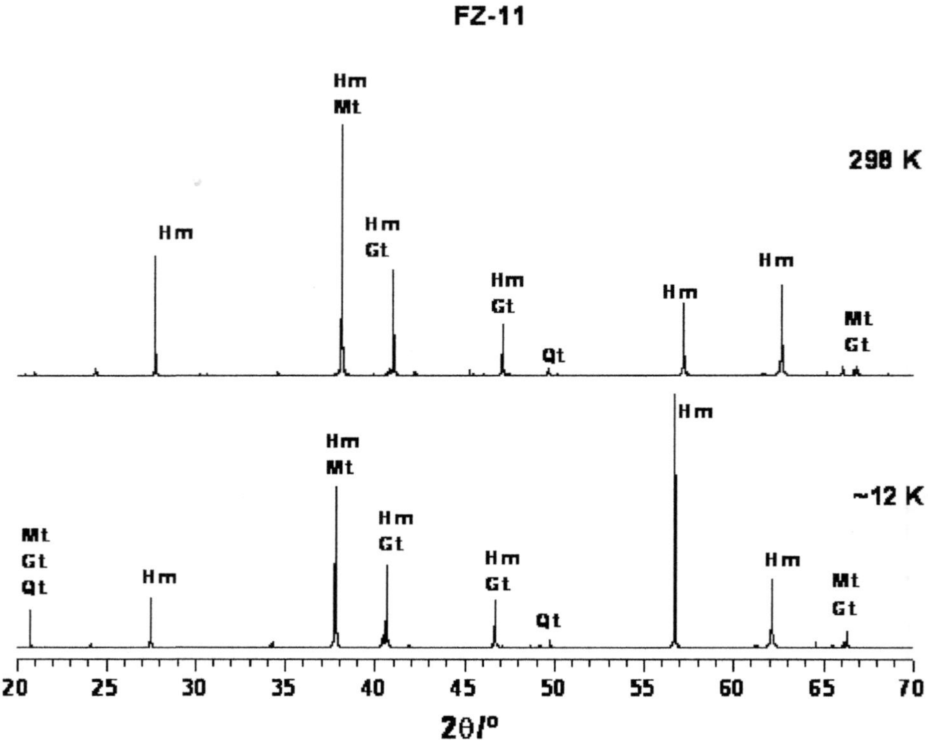

Fig. 3 Synchrotron XRD pattern from FZ-11 sample ($\lambda_{298\ K} = 1.76051$ Å; $\lambda_{12\ K} = 1.74664$ Å) [*Hm* hematite, *Mt* magnetite, *Gt* goethite, *Qt* quartz]

Table 1 Chemical composition as obtained by XRF, expressed on an oxide basis

Elements	Amount/mass%	
	FZ-01	FZ-11
Fe_2O_3	93.3 (7)	97.9 (0)
Al_2O_3	0.7 (3)	0.5 (3)
SiO_2	5.2 (8)	1.1 (3)
TiO_2	0.02 (3)	0.04 (4)
MnO	0.51 (2)	0.08 (0)
Other[a]	0.03 (3)	0.31 (5)

[a] Ca, Mg, K, S and P

and quartz also appear (Fig. 3). Some preferential crystallographic orientation was observed on patterns for both samples, in both temperatures (Fig. 3).

From volumetric determinations, the total iron contents are 65 mass% (FZ-01) and 68 mass% (FZ-11) Fe, and these results agree with previously reported data on related materials of this same area [13, 14]. Corresponding chemical composition obtained with X-ray fluorescence for samples is presented in Table 1. The value of saturation magnetization for sample FZ-01 is nearly zero; for sample FZ-11, $\sigma = 6.90(2)\ J\ T^{-1}\ kg^{-1}$. This result indicates the occurrence of magnetic minerals in sample FZ-11, which is in line with observations from optical petrographic analysis: this magnetic sample unequivocally contains some altered, i.e. partially

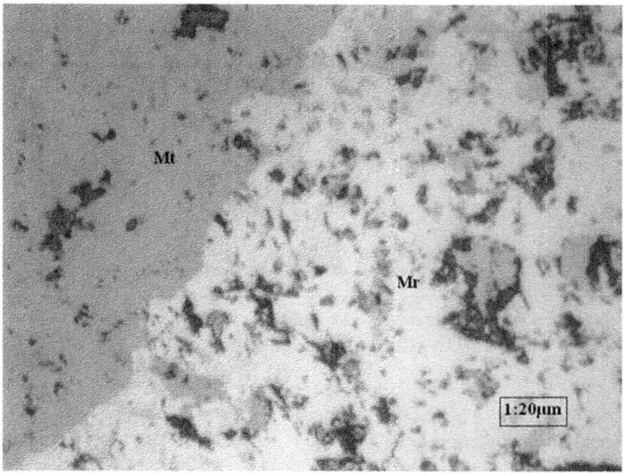

Fig. 4 Reflected light photomicrographics showing alterations in magnetite, sample FZ-11 [*Mt* magnetite, *Mr* martite]

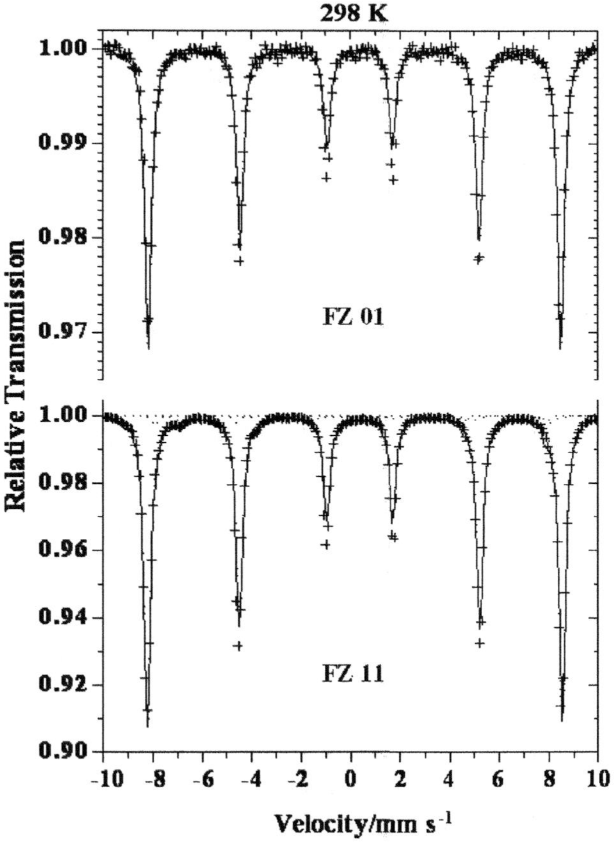

Fig. 5 298 K-Mössbauer spectrum for sample FZ-01 and FZ-11

oxidized magnetite (Fig. 4). 298 K-Mössbauer data also confirmed that hematite is the dominant mineralogical phase, for both samples (Fig. 5), but about 5.7% of the subspectral Mössbauer (Table 2) of sample FZ-11 is assignable to magnetite, which

Table 2 Mössbauer parameters for samples FZ-01 and FZ-11

Sample	Temperature/K	δ/mm s^{-1a}	ε/mm s^{-1}	Bhf/T	Γ/mm s^{-1}	AR/%
FZ-01	298	0.309 (3)	−0.21 (1)	51.7 (5)	0.32 (9)	98.0 (8)
		0.366 (7)	−0.26 (2)	38.2 (1)	0.30 (9)	1.0 (2)
	110	0.48 (7)	0.41 (1)	54.7 (6)	0.37 (7)	100.0 (1)
FZ-11	298	0.375 (3)	−0.20 (6)	51.9 (8)	0.30 (3)	94.2 (5)
		0.144 (1)	0[b]	49.3 (1)	0.31 (1)	2.0 (0)
		0.715 (7)	0[b]	46.5 (7)	0.4 (2)	3.7 (5)
	110	0.468 (6)	0.37 (5)	53.6 (9)	0.33 (2)	83.0 (1)
		0.451 (1)	−0.27 (2)	50.8 (0)	0.6 (8)	16.0 (9)

[a]Isomer shift reported with respect to α-Fe
[b]Fixed value during fitting convergence procedure

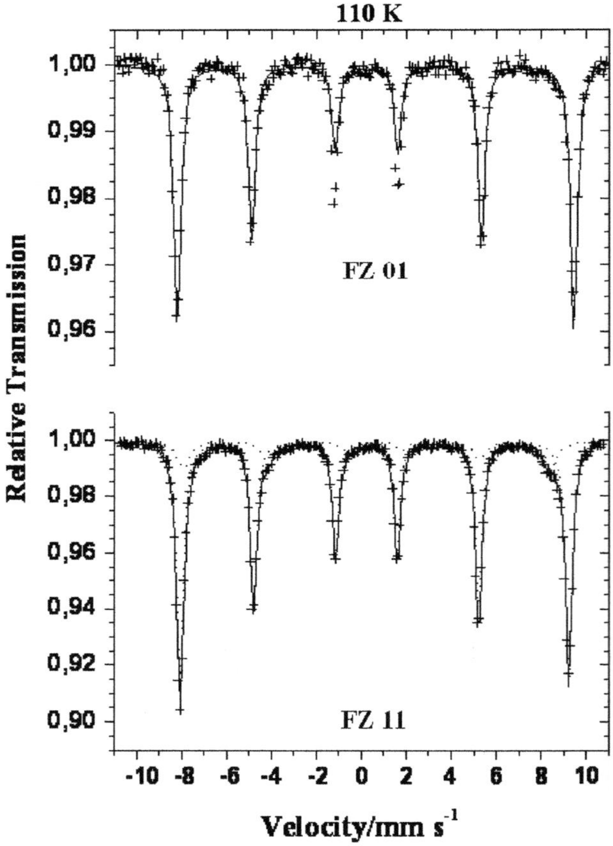

Fig. 6 110 K-Mössbauer spectrum for sample FZ-01 and FZ-11

is fairly consistent with the corresponding saturation magnetization measurement. 110 K-Mössbauer spectra indicate that the hematite in both samples undergoes the Morin transition (characteristic $T_M = 260$ K), as expected for a relatively pure oxide. This is supported at least for the chemical composition of sample FZ-01, with the low proportion of $\sim 1.2 \times 10^2$ mol Fe (mole hematite)$^{-1}$ being isomorphically replaced by Al, as estimated by the equation Al = 35.46 − 70.402a/nm [15] (a =

5.0348 Å, is the trigonal–hexagonal cell dimension, as deduced from XRD data). The 110 K-spectrum for sample FZ-01 evidences a magnetically-ordered component with exceptionally high hyperfine field, $B_{hf} = 54.7$ T (Fig. 6; Mössbauer parameters are presented in Table 2), but this value has already been early reported in the literature [16]. A similarly surprising result was obtained via CEMS measurement with the same sample at 298 K: a magnetically ordered subspectrum corresponding to $B_{hf} = 53.7$ T (figure not shown) contributes to the whole spectrum. This behavior is planned to be now experimentally explored in more detail, as with in-field Mössbauer measurements, in an attempt to seek for a consistent explanation.

4 Conclusions

Hematite is the main mineralogical constituents in both samples. This magnetite–hematite transformation was found to follow a direct pathway, as no maghemite was unequivocally detected, either from X-ray or from Mössbauer data. Results obtained for the non-magnetic sample suggest that this specularite is somehow altered relatively to what would be expected for the characteristic specimen. New numerical analyses, as via Rietveld structural refinement of XRD data, already collected at different temperatures, are now in progress, in a further attempt to have a more consistent crystallographic picture, allowing us to more accurately trace and consistently model the main mineralogical transformation mechanisms involving hematite, either as a direct product from magnetite or as precursor for other iron-bearing species in this geo-system.

Acknowledgements This work was supported by CNPq, FAPEMIG (Brazil) and The Brazilian National Laboratory of Synchrotron Light. CNPq/Prosul (grant no. 490132/2006-5) also supported the international mission by JDF to attend the Latin American Conference on the Applications of the Mössbauer Effect, LACAME'2008, in La Plata, Argentina.

References

1. Fabris, J.D., Coey, J.M.D.: Espectroscopia Mössbauer do ^{57}Fe e Medidas Magnéticas na Análise de Geomateriais. Tópicos em Ciências do Solo. Viçosa, Sociedade Brasileira de Ciência do Solo **2**, 47–102 (2002)
2. Rosière, C.A., Chemale, F. Jr.: Itabiritos e Minérios de Ferro de Alto Teor do Quadrilátero Ferrífero—Uma Visão Geral. Geonomos **8**(2), 27–43 (2000)
3. Fuzikawa, K.: Inclusões Fluidas: Métodos Usuais de estudo e aplicações. Contribuições à Geologia e à Petrologia—SBGM, pp. 29–44 (1985)
4. Goulart, A.T., et al.: Multiple iron-rich spinel phases and hematite in a magnetic soil developing on tuffite. Phys. Chem. Miner. **25**(1), 63–69 (1997)
5. Ríos, F.J., Alves, J.V., Pérez, C.A., Rosière, C.A., Fuzikawa, K., Costa, E.C., Chaves, A.O., Prates, S.S., de Barrio, R.: Combined investigations of fluid inclusions in opaque ore minerals by NIR/SWIR Microscopy and Microthermometry and Synchrotron Radiation X-ray Fluorescence. Applied Geochemistry. Special Issue Frontiers in Analytical Geochemistry **21**(5), 813–819 (2006)
6. Ríos, F.J., Pérez, C.A., Fuzikawa, K., Alves, J.V., Gonçalves, R., Chaves, A.O., Mattos, E., Souza, A., Gonzalez Guillot, M., Chaves, A.M., Pereira, A.R.P., de Lima, T.F., Rosiére, C.A., Correia Neves, J.M.: Estudos por μXRF-Sincrotron de fluidos mineralizadores detectados em minérios uraníferos, ferríferos e depósitos epitermais. REM **59**(4), 373–377 (2006)
7. Rosière, C.A., Rios, F.J.: The origin of hematite in high-grade iron base don infrared microscopy and fluid inclusion studies: the example of the Conceição Mine, Quadrilátero Ferrífero, Brazil. Econ. Geol. **99**, 611–624 (2004)

8. Doriguetto, A.C., Fernandes, N.G., Persiano, A.I.C., Nunes Filho, E., Grenèche, J.M., Fabris, J.D.: Characterization of a natural magnetite. Phys. Chem. Miner. **30**, 249–255 (2003)
9. Santana, G.P., Fabris, J.D., Goulart, A.T., Santana, D.P.: Magnetite and its transformation to hematite in a soil derived from steatite. Rev. Bras. Cienc. Solo **25**, 33–42 (2001)
10. Moura, C.S., Fabris, J.D., Mussel, W.N., Rosière, C.A.: Iron oxides of a magnetic soil derived from dolomitic itabirite. Hyperfine Interact. (C) **4**, 17–23 (1999)
11. Lagoeiro, L.E.: Transformation of magnetite to hematite and its influence on the dissolution of iron oxide minerals. J. Metamorph. Geol. **16**, 415–423 (1998)
12. Coey, J.M.D., Cugat, O., McCauley, J., Fabris, J.D.: A portable soil magnetometer. Rev. Fis. Apl. Instrum. **7**(1), 25–30 (1992)
13. Silva, A.M., Gibotti, M. Jr.: Geologia das Minas de Fazendão e Adjacências. Monografia. UFOP (1989)
14. Rosière, C.A., Chemale, F. Jr., Guimarães, M.L.V.: Um modelo para a Evolução microestrutural dos minérios de ferro do Quadrilátero Ferrífero. Parte I—Estruturas e Recristalização. Geonomos **1**(1), 65–84 (1993)
15. Kämpf, N., Schwertmann, U.: Avaliação da Estimativa de Substituição de Fe por Al em Hematitas de Solos. Rev. Bras. Cienc. Solo **22**, 209–213 (1998)
16. Spier, C.A., Oliveira, S.M.B., Rosière, C.A., Ardisson, J.D.: Mineralogy and trace-element geochemistry of the high-grade iron ores of the Águas Claras Mine and comparison with the Capão Xavier and Tamanduá iron ore deposits, Quadrilátero Ferrífero, Brazil. Miner. Depos. **43**, 229–254 (2008)

Hyperfine Interact (2010) 195:77–83
DOI 10.1007/s10751-009-0104-8

Mössbauer studies of one representative hydromorphic soil of the coastal area of the Rio de La Plata

B. A. Guichon · J. Desimoni · R. C. Mercader · P. A. Imbellone

Published online: 2 October 2009

Abstract In the present work, we have applied Mössbauer spectroscopy as well as the traditional chemical analyses to assess the contents of different states of Fe in oxides and hydroxides in an Entisol soil from the Argentine coastal plain of the Rio de La Plata. Tentative assignments for the different Fe ion sites are proposed. Our findings show that the isomer shift and quadrupole splitting are sensitive to the changes detected in the Fe^{2+} contents of the soluble phase while others remain constant within the experimental uncertainties.

Keywords Hydromorphic soils · Mössbauer spectroscopy · Fe oxidation states

1 Introduction

The geomorphologic locations of some areas render flooding conditions that differ along the year and give rise to soils that are generally named *hydromorphic* soils. A large part of the Argentine coastal plain of the Rio de La Plata is covered with wetland soils whose geochemical properties and the influence of the different components of the environment are not totally understood. In a recent paper [1] the evolution of some hydromorphic parameters, such as Eh and concentration of soluble Fe^{2+} and Mn^{2+} and moisture, formed on different parent materials and landforms, has been analyzed in an attempt to establish the relationships of these parameters with external factors, including floods, real evapotranspiration or water table depth, including also the expression of redoximorphic features, in poorly drained soils located along the Argentine coastal plain of the Rio de La Plata.

B. A. Guichon · P. A. Imbellone
Instituto de Geomorfología y Suelos, Universidad Nacional de La Plata, 3 No.584, 1900 La Plata, Argentina

J. Desimoni · R. C. Mercader (✉)
Departamento de Física, IFLP-CONICET, Facultad de Ciencias Exactas, Universidad Nacional de La Plata, CC. 67, 1900 La Plata, Argentina
e-mail: mercader@fisica.unlp.edu.ar

Alternating dry and wet periods along the year bring on oxidized or reduced states of the soil profile horizons that thus, for the Fe and Mn compounds in particular, have a much more active geochemistry than for other types of soil. When flooding occurs, the soil-iron in the soil tends to be reduced to soluble Fe^{2+}. Under subsequent drier season, the Fe^{2+} in the profile gradually oxidizes to Fe^{3+} and precipitates as oxides or hydroxides with variable degree of crystallinity. The degree of segregation of the oxides in the profile and the assessment of crystalline iron species helps toward a deeper understanding of the pedogenesis process associated with the hydromorphic states. Changes between different oxidation states are faster for iron than for other ions, but in soils they are not immediate, as they are strongly influenced by different local parameters like the presence of organic matter, pH, Eh, or the granulometric distribution of the soil material.

In principle, because of the sensitivity of the hyperfine parameters obtained by ^{57}Fe Mössbauer analysis, it should be possible to study in more detail than with other methods the oxidized Fe formed by reactions taking place during pedogenesis [2]. However, in practice, although it is not impossible, a thorough study is extremely time-consuming and yields results difficult to assess accurately. This is because of the complexity of the soil system and the great overlapping that the Mössbauer signals have for the different iron species present in the soil [3].

In this work, we attempt to relate measured room temperature Mössbauer parameters with some earlier reported chemical data [1] for the soluble phase.

2 Experimental materials and methods

Soil materials from a Fluvaquent [4] (sampling site, at geographical coordinates 34°47′10″ S 58°00′50″ W) were sampled along the year. Twenty-four monthly in situ determinations of pH and Eh (by the method of Vizier [5]) were made in the three: A, 2Cxg, and 2Cg horizons. Table 1 displays selected properties of the investigated soil. The Pt electrodes were calibrated with the Zobell solution, using the equation corrected for the calomel electrode [6].

The Fe^{2+} extraction was made with a 3% $AlCl_3$ solution (according to ref. [7]). Fe^{2+} was analyzed by colorimetry with *o*-phenantroline. The samples were put in air-tight plastic jars immediately after collection, refrigerated during transport and stored for up to 24 h in a refrigerator at 4°C until analyzed.

The Mössbauer spectra were taken for samples selected according to the maximum and minimum states of reduction so found. Therefore, each horizon was analyzed through samples that had yielded the maxima and minima values for the contents of Fe^{2+} determined by chemical methods. Mössbauer spectra (MS) of the total fraction (crushed and sieved through a 250 μm mesh) without previous separation were taken in transmission geometry using a 5mCi ^{57}Co in a Rh matrix source and recorded in a standard constant acceleration spectrometer in the velocity range of about ±12 mm/s. Due the different iron hyperfine sites, the MS were numerically analyzed using hyperfine field distributions [8].

3 Results and discussion

Figure 1 shows the Mössbauer spectra recorded for these samples. Two main contributions can be separated: (1) a magnetic sextet and (2) a central set of doublets.

Table 1 pH, electrical conductivity, contents of organic matter, sand, silt and clay, matrix color, structural type and mottles/concretions of Fe–Mn of the three horizons of the Fluvaquent soil at location: 34°47′10″ S; 58°00′50″ W, altitude 2.00 m above sea level

Horizon	Depth (cm)	pH paste	EC (dS m^{-1})	OM (%)	Sand (%)	Silt (%)	Clay (%)	Color (matrix)		Struct (type)	Mottles/concr. of Fe–Mn
								Dry	Moist		
A	0–10	5.5	<1.0	6.99	64.9	13.3	21.8	10YR 5/2	10YR 3/2	Granular	–/–
2Cxg	10–27	6.3	<1.0	0.26	94.3	3.3	2.4	10YR 6/1	10YR 4/1	Massive	***/***
2Cg	27–48+	5.6	<1.0	0.43	96.9	1.8	1.3	2.5Y 5/3	10Y 4/1	Massive	**/**

EC electrical conductivity, *OM* organic matter, *** abundant, **common, * few, – none

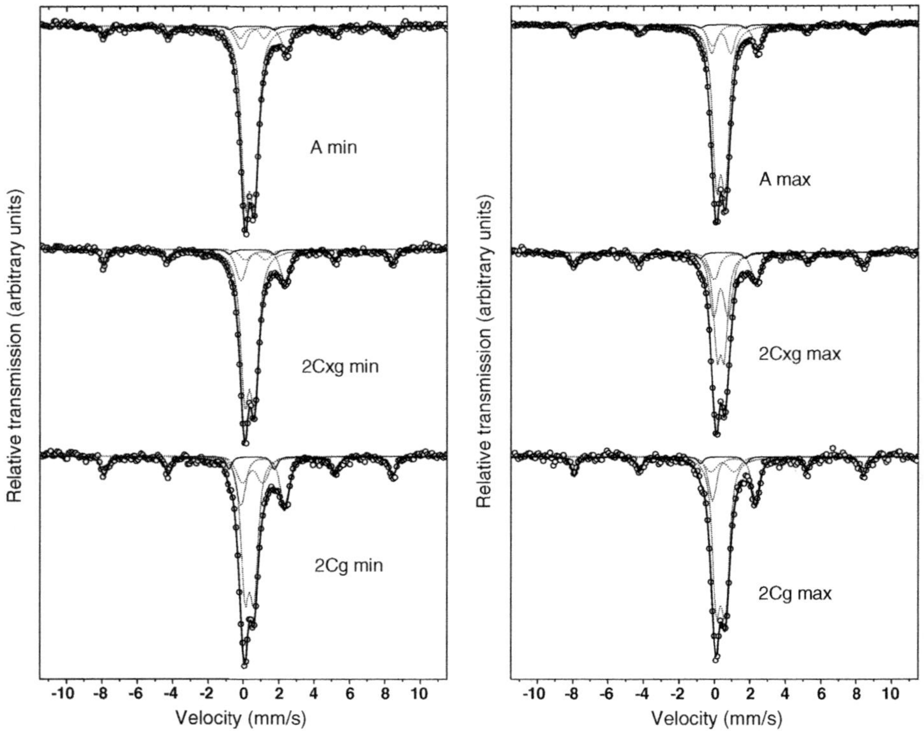

Fig. 1 Room temperature Mössbauer spectra of two sets of samples with different Fe^{2+} soluble contents (labeled 'min' and 'max') for different horizons of an Entisol of the Rio de La Plata coastal plain in Argentina. The *solid line* corresponds to the least-squares-fitted spectra. Parameters are shown in Table 2

The sextet is assignable to α-Fe_2O_3, whereas the central part was fitted by assuming three quadrupole interactions, two of them corresponding to Fe^{3+} and the other one to Fe^{2+}. The resulting fitting parameters are quoted in Table 2.

While the sextets are unequivocally assignable to hematite, the assessment of Mössbauer parameters for the central doublets is more ambiguous because of the different Fe species that have similar parameters that overlap in the same region of the spectra [3]. We have found broadened contributions from two Fe^{3+} and one Fe^{2+} doublets, which may belong to minerals, to Fe hydroxides or to finely divided Fe oxides under superparamagnetic relaxation processes that, in addition, can have some Fe ions partially substituted for other ions, like Al or Mn. Therefore, we have followed the assignments performed in a recent paper by Paduani et al. [9].

The doublet noted by Fe^{3+}(doublet I) in Table 2 comprises signals arising from iron hydroxides and from Fe^{3+} ions in octahedral coordination. Therefore, all signals originating in minority Fe minerals as well as the different Fe hydroxides that may be present in the sample will contribute to the spectral area assigned to this fraction.

The iron fraction noted with Fe^{3+}(doublet II) in the table, which can be attributed to Fe^{3+} ions located at tetrahedral sites [9], displays relative spectral areas that do not change much, within experimental uncertainties, for the different samples collected at diverse times and depths. We assign this fraction to detritic Fe minerals.

Table 2 Room temperature-Mössbauer parameters for the two sets of soil materials

Sample/depth	Fe^{3+}(doublet I)				Fe^{2+}				Fe^{3+}(doublet II)					α-Fe_2O_3			
	Δ	σ	δ	F(%)	Δ	σ	δ	F(%)	Δ	σ	δ	F(%)	B	σ	δ	2ε	F(%)
A min/5 cm	0.53	0.20	0.37	71_2	2.52	0.38	1.13	13_1	1.38	0.20	0.49	6_2	51	0.47	0.36	−0.21	11_1
2Cxgmin/19 cm	0.56	0.22	0.36	65_4	2.47	0.32	1.10	17_2	1.16	0.35	0.68	6_4	51	0.00	0.36	−0.21	13_2
2Cgmin/35 cm	0.52	0.26	0.36	49_7	2.46	0.36	1.09	21_3	1.14	0.00	0.64	13_9	51	0.64	0.37	−0.19	17_3
A max/5 cm	0.51	0.20	0.36	66_2	2.55	0.38	1.16	12_1	1.08	0.20	0.38	13_1	51	0.62	0.35	−0.22	9_1
2Cxgmax/19 cm	0.53	0.26	0.36	61_2	2.48	0.34	1.11	15_1	1.08	0.47	0.49	10_3	51	0.98	0.36	−0.24	14_2
2Cgmax/35 cm	0.53	0.26	0.37	59_3	2.44	0.26	1.10	19_2	1.32	0.52	0.45	8_4	51	0.28	0.37	−0.23	14_2

Δ (mm/s) = quadrupole splitting, σ (mm/s) = Gaussian line-width, δ (mm/s) = isomer shift relative to α-Fe, B (T) = hyperfine field, 2ε (mm/s) = quadrupole shift, and F (%) = relative subspectral areas. Spectra were fitted with an extended Voigt-based fitting method. Estimated uncertainties for subspectral areas are quoted as sub-indexes

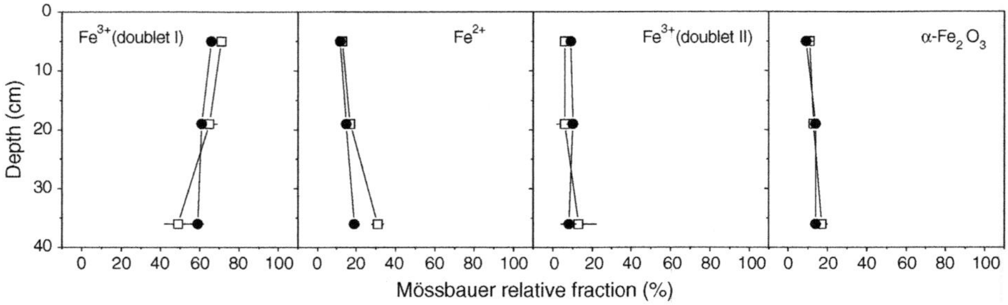

Fig. 2 Variation of the Fe contents for soil samples, according to Mössbauer subspectral areas, for Fe^{2+} and Fe^{3+}, after the fittings of the 'max' and 'min' samples. The *solid circles* correspond to samples with the maximum and the *open squares* to those with the minimum Fe^{2+} soluble contents

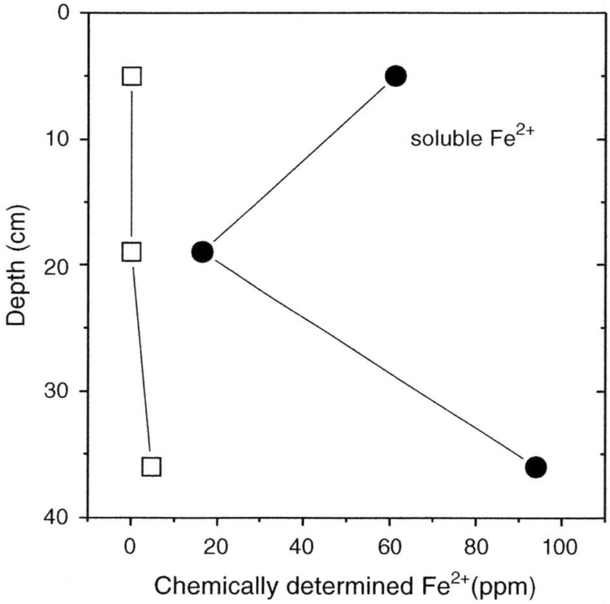

Fig. 3 Variation of Fe^{2+} ions content in the soil materials determined by chemical methods. **a** *circle*: maximum and **b** *square*: minimum reduced state

The Fe^{2+} fraction can be partially due to Fe ions located in tri-octahedral coordination [9], but determining the Fe species to which it belongs, would require performing measurements at different temperatures, which is the usual procedure in Mössbauer studies of soils [2, 3].

Figure 2 shows the relative area variations for samples from different soil horizons. The fraction belonging to the Fe^{3+}(doublet I) species has the higher relative fraction for all horizons and displays variations that can be distinguished from one horizon to the next. The other fractions exhibit differences that are not so clearly defined and lie within the experimental uncertainties. In both the maxima and the minima samples of all horizons, the Fe^{3+}(doublet I) fraction displays greater variations than the other fractions do, with the exception of the Fe^{2+} fraction of the deepest horizon.

Figure 3 shows the variation of the soluble Fe^{2+} ions fraction determined by the chemical methods described in [1]. The greater changes of the Fe^{2+} contents between the soil horizons are observed for the maximum values of the reduced state. The higher value of Fe^{2+} in the soil solution belongs to the deepest horizon, which remains saturated for most of the year. Currently, to disclose the system dynamics, we are carrying out detailed correlation analyses of the different parameters to determine the degree in which the observed changes in the solid phase are due to the different oxidation states of the soil.

4 Conclusions

Although this study has been carried out by taking the Mössbauer spectra in very simple and straightforward manner, it can be seen that some hyperfine parameters are sensitive to the changes detected in the Fe^{2+} contents of the soluble phase. The first relations found in this study between the Fe^{2+} contents are clearer defined for different redoximorphic conditions of Entisol. Further studies are in progress.

References

1. Imbellone P.A., Guichon B.A., Giménez J.E.: Hydromorphic soils of the River Plate coastal plain, Argentina. Latin American Journal of Sedimentology and Basin Analysis, (2009, in press)
2. Otero, X.L., Ferreira, T.O., Huerta-Díaz, M.A., Partiti, C.S.M., Souza, V. Jr., Vidal-Torrado, P., Macías, F.: Geoderma **148**, 318 (2009)
3. Fabris, J.D., Coey, J.M.D.: Topicos em Ciencias do Solo, Viçosa, Sociedade Brasileira de Ciencia do Solo **2**, 47 (2002)
4. Soil Survey Staff, Keys to Soil Taxonomy, 10th edn. USDA—Natural Resources Conservation Service, p. 668. Washington, DC, Pocahontas, Blacksburg, VA (2006)
5. Vizier, J.F.: Pédologie **8**, 33 (1970)
6. Lévy G., Toutain F.: Pédologie. In: Bonneau, M., Souchier, B. (eds.), Masson, Paris, vol. 2, p. 313 (1979)
7. Ignatieff, V.: Soil Sci. **51**, 249 (1941)
8. Lagarec K., Rancourt D.G.: Mössbauer Spectral Analysis Software, Version 1.0. Department of Physics, University of Ottawa (1998)
9. Paduani, C., Samudio Pérez, C.A., Gobbi, D., Ardisson, J.D.: Appl. Clay Sci. **42**, 559 (2009)

Hyperfine Interact (2010) 195:85–91
DOI 10.1007/s10751-009-0105-7

^{57}Fe Mössbauer spectroscopy of radiation damaged samarskites and gadolinites

Dariusz Malczewski · Agnieszka Grabias · Grzegorz Dercz

Published online: 26 September 2009

Abstract We report the results of ^{57}Fe Mössbauer spectroscopy, gamma-ray spectrometry and X-ray diffraction of two fully metamict samarskites and two partially metamict gadolinites. The absorbed α-dose for these minerals are found to range from 3.6×10^{15} α-decay/mg for one of the gadolinite samples to 7.7×10^{17} α-decay/mg for one of the samarskite samples. The Mössbauer spectra of samarskites and gadolinites show increasing line widths of the Fe^{2+} doublets with absorbed α-dose. We also observe that the increase in average quadrupole splitting of the Fe^{2+} components correlates better with absorbed α-dose from ^{232}Th than with total α-dose.

Keywords Samarskite · Gadolinite · Mössbauer spectroscopy · Metamict minerals

1 Introduction

Metamict minerals such as samarskites and gadolinites are a class of natural amorphous materials that were initially crystalline but have become amorphous. These minerals contain radioactive elements that cause structural damage mainly by progressive overlapping nuclear recoil nuclei collision cascades from α-decays of

D. Malczewski (✉)
Faculty of Earth Sciences, University of Silesia,
Bedzińska 60, 41-200 Sosnowiec, Poland
e-mail: dariusz.malczewski@us.edu.pl

A. Grabias
Institute of Electronic Materials Technology,
Wolczyńska 133, 01-919 Warszawa, Poland

G. Dercz
Institute of Materials Science, University of Silesia,
Bankowa 12, 40-007 Katowice, Poland

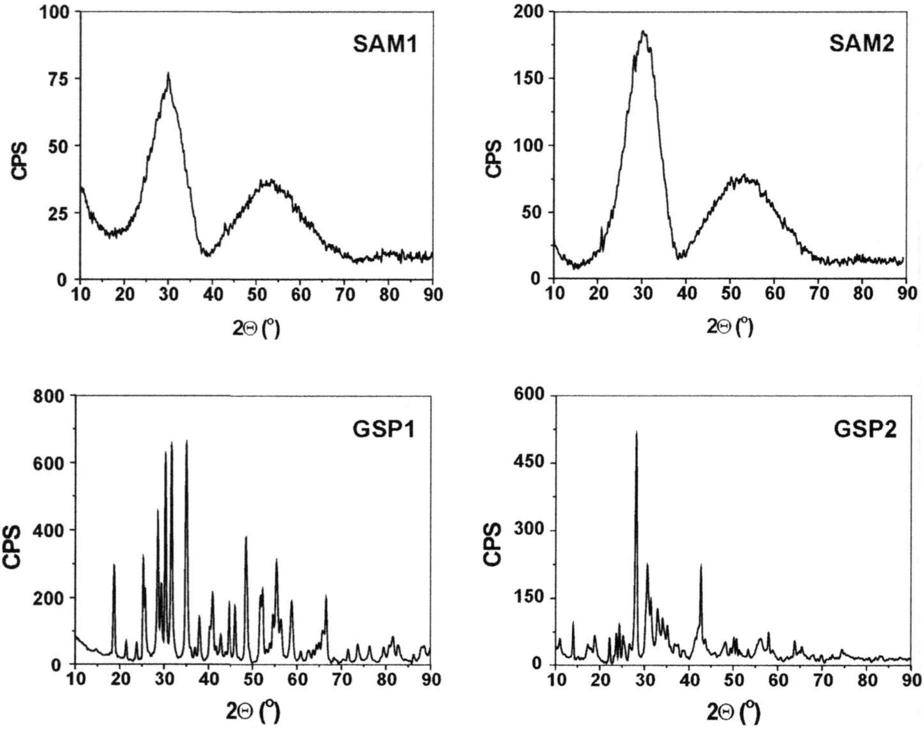

Fig. 1 XRD patterns of investigated metamict samples

^{238}U, ^{232}Th, ^{235}U and their daughter products. Because of the natural occurrence of these actinide elements, they serve as analogs for radiation effects in high-level nuclear waste [1].

Samarskite is a complex niobium-tantalum-titanium oxide. Samarskite has the general structural formula $A^{3+}B^{5+}O_4$ where A = Ca, Ti, Fe^{2+}, Fe^{3+}, rare earth elements (REE), U and Th, and B = Nb and Ta [2]. Analysis of average site charges and cation radii indicate that both A and B sites have octahedral coordination [2]. Gadolinite has a formula $REE_2Fe^{2+}Be_2Si_2O_{10}$, where REE means both rare earth elements and yttrium. Structurally, gadolinites consist of sheets of SiO_4 and BeO_4 tetrahedra interconnected by layers of distorted octahedral Fe and 8-coordinate Y, REE and actinides (U and Th) [3].

The purpose of this work is to show that certain changes in hyperfine parameters are associated with the absorbed α-dose and are common for the investigated complex metamict oxides and silicates.

2 Samples and experimental procedures

A large, massive brownish black sample of samarskite, SAM1, was collected in pegmatites in Centennial Cone, Jefferson Co., Colorado. A massive chunk of dark brown samarskite, sample SAM2, was collected from the Ross Mine, Yancy Co., North

Table 1 Ages of the samples, Fe, U and Th concentrations and calculated α-doses

	Samarskite (SAM1)	Samarskite (SAM2)	Gadolinite (GSP1)	Gadolinite (GSP2)
Age (Ma)	1,400–1,700[a]	1,000–1,200[b]	328(12)[c]	328(12)[c]
Fe (wt.%)	5.9	5.4	8.5	12.1
U (wt.%)	11.4	19.0	0.22	0.28
Th (wt.%)	1.7	0.41	0.43	0.63
Calculated total dose (α-decay/mg)[d]	$7.1(8) \times 10^{17}$	$7.7(8) \times 10^{17}$	$3.6(2) \times 10^{15}$	$4.7(2) \times 10^{15}$
Calculated dose from ^{232}Th (α-decay/mg)	$2.1(2) \times 10^{16}$	$3.6(3) \times 10^{15}$	$1.12(4) \times 10^{15}$	$1.62(6) \times 10^{15}$

[a] [4]

[b] Mesoproterozoic rocks

[c] [5]

[d] Dose has been calculated from the equation: $D = 8 \times N_{238}\left(e^{t\lambda 238} - 1\right) + 7 \times N_{235}\left(e^{t\lambda 235} - 1\right) + 6 \times N_{232}\left(e^{t\lambda 232} - 1\right)$, where N_{238}, N_{235} and N_{232} are the present number of atoms of ^{238}U, ^{235}U and ^{232}Th per milligram, λ_{238}, λ_{235} and λ_{232} are the decay constants of ^{238}U, ^{235}U and ^{232}Th, and t is the geologic age

Carolina. Figure 1 shows XRD patterns of each of the samples. The lack of crystalline peaks for SAM1 and SAM2 indicates complete metamictization (amorphization) of the samples. The gadolinite samples (GSP1 and GSP2) are from pegmatites of Szklarska Poręba in southwestern Poland. The sample GSP1 is dark grayish matte while the sample GSP2 is glassy black. Despite being from the same location, the two samples exhibit different degrees of metamictization, as indicated by differing numbers of crystalline peaks as seen in Fig. 1. This discrepancy is due to different U and Th concentrations. Ages, Fe, U and Th concentrations and calculated α-doses of the four samples are given in Table 1.

The concentrations of ^{232}Th, ^{238}U and ^{235}U were calculated for each sample based on the gamma-ray activities of ^{228}Ac (^{232}Th), ^{226}Ra, ^{214}Pb, ^{214}Bi (^{238}U) and ^{235}U. Gamma-ray spectra were collected using an HPGe detector (32% efficiency with energy resolution of 0.86 keV at 122 keV) and analyzed using the Genie 2000 v.3 software package. Iron concentrations were obtained using a JEOL JSM-6480 Scanning Electron Microscope. The samples were ground into powder and prepared in the shape of a thin disc absorber. The Mössbauer transmission spectra were recorded at room temperature using a constant acceleration spectrometer, a multichannel analyzer with 512 channels and a linear arrangement of a ^{57}Co/Rh source (=50 mCi) absorber and detector. All Mössbauer spectra were numerically analyzed by the fitting software programs Recoil and MEP. X-ray powder diffraction (XRD) patterns were obtained using a PHILIPS X'Pert diffractometer in the Θ–Θ system and Cu $K\alpha$ radiation in scan mode with step size 0.02°.

3 Results and discussion

The Mössbauer spectra of the samarskite and gadolinite samples with the corresponding quadrupole splitting distributions (QSD) are shown in Fig. 2. The hyperfine parameters derived from the fitting procedure are summarized in Table 2.

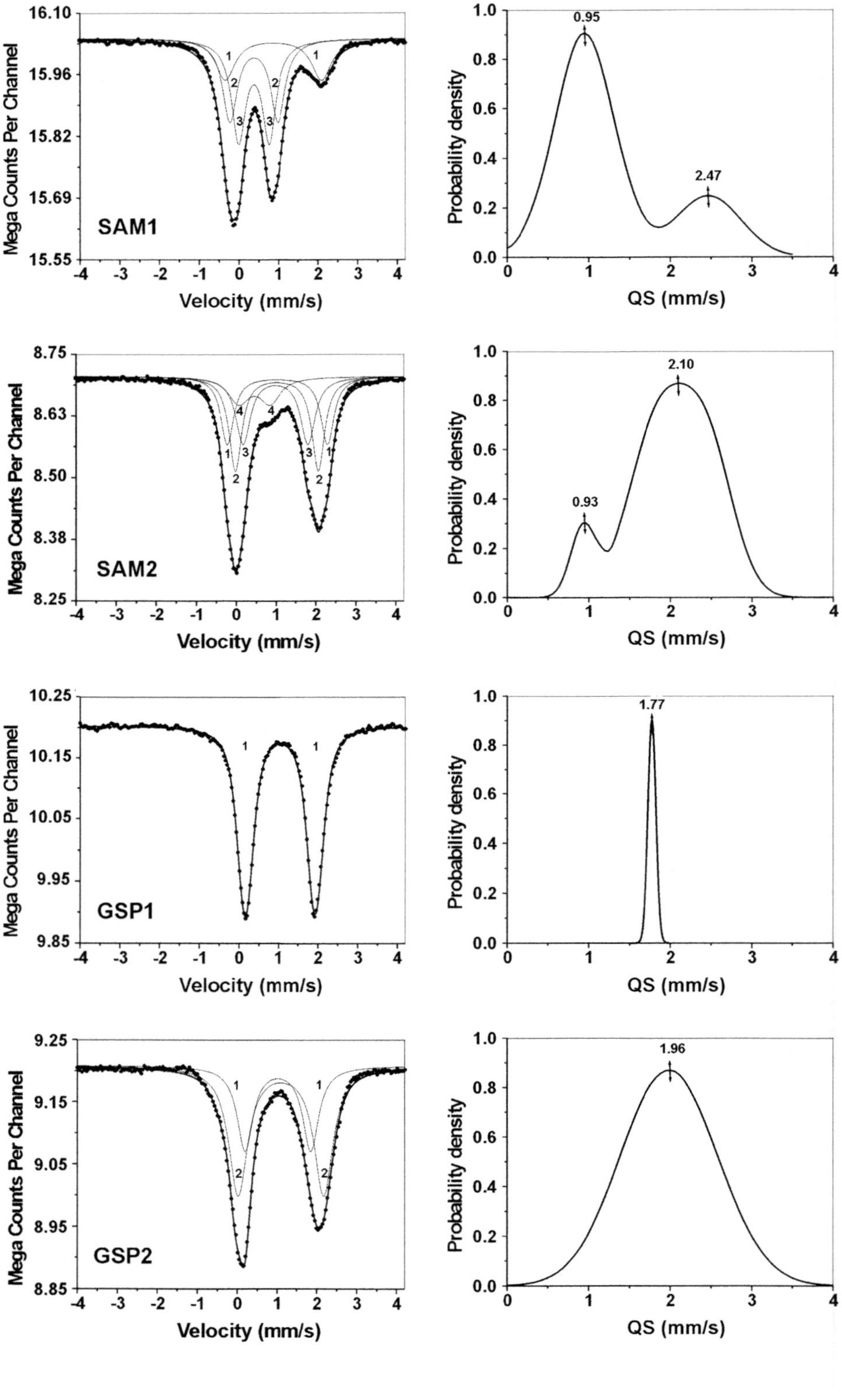
SAM1
Mega Counts Per Channel
Velocity (mm/s)
Probability density
QS (mm/s)
0.95
2.47
SAM2
Mega Counts Per Channel
Velocity (mm/s)
Probability density
QS (mm/s)
2.10
0.93
GSP1
Mega Counts Per Channel
Velocity (mm/s)
Probability density
QS (mm/s)
1.77
GSP2
Mega Counts Per Channel
Velocity (mm/s)
Probability density
QS (mm/s)
1.96

Fig. 2 (*Lefthand plots*) ^{57}Fe Mössbauer spectra at room temperature of metamict samarskite and gadolinite samples. *Solid dots* experimental data, *thick solid line* fitted curve, *thin solid line* fitted doublets. (*Righthand plots*) Corresponding QSD

Table 2 Parameters for ^{57}Fe Mössbauer spectra (shown in Fig. 2) for investigated metamict samples

Sample	Doublet no.	χ^2	δ (mm/s)	Δ (mm/s)	Γ (mm/s)	Assignment (CN)[a]	Intensity
Samarskite (SAM1)		1.6					
	1		0.88 (2)	2.43 (2)	0.30 (2)	Fe^{2+} (6)	0.24 (2)
	2		0.382 (5)	1.21 (3)	0.22 (2)	Fe^{3+} (6)	0.35 (3)
	3		0.379 (3)	0.77 (3)	0.21 (1)	Fe^{3+} (6)	0.41 (3)
Samarskite (SAM2)		2.3					
	1		1.038 (4)	2.51 (3)	0.19 (1)	Fe^{2+} (6)	0.23 (3)
	2		1.031 (3)	2.08 (2)	0.19 (3)	Fe^{2+} (6)	0.34 (4)
	3		0.994 (6)	1.63 (2)	0.22 (1)	Fe^{2+} (6)	0.28 (3)
	4		0.39 (2)	0.92 (3)	0.31 (2)	Fe^{3+} (6)	0.15 (1)
Gadolinite (GSP1)		1.6					
	1		1.045 (1)	1.775 (3)	0.229 (2)	Fe^{2+} (6)	1.0
Gadolinite (GSP2)		2.5					
	1		1.025 (5)	1.66 (2)	0.24 (1)	Fe^{2+} (6)	0.35 (4)
	2		1.100 (3)	2.15 (2)	0.29 (1)	Fe^{2+} (6)	0.65 (4)

Isomer shift values (δ) are given relative to the α-Fe standard at room temperature

[a]Coordination number

The Mössbauer spectrum of the sample SAM1 is fitted to three quadrupole doublets assigned to Fe^{2+} (labeled with a number 1 in Fig. 2; details given in Table 2) and Fe^{3+} (labeled 2 and 3) in octahedral positions with a total relative contribution of 0.24 Fe^{2+}. Based on annealing experiments recently performed in our laboratory, it appears that nearly the entire content of Fe^{2+} in SAM1 is a result of the metamictization process. The main features seen in the spectrum of SAM2 are two broad, asymmetric peaks from divalent iron components. The corresponding QSD has also two maxima at 0.93 and 2.10 mms^{-1}. Based on this QSD, the Mössbauer spectrum can be fitted to three Fe^{2+} doublets (nos. 1, 2 and 3) and one Fe^{3+} doublet (no. 4). Unlike in SAM1, the total relative content of Fe^{2+} is as high as 0.85. Most likely, the very high content of Fe^{2+} in samarskite from Ross Mine reflects the high initial ratio of Fe^{2+} to total Fe at the time of its formation rather than a result of the metamictization process. It is known that Fe^{2+}/Σ Fe for samarskites often differ from sample to sample [6].

The Mössbauer spectrum of the gadolinite sample GSP1 is fitted to one doublet (no. 1) assigned to Fe^{2+} in a uniform octahedral site. The corresponding QSD is very narrow with maximum at 1.77 mm s^{-1}. The sample represents an intermediate stage between the crystalline and metamict state of gadolinite [7]. Similar to the SAM2 spectrum, the spectrum of gadolinite sample GSP2 is characterized by two broad, asymmetric peaks from solely Fe^{2+} components. The corresponding QSD is continuous with central maximum at 1.96 mm s^{-1}. Based on this QSD, the Mössbauer spectrum of GSP2 can be fitted to two Fe^{2+} doublets (nos. 1 and 2). The doublet no. 1 represents the Fe^{2+} octahedra that have undergone a contraction during

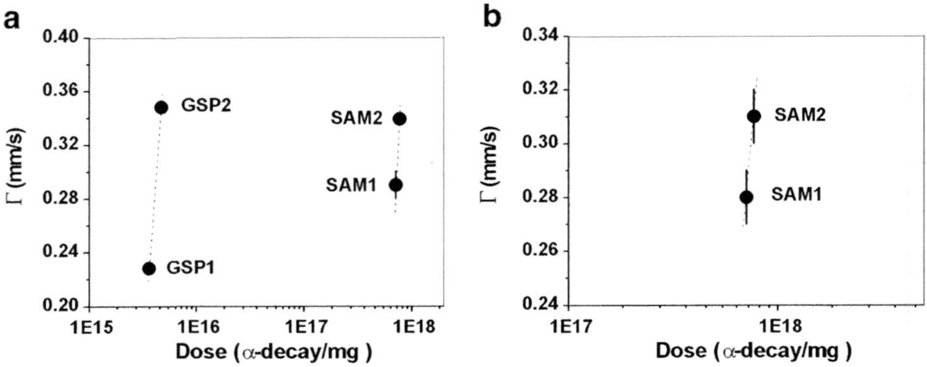

Fig. 3 The total line widths of high energy peaks of **a** Fe^{2+} and **b** Fe^{3+} components vs. total absorbed α-dose

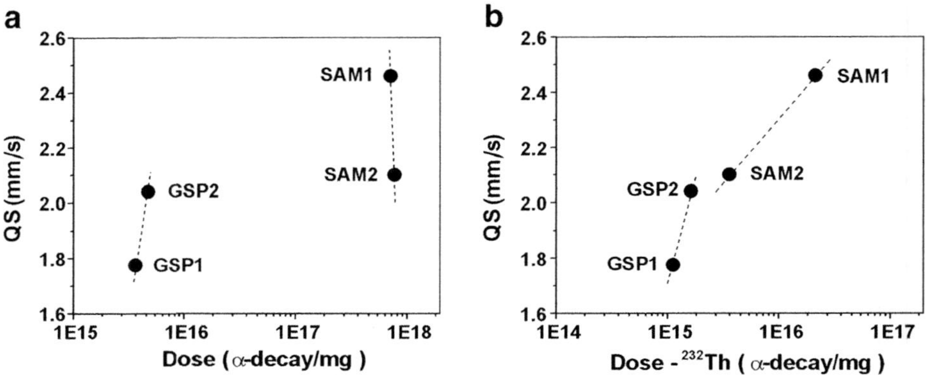

Fig. 4 Average quadrupole splittings for Fe^{2+} doublets **a** vs. total absorbed α-dose and **b** vs. absorbed α-dose from the ^{232}Th series

metamictization, while the doublet no. 2 represents the Fe^{2+} octahedra that have undergone an expansion during metamictization.

The total line widths of the high energy peaks of Fe^{2+} and Fe^{3+} components vs. total α-dose are shown in Fig. 3a, b, respectively. Figure 3a shows that the line widths of the Fe^{2+} peaks increase with α-dose both for gadolinites and samarskites. Figure 3b shows that the Fe^{3+} line widths increase with α-dose for samarskites.

Figure 4a, b show the average of quadrupole splittings for Fe^{2+} doublets vs. total absorbed α-dose and vs. absorbed α-dose from the ^{232}Th series, respectively. As can be seen in Fig. 4, plotting the quadrupole splitting as a function of the α-dose from ^{232}Th gives the same increasing trend both for gadolinites and samarskites, unlike plotting as a function of total α-dose. The average energy of α particles from the ^{232}Th series is 6.14 MeV and the average energy of recoil nuclei is about 105 keV. These values are higher than for the ^{238}U series, where the average energy of α particles and of recoil nuclei is 5.34 MeV and 89 keV, respectively. The α-decays from the ^{232}Th series cause larger radiation damage than α-decays from ^{238}U. From the ^{232}Th decay, SAM1 has obtained an α-dose six times higher than that of SAM2, whereas this ratio for gadolinites (GSP2/GSP1) is about 1.4 (Table 1). Thus, for metamict minerals, the increase of average quadrupole splittings of Fe^{2+} doublets in the high degree may be controlled by the α-dose from ^{232}Th.

4 Conclusions

The line widths of the Fe^{2+} components for gadolinites and samarskites and the widths of the Fe^{3+} components for samarskites increase with total absorbed α-dose. The increase in average quadrupole splitting for Fe^{2+} doublets correlates with absorbed α-dose from ^{232}Th for both gadolinites and samarskites.

Acknowledgement This work was supported by the State Committee for Scientific Research, Poland, through grant no. 2P04D06229.

References

1. Weber, W.J., Ewing, R.C., Catlow, C.R.A., Diaz de la Rubia, T., Hoobs, L.W., Kinoshita, C., Matzke, H., Motta, A.T., Nastas, M., Salje, E.K.H., Vance, E.R., Zinkle, S.J.: Radiation effects in crystalline ceramics for the immobilization of high-level nuclear waste and plutonium. J. Mater. Res. **13**, 1434–1484 (1998)
2. Warner, J.K., Ewing, R.C.: Crystal chemistry of samarskite. Am. Mineral. **78**, 419–424 (1993)
3. Miyawaki, R., Nakai, I., Nagashima, K.: A refinement of the crystal structure of gadolinite. Am. Mineral. **69**, 984–953 (1984)
4. Bryant, B., McGrew, L.W., Wobus, R.A.: Geologic map of the Denver 1 × 2 degree quadrangle, north-central Colorado: U.S. Geological Survey Miscellaneous Investigations Map I-1163, scale 1:250,000 (1981)
5. Pin, H., Mierzejewski, M.P., Duthou, J.L.: Age of Karkonosze Mts. Granite dated by isochrome Rb/Sr and its initial $^{87}Sr/^{86}Sr$ value. Prz. Geol. **10**, 512–517 (1987)
6. Nakai, I., Akimoto, J., Imafuku, M., Miyawaki, R., Sugitani, Y., Koto, K.: Characterization of the amorphous state in metamict silicates and niobatesby EXAFS and XANES analyses. Phys. Chem. Miner. **15**, 113–124 (1987)
7. Malczewski, D., Janeczek, J.: Activation energy of annealed metamict gadolinite from ^{57}Fe Mössbauer spectroscopy. Phys. Chem. Miner. **29**, 226–232 (2002)

Hyperfine Interact (2010) 195:93–98
DOI 10.1007/s10751-009-0106-6

Synthesis of hematite nanowires using a mesoporous hard template

L. A. Cano · J. F. Bengoa · S. J. Stewart · R. C. Mercader · S. G. Marchetti

Published online: 30 September 2009

Abstract The influence of the impregnation media (ethanol or water) and the calcination atmosphere (air and NO/He) on the hematite nanowires production embedded on a hard template (MCM-41) was studied. The solids were characterized by X-ray diffraction, Mössbauer spectroscopy and magnetic measurements. The results obtained indicate that the more appropriate conditions for the iron oxide nanowires to get inside the MCM-41 hard template seem to be reached using water as a solvent and air as calcination atmosphere.

Keywords Nanoarrays · Nanoparticles · Hematite · Fe/MCM-41

1 Introduction

The properties exhibited by arrays of nanowires and nanotubes are currently intensively investigated because of the new phenomena that they give rise and the wealth of applications that these systems offer, like potential applications in high-density magnetic recording media and other nanodevices [1]. Several methods have been used to obtain nanowires with specific physical properties, for example: laser ablation [2], electrodeposition [3] or supercritical fluid phase inclusion [4]. However, these techniques have the common disadvantage of their complexity.

Aiming at overcoming this difficulty, a highly ordered mesoporous template could be used as a host, where the material of interest can be embedded. MCM-41 is an almost perfectly ordered mesoporous material suitable for this purpose. It has been reported that conveniently magnetic compounds [5, 6] can be confined in a quasi

L. A. Cano · J. F. Bengoa · S. G. Marchetti (✉)
Cindeca, Facultad De Ciencias Exactas, UNLP, CICPBA, CONICET, La Plata, Argentina
e-mail: march@quimica.unlp.edu.ar

S. J. Stewart · R. C. Mercader
IFLP-CONICET, Facultad De Ciencias Exactas, UNLP, La Plata, Argentina

one-dimensional array due to its particular topology of regular hexagonal parallel channels with usual average diameters of less than 5 nm.

In a previous work [7] we have obtained nanowire geometry by embedding hematite nanoparticles in the hexagonal array of the very narrow channels of MCM-41. Hematite was chosen as the first step of the preparative methods because its handling is easier than the magnetic iron oxides, maghemite and magnetite. Notwithstanding, the magnetic oxides are more interesting due to their potential applications [1]. Currently we are carrying out alternative experiments to induce the structural transformations that lead from hematite nanowires to maghemite and magnetite nanowires.

In this work, we report the synthesis of three samples prepared in different conditions by incipient wetness impregnation of MCM-41 with iron nitrate to yield an iron nominal concentration of 15 wt.%. The aim of the present work is to study the effect of the nature of the impregnating solution and the calcination atmosphere on the structural properties of the iron oxides nanowires.

2 Experimental section

The MCM-41 matrix was prepared according to the methodology proposed by Ryoo and Kim [8]; firstly 40 g of sodium silicate (26.1% SiO_2) is dissolved into 74 g of water. The solution was then slowly added to 38 ml of cetyltrimethylammonium chloride and 0.65 ml of NH_3 with vigorous stirring at room temperature. This mixture was heated in a polypropylene bottle, without stirring to 373 K for 24 h, afterwards, it was cooled to room temperature. The pH was adjusted to approximately 11 by drop wise addition of acetic acid with vigorous stirring. The reaction mixture was heated again to 373 K for 24 h. This procedure for pH adjustment and subsequent heating was repeated twice. The resulting solid was filtered, washed and dried in air at room temperature. It was then calcined at 813 K for 1 h in flowing N_2 (150 cm^3/min) followed by 6 h in flowing air (150 cm^3/min).

Three solids were obtained treating MCM-41 in the following ways in order to obtain iron concentrations of about 15 wt.%:

Fe/MCM-41 (alc-air) incipient wetness impregnation with $Fe(NO_3)_3{\cdot}9H_2O$ in ethanolic solution. The sample was dried at room temperature in atmospheric air, calcined in dry air stream (100 cm^3/min) from 298 to 723 K at 1 K/min and kept at 723 K for 4 h.

Fe/MCM-41 (alc-NO/He) impregnated in the same way that Fe/MCM-41 (alc-air). The sample was dried at room temperature in atmospheric air, calcined in a dry mixture of 1% (*V*/*V*) of NO in He (100 cm^3/min) from 298 to 723 K at 1 K/min and kept at 723 K for 4 h following the methodology proposed by Sietsma et al. [9].

Fe/MCM-41 (aq-air) incipient wetness impregnation with $Fe(NO_3)_3{\cdot}9H_2O$ aqueous solution. The sample was dried at room temperature in atmospheric air, and calcined in the same way than Fe/MCM-41 (alc-air).

The resulting solids were characterized by X-ray diffraction (XRD), Nitrogen adsorption (BET), Mössbauer spectroscopy and magnetic measurements.

All XRD patterns were measured using a standard automated powder X-ray diffraction system Philips PW 1710 with diffracted-beam graphite monochromator using Cu Kα radiation ($\lambda = 1.5406$ Å). Nitrogen adsorption isotherms were recorded at the temperature of liquid nitrogen (77 K) using a Micromeritics ASAP 2020 apparatus.

The Mössbauer spectra at 298 and 30 K were taken in transmission geometry with a constant acceleration spectrometer. A source of ^{57}Co in Rh matrix of nominally 50 mCi was used. Velocity calibration was performed against a metallic 12 μm-thick α-Fe foil at room temperature. All isomer shifts (δ) are referred to this standard. The temperature between 30 and 298 K was varied using a Displex DE-202 Closed Cycle Cryogenic System. The Mössbauer spectra were fitted using the Recoil program [10]. At 30 K the Blume and Tjon's dynamic line shape site analysis option was used [10]. Lorentzian lines with equal linewidths were considered for each spectrum component in a complete superparamagnetic regime.

Magnetic measurements of Zero Field Cooling (ZFC) and Field Cooling (FC), recorded between 5 and 300 K, were carried out using a commercial superconducting quantum interference device magnetometer from Quantum Design with a field of $H_{FC} = 500$ Oe.

3 Results and discussion

The typical structural properties of an ordered mesoporous solid MCM-41, used as template, were confirmed by XRD and by their textural properties measured by N_2 adsorption using the BET method. Because of the lack of space, the results are not shown here. The XRD measurements allowed verifying that the impregnation and calcination treatments did not modify the solid structure.

The Mössbauer spectra of all samples at 298 K only display a doublet (Fig. 1). The hyperfine parameters obtained by the fitting procedure (Table 1) are identical, within the experimental errors, for all solids. They can be assigned to superparamagnetic α-Fe_2O_3 or paramagnetic Fe^{3+} ions. Taking into account our previous results, the absence of a sextuplet at room temperature, is an indication that nearly all the iron species are inside the MCM-41 channels [7].

When the temperature decreases down to 30 K, the spectrum of Fe/MCM-41 (aq-air) shows a magnetic splitting with a very intense central peak and four additional peaks (Fig. 1). One relaxing sextet and a very broad singlet were used in the fittings. The hyperfine parameters of the sextuplet can be assigned to small α-Fe_2O_3 crystals [11] (Table 1). Therefore, from the thermal evolution of the spectra we conclude that iron species in Fe/MCM-41 (aq-air) consists of α-Fe_2O_3constituted by crystallites of very small size, which are in a superparamagnetic relaxation regime at room temperature. Since at 30 K the magnetic signal area is about 58%, we can consider that the blocking temperature is not far from this value.

The Fe/MCM-41 (alc-air) spectrum at 30 K displayed a central doublet and a curved background typical of a superparamagnetic signal belonging to a magnetic system that is close to its blocking temperature (Fig. 1). It was fitted with a doublet and a relaxing sextet. Following the same reasoning that in the sample impregnated

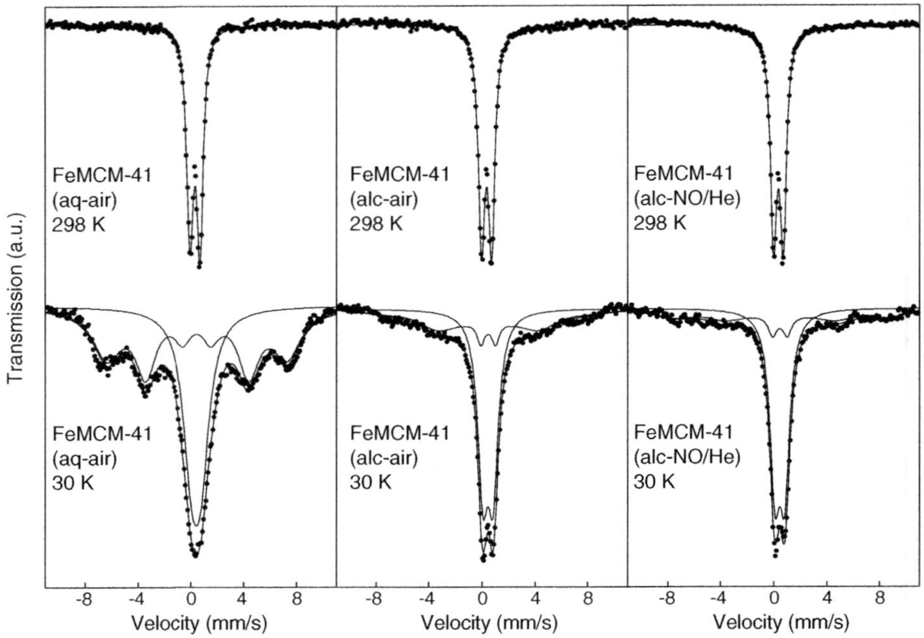

Fig. 1 Mössbauer spectra of all solids at 298 and 30 K

Table 1 Hyperfine parameters and blocking temperatures of the samples

Temperature	Parameters	Fe/MCM-41 (aq-air)	Fe/MCM-41 (alc-air)	Fe/MCM-41 (alc-NO/He)
298 K	Δ (mm/s)	0.75 ± 0.01	0.75 ± 0.01	0.72 ± 0.01
	δ (mm/s)	0.35 ± 0.01	0.35 ± 0.01	0.36 ± 0.01
	%	100 ± 2	100 ± 2	100 ± 2
30 K	H (T)	54.2 ± 0.3	53[a]	53[a]
	δ (mm/s)	0.46 ± 0.03	0.47[a]	0.47[a]
	2ε (mm/s)	-0.02 ± 0.05	0[a]	0[a]
	%	58 ± 2	39 ± 2	36 ± 5
	δ (mm/s)	0.44 ± 0.01	–	–
	%	42 ± 1	–	–
	Δ (mm/s)	–	0.77 ± 0.01	0.73 ± 0.03
	δ (mm/s)	–	0.46 ± 0.01	0.46 ± 0.01
	%	–	61 ± 1	64 ± 1
ZFC-FC blocking temperature (K)		40 ± 2	15 ± 2	< 6

H hyperfine magnetic field in Tesla, *δ* isomer shift (all the isomer shifts are referred to α-Fe at 298 K), 2*ε* quadrupole shift, *Δ* quadrupole splitting
[a] Parameter held fixed in fitting

with aqueous solution, the only species present would be α-Fe_2O_3. However, in this sample the blocking temperature is lower than 30 K, since only 39% is magnetically blocked at that temperature. Therefore, the nature of the solvent has produced a difference on the α-Fe_2O_3 crystallite sizes. Perhaps, the lower surface tension of

Fig. 2 ZFC-FC curves for Fe/MCM-41 (aq-air)

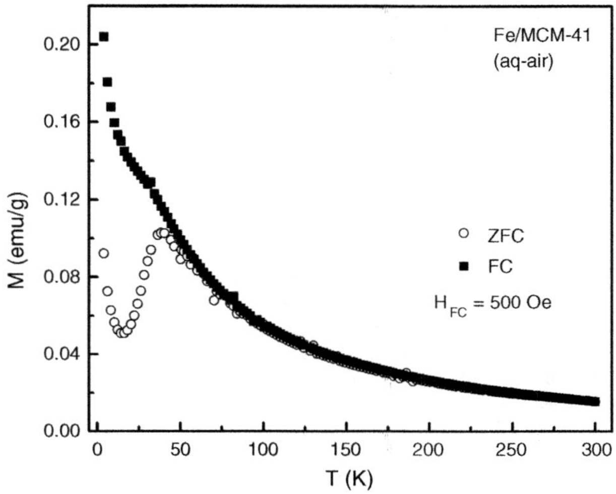

the ethanolic solution has favoured the "spreading" of the iron species on the wall channels of the MCM-41 template leading to smaller crystallite sizes.

To study the influence of the calcination atmosphere, a sample impregnated in ethanolic solution was calcined in NO/He flow, as proposed by Sietsma et al. [9]. According to these authors, this procedure enables the preparation of high loading of Co_3O_4 (15–18 wt.%) inside the pores of SBA-15 with high dispersion. In our system (Fe/MCM-41 (alc-NO/He)) the Mössbauer spectrum at 30 K is very similar to that obtained by Fe/MCM-41 (alc-air; Fig. 1). There is only a slight difference between the percentages of the magnetically blocked signal (Table 1). Therefore, the blocking temperatures of both samples must be very similar.

The values of the blocking temperatures obtained from ZFC-FC magnetization results are in good agreement with the estimations from Mössbauer spectroscopy, taking into account the different sampling times of both techniques. These values are displayed in Table 1. Besides, at temperatures higher than the corresponding blocking temperatures, the ZFC-FC curves are coincident, confirming the superparamagnetic state of the iron species. As an example, in Fig. 2 it can be seen the ZFC-FC curves for Fe/MCM-41 (aq-air).

4 Conclusions

In the present work, we have demonstrated that in Fe/MCM-41 (15% wt of Fe) system, the effect of the solvent on the structural properties of the iron species is more important than that of the calcination atmosphere. Considering that in all samples the maximum diameter of the iron oxide particles is determined by the channel diameter of the MCM-41, the higher blocking temperature in Fe/MCM-41 (aq-air) would indicate a higher shape anisotropy magnetic constant. Therefore, the more appropriate conditions for the iron oxide nanowires to get inside the MCM-41 hard template seem to be reached using water as a solvent and air as calcination atmosphere.

Acknowledgements The authors acknowledge support of this work by Consejo Nacional de Investigaciones Científicas y Técnicas, Comisión de Investigaciones Científicas de la Provincia de Buenos Aires, Universidad Nacional de La Plata and Agencia Nacional de Promoción Científica y Tecnológica (PICT no. 14-11267).

References

1. Pankhurst, Q.A., Pollard, R.J.: J. Phys. Condens. Matter. **5**, 8487 (1993)
2. Morales, A.M., Lieber, C.M.: Science **279**, 208 (1998)
3. Lederman, M., O'Barr, R., Schultz, S.: Trans. Magn. **31**, 3793 (1995)
4. Crowley, T.A., Ziegler, K.J., Lyons, D.M., Erts, D., Olin, H., Morris, M.A., Holmes, J.D.: Chem. Mater. **15**, 3518 (2003)
5. Liu, S., Wang, Q., Van der Boort, P., Cool, P., Vansant, E.F., Jiang, M.: J. Magn. Magn. Mater. **280**, 31 (2004)
6. Jung, J.S., Choi, K.H., Chae, W.S., Kim, Y.R., Jun, J.H., Malkinski, L., Kodenkandath, T., Zhou, W., Wiley, J.B., O'Connor, C.J.: J. Phys. Chem. Solids **64**, 385 (2003)
7. Cagnoli, M.V., Gallegos, N.G., Bengoa, J.F., Alvarez, A.M., Moreno, M.S., Roig, A., Marchetti, S.G., Mercader, R.C.: In: American Institute of Physics Conference Proceedings, vol. 765, pp. 13–22 (2005)
8. Ryoo, R., Kim, J.M.: J. Chem. Soc. Chem. Commun. **711** (1995)
9. Sietsma, J.R.A., den Breejen, J.P., de Jongh, P.E., van Dillen, A.J., Bitter, J.H., de Jong, K.P.: Stud. Surf. Sci. Catal. **167**, 55–60 (2007)
10. Lagarec, K., Rancourt, D.G.: Mossbauer Spectral Analysis Software, Version 1.0. Department of Physics, University of Ottawa (1998)
11. Răileanu, M., Crişan, M., Crişan, D., Brăileanu, A., Drăgan, N., Zaharescu, M., Stan, C., Predoi, D., Kuncser, V., Marinescu, V.E., Hodorogea, S.M.: J. Non-Cryst. Solids **354**, 624–631 (2008)

Hyperfine Interact (2010) 195:99–104
DOI 10.1007/s10751-009-0107-5

Mössbauer characterization of surface-coated magnetic nanoparticles for applications in transformers

**E. S. Leite · A. C. Oliveira · V. K. Garg ·
P. P. C. Sartoratto · W. R. Viali · P. C. Morais**

Published online: 25 September 2009

Abstract In this study the synthesis of very small magnetite nanoparticles is reported. Analysis of the (311) X-ray diffraction line indicates nanoparticle with 2.2 nm average diameter. The as-synthesized nanosized magnetite sample was submitted to a chemical oxidation process for conversion to maghemite. However, the yielding of the chemical oxidation process, as indicated by the analysis of the Mössbauer data, was about 20%. This finding indicates a strong size-dependence of the oxidation process, reducing the yielding of the chemical process used as the nanoparticle size reduces down to extremely low values.

Keywords Mössbauer · Magnetite nanoparticle · Oxidation process · Size-dependence · Magnetic fluid · Transformer

1 Introduction

Development of oil-based magnetic fluid (MF) samples for use in high voltage transformers technology requires highly-stable MFs, displaying specific electrical properties and able to operate at temperatures above 100°C and under high voltages for long periods of time [1]. Colloidal stability of organic-based MF samples can be achieved via balancing the Van der Waals, magnetic dipole and steric interactions,

E. S. Leite (✉) · A. C. Oliveira · V. K. Garg · P. C. Morais
Instituto de Física, Núcleo de Física Aplicada, Universidade de Brasília,
Brasília 70910-900, Brazil
e-mail: esilva2308@hotmail.com

P. P. C. Sartoratto · W. R. Viali
Instituto de Química, Universidade Federal de Goiás, Goiânia, Goiâs 74001-970, Brazil

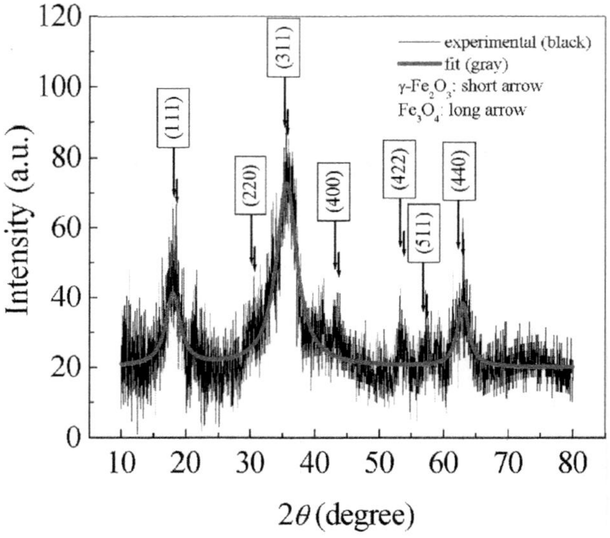

Fig. 1 X-ray diffractogram of the sample after the oxidation process. The *gray solid line* is the curve-fitting of the three main features using Lorentzian shaped curves. The standard X-ray lines of magnetite and maghemite are indicated with long and short arrows, respectively

the latter due to the presence of long chain hydrocarbon molecules comprising the molecular shell dressing the suspended magnetic core nanoparticle. Phase stability of the magnetic nanosized core is also a key issue in producing highly-stable MF samples for applications in high voltage transformers. In this study, Mössbauer spectroscopy was used as the main technique in the investigation of nanosized magnetic particles surface-coated with oleic acid (OA). The nanosized magnetic powder used for OA surface-coating was obtained by oxidation of native nanosized magnetite. Nevertheless, the oxidation process aimed to oxidize magnetite (Fe_3O_4) up to maghemite (γ-Fe_2O_3) seems to be size-dependent. Therefore, starting with a polydispersed magnetite sample one has the possibility to obtain a partially-oxidized sample, consisting of a mixture of magnetite and maghemite. In order to investigate this aspect of the nanoparticulated powder sample Mössbauer spectra were recorded at 300 and 77 K. X-ray diffraction (XRD) data of the investigated sample will be used to provide information in regard to the average size of the nanoparticulated material and to support the discussion of the Mössbauer data *regarding the material phases*.

2 Experimental

Preparation of native maghemite nanoparticle was accomplished in a two-step procedure whereas its surface-coating with oleic acid (OA) was realized in a third step [2]. In the first step magnetite nanoparticle was precipitated after pouring NaOH aqueous solution under vigorous stirring into an aqueous solution containing both Fe(II) and Fe(III) ions. In the second step the oxidation of the as-produced magnetite was performed by adding HCl aqueous solution to the fresh black sediment (setting the pH at 3.5) and the resulting suspension was heated at 97°C under stirring, for 3 h, while oxygen was bubbled throughout the aqueous suspension. In the third step pure oleic acid was added under stirring to the previously-oxidized suspension

Table 1 Standard XRD lines (2θ) of maghemite (JCPDS-ICDD 1346 chart), magnetite (JCPDS-ICDD 9-629 chart) and the approximated values of the XRD lines (2θ) identified in the sample investigated in the present study

Hkl	111	220	311	400	422	511	440
$\gamma-Fe_2O_3$ (2θ)	18.38	30.24	35.63	43.28	53.73	57.27	62.92
Fe_3O_4 (2θ)	18.27	30.09	35.42	43.05	53.39	56.94	62.51
This study	18.02	30.44	35.66	43.26	53.06	58.34	63.02

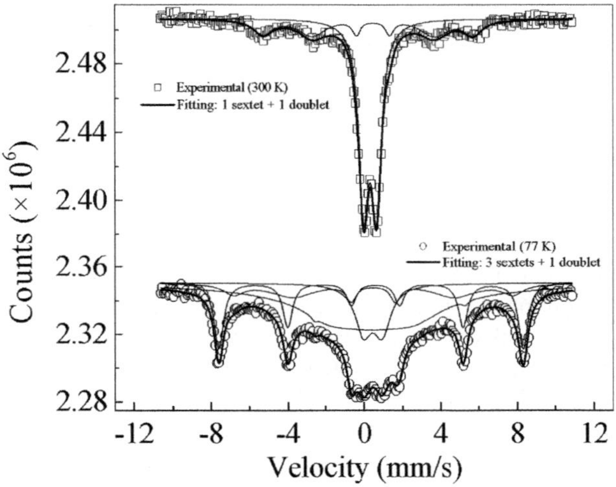

Fig. 2 Three hundred and 77 K Mössbauer spectra of the sample after the oxidation process. *Open squares* and *circles* are experimental data whereas *thin solid lines* represent the sub-spectra obtained from the curve-fitting using combination of sextets and doublets. The *thicker solid line* represents the convolution of the sub-spectra

whereas the pH was set at 6 by adding NH_4OH aqueous solution. The suspension was stirred for 30 min. At the end of the third step the OA-coated nanoparticles were captured in organic phase which was separated from the aqueous phase by decantation and further washed with ethanol until the excess of free oleic acid was completely removed. The OA-coated nanoparticles were then dried out (powder sample), removing the ethanol for further suspension in insulating mineral oil to produce the stock diluted MF sample (4 g of the OA-coated nanoparticles in 100 mL of insulating oil). The powder sample produced before the surface-coating step (AO-coating) was investigated using XRD and Mössbauer spectroscopy. Data in Fig. 1 represents the XRD pattern of the OA-coated nanosized sample in the 2θ range of 10° to 80°, showing the typical reflection lines associated to both magnetite and maghemite. Long (short) arrows represent the standard XRD line positions of magnetite (maghemite), as quoted in the JCPDS-ICDD charts (9-629 and 1346 for magnetite and maghemite, respectively) and collected in Table 1. The peak positions observed in the XRD of the sample investigated are also included in Table 1. The 300 and 77 K Mössbauer spectra of the sample are presented in Fig. 2. Symbols in Fig. 2 represent the experimental points, thin solid lines represent the Mössbauer sub-spectra and the thick solid lines are the best curve fitting for the data. The corresponding hyperfine parameters are collected in Table 2.

Table 2 Mössbauer parameters associated to the 77 K sub-spectrum obtained from the present study and from the magnetite sample reported on [4]

Spectrum component	IS (mm/s)	QS (mm/s)	IS (mm/s)	QS (mm/s)	HF (T)	Area %
Doublet (Fe_3O_4)[a]	0.46	0.97				13.2
Sextet (Fe_3O_4)[a]			0.48	−0.0003	45.4	16.1
Sextet (Fe_3O_4)[a]			0.89	1.86	24.5	50.2
Sextet (γ-Fe_2O_3)[a]			0.47	−0.23	49.4	20.5
Sextet (Fe_3O_4)[b]			0.40	0.00	50.7	42
Sextet (Fe_3O_4)[b]			0.54	−0.04	52.8	14
Sextet (Fe_3O_4)[b]			0.78	−0.03	49.9	25
Sextet (Fe_3O_4)[b]			0.86	0.02	46.0	15
Sextet (Fe_3O_4)[b]			1.15	2.19	34.5	4

[a] Present study

[b] [4]

3 Data analysis

Though the XRD data presented in Fig. 1 does not allows the estimation of the magnetite-to-maghemite content ratio precisely it was used for both identifying the XRD peaks associated with the two main phases (magnetite and maghemite) as well as to estimate the average nanoparticle diameter (2.2 nm), the later using the Scherrer's relation [3]. The standard XRD lines associated to magnetite and maghemite (see Table 1) are indicated in Fig. 1 by long and short arrows, respectively. The XRD features observed in Fig. 1 can be quickly identified with the typical XRD lines of magnetite and maghemite, indicating the presence of both phases. The solid gray line in Fig. 1 represents the best curve fitting of the three main XRD features, namely (111), (311) and (440) using Lorentzian lines. The corrected linewidth at half-height of the (311) reflection was used to estimate the average nanoparticle diameter of about 2.2 nm. Analysis of the Mössbauer spectra, however, was used to provide the information in regard to the magnetite-to-maghemite content ratio. The room-temperature Mössbauer spectrum was curve-fitted using one sextet and one doublet, as shown in Fig. 2 (upper panel). Differently, the liquid-nitrogen Mössbauer spectrum was curve-fitted using three sextets and one doublet, as indicated in Fig. 2 (lower panel). Table 2 summarizes the hyperfine parameters obtained from the 77 K Mössbauer sub-spectra. Included in Table 2 are typical hyperfine parameters obtained from the analysis of a nanosized magnetite sample at 77 K, whose Mössbauer spectrum was curve-fitted using five sextets [4].

The Mössbauer spectra presented in Fig. 2, including the spectrum evolution from 77 to 300 K, is typical of superparamagnetic particles [5, 6]. The values we found for the hyperfine parameters of the sub-spectra indicate a mixture of magnetite and maghemite. From the Mössbauer analysis about 20% of the sample is composed by nanosized maghemite whereas about 80% consists of nanosized magnetite, as indicated in Table 2. This assumption is supported by the analysis of the 77 K Mössbauer spectrum as follows. As presented in Table 2 the hyperfine parameters obtained from the analysis of the sextet, here claimed to be associated to nanosized maghemite, is in agreement with the literature [7]. Likewise, the hyperfine parameters obtained from the analysis of the two sextets, here claimed to be associated to nanosized magnetite (sites A and B), is in agreement with the literature [4, 8]. In addition, the hyperfine parameters obtained from the analysis of the doublet of the 77 K Mössbauer spectrum, here claimed to be associated to nanosized magnetite, is in good agreement with the literature [9]. This conclusion is supported by the XRD features observed in Fig. 1 and quoted in Table 1. As far as the two-phase composition of the sample is concerned the analysis of the Mössbauer spectrum performed at 77 K is consistent with the analysis of the XRD data. Indeed, Mössbauer parameters obtained from measurements performed on nanosized oxide-based magnetic particles, as for instance magnetite, may reveal a significant variability which could be explained by the preparation route employed, including oxidation/reduction steps, average size, size dispersion, and surface functionalization [4].

What is quite surprising from the analysis of the 77 K Mössbauer spectrum is the high content of magnetite, despite the oxidation chemical process the sample was submitted to. This finding indicates the strong dependence of the chemical oxidation yielding upon the particle size, for this particular oxidation route of magnetite. Our findings indicate that the oxidation yielding reduces as the average magnetite nanoparticle size reduces. The literature reports high oxidation yielding of magnetite using this chemical route, but the average particle diameter used by van Ewijk et al. [2] was about 10 nm. In addition, as far as the particle size dispersion is concerned the literature reports wide distributions [2]. This characteristic supports the hypothesis of partial oxidation of magnetite to maghemite, taking place only in the higher particle size end of the size distribution. We then hypothesize that the oxidation process possibly creates a selective (size-dependent) fraction of core-magnetite structure, extremely small, responsible for the doublet observed at 77 K. Within this picture the oxidized sample may be best represented by a bimodal distribution of nanosized magnetite, the bigger component being responsible for the sextets observed at 77 K. Note that the (311) X-ray linewidth may probe the smallest component of this bimodal distribution with an average diameter of 2.2 nm. Actually, further analysis taking into account samples with quite different average particle diameter is presently under progress to elucidate the size-dependence of the chemical oxidation route used in this study.

4 Conclusions

In conclusion, nanosized magnetite particles was synthesized and further oxidized to maghemite. However, the oxidation chemical route used in the present study seems to be strongly size-dependent. In the very small size limit the oxidation process used here presented a reduced chemical yielding, as indicated by the analysis of the Mössbauer spectra. These findings were also supported by X-ray diffraction analysis.

Acknowledgements This work was supported by the Brazilian agencies CNPq, ELETRONORTE, and FINATEC.

References

1. Sartoratto, P.P.C., Neto, A.V.S., Lima, E.C.D., Rodrigues de Sá, A.L.C., Morais, P.C.J.: Appl. Phys. **97**, 10Q917 (2005)

2. van Ewijk, G.A., Vroege, F.J., Philipse, A.P.: J. Magn. Magn. Mater. **201**, 31 (1999)
3. Cullity, B.D.: Elements of X-ray Diffraction, pp. 101. Addison Wesley, Reading (1978)
4. Daou, T.J., Pourroy, G., Bégin-Colin, S., Grenèche, J.M., Ulhaq-Bouillet, C., Legaré, P., Bernhardt, P., Leuvrey, C., Rogez, G.: Chem. Mater. **18**, 4399 (2006)
5. Morais, P.C., Skeff Neto, K.J.: Appl. Phys. **54**, 307 (1983)
6. Li, X., Kutal, C.J.: Alloys Comp. **349**, 264 (2003)
7. Sprenkel-Segel, E.L.: J. Geophys. Res. **75**, 6618 (1970)
8. Greenwood, N.N., Gibb, T.C.: Mössbauer Spectroscopy, pp. 241. Chapman and Hall, London (1971)
9. Golden, D.C., Ming, D.W., Bowen, L.H., Morris, R.V., Lauer, H.V. Jr.: Clays Clay Miner. **42**, 53 (1994)

Hyperfine Interact (2010) 195:105–109
DOI 10.1007/s10751-009-0131-5

Formation of nanostructured $Al_{64}Cu_{23}Fe_{13}$ icosahedral quasicrystal by the ball milling technique

J. Quispe Marcatoma · C. V. Landauro · M. Taquire · Ch. Rojas Ayala · M. Yaro · V. A. Peña Rodriguez

Published online: 1 October 2009

Abstract Nanocrystalline $Al_{64}Cu_{23}Fe_{13}$ icosahedral quasicrystal has been obtained by milling of solid quasicrystal precursors prepared by arc-melt. The local structure around Fe atoms was studied by Mössbauer spectroscopy using a quadrupole splitting distribution method. Mössbauer results of annealed and milled samples show the existence of a broadened distribution of Fe sites which is associated to intrinsic disorder. The structural characterization was determined using x-ray diffraction. The average grain-size of the nanostructured quasicrystal, obtained from the line broadening of the X-ray diffraction peaks, was estimated to be of the order of 10 nm for a sample milled by 5 h.

Keywords Quasicrystals · Mechanical milling · Nanostructures

1 Introduction

Since the discovery of quasicrystals (QCs) [1] the interest to employ these materials for diverse technological applications was growing. Specially exciting properties for industrial applications are low coefficient of friction [2, 3], high hardness [4], low surface energy [5], and good wear-resistance [6]. QCs are also potential candidates for applications as thermoelements [7–10] and in catalysis [11]. In particular, the stable icosahedral *i*-Al-Cu-Fe QC, discovered in 1987 by Tsai et al. [12], is currently one of the best studied icosahedral QCs due to its excellent thermal stability.

J. Quispe Marcatoma (✉) · C. V. Landauro · M. Taquire · Ch. Rojas Ayala · M. Yaro · V. A. Peña Rodriguez
Facultad de Ciencias Físicas, Universidad Nacional Mayor de San Marcos, P.O.Box 14-0149, Lima 14, Peru
e-mail: jquispem@unmsm.edu.pe

M. Yaro
Universidade Federal do ABC, Rua Santa Adélia 166-Bairro-Bangu, Santo André, Sao Paulo, Brazil

Moreover, the nominal composition $Al_{64}Cu_{23}Fe_{13}$ is considered to represent the ideal icosahedral composition [13].

In the last years nanostructured materials are intensively studied because they provide the possibility to improve the physical properties of their solid counterparts. The challenge lies in controlling these properties as a function of the grain size. Hence, the study of the nanostructured counterparts of this promising new material is of relevant importance. In this sense, ball milling has received special attention since it was firstly used by Benjamin and Koch [14, 15] as one of the powerful methods for material processing, and specially for reducing the grain size (to the nanometer scale) of the sample under study [15–18].

The main goal of the present work is to study the possibility to obtain nano-quasicrystalline *i*-Al-Cu-Fe starting from a solid sample produced by arc furnace that should be then nanostructured employing a high energy vibratory ball mill equipment. The experimental details are described in the next section. Section 3 is devoted to the results and discussion. In Section 4 we summarize the main results of the present work.

2 Experimental details

Alloy ingots of the precursor icosahedral solid samples of nominal composition $Al_{64}Cu_{23}Fe_{13}$ were prepared from a mixture of the high-purity chemical elemental powders of Al (99.99+%), Cu (99.99+%) and Fe (99.9+%) using arc furnace in argon atmosphere. Then, the samples were annealed at 700°C in evacuated quartz tubes for up to 168 h with a subsequent quenching in liquid nitrogen. In order to obtain a nanostructured quasicrystal these samples were milled for 5 h in a high energy ball milling SPEX 8000 employing a 7:1 charge ratio. The milling was made under argon atmosphere.

The phase purity of the samples was confirmed by powder x-ray diffraction (XRD) using Cu-Kα radiation ($\lambda = 1.5406$Å). The local structure around of Fe sites was analyzed by transmission Mössbauer spectroscopy (TMS) employing a 25 mCi $^{57}Co/Rh$ radioactive source. Mössbauer spectra, recorded at room temperature, were fitted using the NORMOS fitting program. The isomer shift values are given relative to the α-Fe value, obtained at room temperature.

3 Results and discussion

Figure 1 shows the XRD patterns of the $Al_{64}Cu_{23}Fe_{13}$ sample (a) as prepared in the arc furnace, (b) after heat treatment, and (c) after 5 h milling. As can be seen in Fig. 1a, b the quasicrystalline sample needs to be heat treated in order to remove other metastable phases (indicated by arrows in Fig. 1a), which are generally associated to the $Al_{13}Fe_4$ and ε-Al_2Cu_3 phases [19]. From our analysis we confirm the presence of the ε-Al_2Cu_3 phase, however, the presence of the $Al_{13}Fe_4$ phase could not be clearly established. The presence of the quasicrystalline phase is evident from the XRD pattern shown in Fig. 1b where all the diffraction peaks can be indexed according to the icosahedral scheme proposed by Cahn et al. [20], in good agreement with previous works [21, 22]. Following Weisbecker et al. [23] we determine the six-

Fig. 1 X-ray diffraction patterns of the *i*-$Al_{64}Cu_{23}Fe_{13}$ sample *(a)* as prepared by arc furnace, *(b)* after heat treatment, and *(c)* after 5 h milling

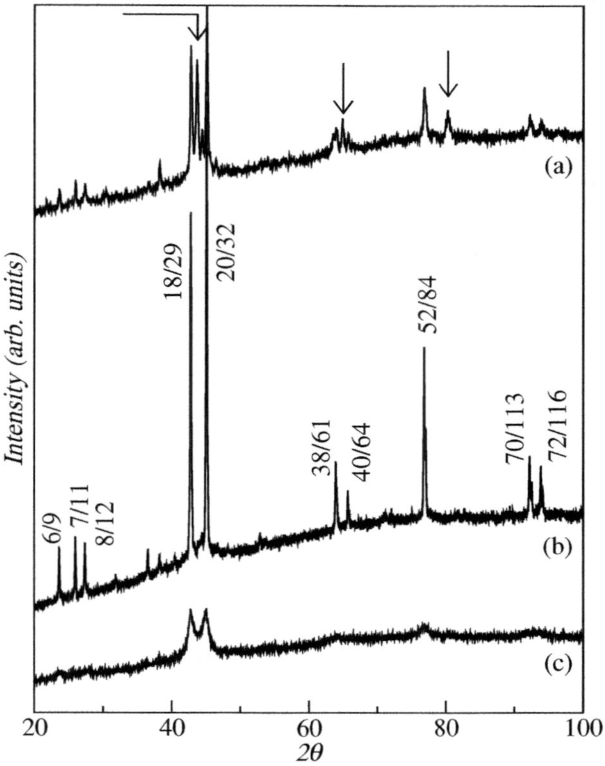

dimensional lattice constant obtaining a value of $a_{6D} = 6.309$Å which is very close to values reported previously [23, 24] (for more details see [25]). After ball-milling, a nanostructured quasicrystal is formed, which is indicated by a reduction of the peak intensity and the broadening of the diffraction peaks. To determine the grain-size of the sample we employ the Scherrer equation. Thus, after 5 h milling the average grain-size has been estimated to be ~10 nm. It is worth mention that there is not evidence of the formation of other phases during the milling process, as can be seen in Fig. 1c.

The local order around the Fe atoms was studied from TMS measurements of the solid sample (before and after heat treatment) and its nanostructured counterpart (for 5 h milling). In order to fit the Mössbauer spectra we employ a quadrupole splitting distribution associated to a non-equivalent paramagnetic Fe-sites distribution. A distribution of Fe-sites could be expected as a result of the temperature of measurements (room temperature) and/or the presence of intrinsic disorder, as suggested in previous works [26–29], which should be more significant in the nanostructured sample. For our fit we fixed the full width at half maximum $\Gamma = 0.248$ mm/s for all sites. We obtain $\langle\delta\rangle = 0.222(2)$ mm/s, $\langle\Delta\rangle = 0.365(2)$ mm/s and $\chi^2 = 1.07$ for the as-cast sample; $\langle\delta\rangle = 0.230(2)$ mm/s, $\langle\Delta\rangle = 0.390(1)$ mm/s and $\chi^2 = 1.35$ for the heat-treated sample; and $\langle\delta\rangle = 0.238(1)$ mm/s, $\langle\Delta\rangle = 0.441(3)$ mm/s and $\chi^2 = 1.09$ for the 5 h milled sample. These values are very close to those obtained in the literature [26–29]. The results are shown in Fig. 2.

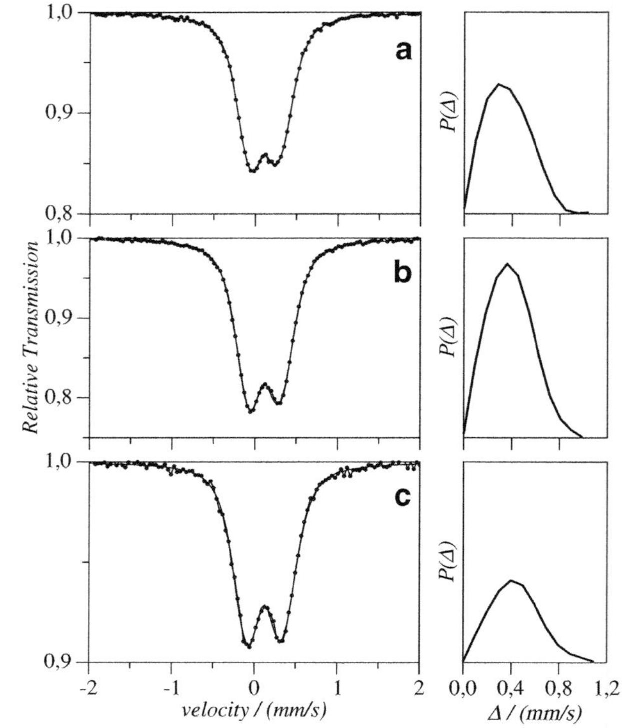

Fig. 2 Mössbauer spectra of the *i*-$Al_{64}Cu_{23}Fe_{13}$ sample **a** as prepared by arc furnace, **b** after heat treatment, and **c** after 5 h milling. The *right panels* correspond to the distributions $P(\Delta)$ for each case of the *left* (see text for details)

From the XRD y TMS measurements we conclude the following: (i) for the initial sample (before heat treatment) there is an extended distribution of non-equivalent Fe-sites which can be associated to the inhomogeneities of the sample, as confirmed by XRD. (ii) The homogeneity of the sample is achieved after heat treatment (see Fig. 1b) however TMS indicates the presence of intrinsic disorder (a distribution of Fe-sites is necessary to obtain a good fit to the experiments, see below). (iii) The 5 h milled sample does not present remarkable changes in the distribution of the Fe-sites. We observe only a reduction of the relative transmission of the Mössbauer spectrum which is expected in nanostructured materials where only the long-range order of the solid sample should be affected (preserving the local order) during the milling process.

Finally, it is worth mention that because of the good structural quality of the thermally treated icosahedral QC (as can be seen from Fig. 1b) and considering the Cockayne model [30] we could expect that a two Fe-sites model is the most recommended one to fit the experiments. Although the fits reproduce the spectra well, the values of Γ differ significantly from 0.25 mm/s conducing to parameter values without physical meaning, as detailed discussed in [26]. On the other hand, a distribution of Fe-sites model seems to be more pertinent to describe the Mössbauer spectra because of the smaller value of χ^2 and the physically justifiable value of Γ, in agreement with other experimental works [26–29].

4 Summary

In summary, we have shown that a nanostructured icosahedral quasicrystal can be obtained after milling its solid counterpart. After 5 h milling the quasicrystalline sample reduces its average grain-size (preserving the local order) to $\sim$ 10 nm. The structure and micro-structure of the studied samples were controlled by means of x-ray diffraction and Mössbauer spectroscopy, respectively. There is also evident the presence of intrinsic disorder in the studied samples. The next step is to obtain the dependence of the milling time with the grain-size in order to study the physical properties of these nanostructured quasicrystals as a function of the grain-size.

Acknowledgements We are grateful to J. Bravo for fruitful discussions. M. Yaro is thankful to the National University of San Marcos for partial financial support through its scientific initiation program (PIC-2006).

References

1. Shechtman, D., Blech, I., Gratias, D., Cahn, J.W.: Phys. Rev. Lett. **53**, 1951 (1984)
2. Dubois, J.M.: Mater. Sci. Eng. A **294–296**, 4 (2000)
3. Brunet, P., Zhang, L.-M., Sordelet, D.J., Besser, M., Dubois, J.M.: Mater. Sci. Eng. A **294–296**, 74 (2000)
4. Wolf, B., Bambauer, K.-O., Paufler, P.: Mater. Sci. Eng. A **298**, 284 (2001)
5. Belin-Ferré, E., Dubois, J.M., Fournée, V., Brunet, P., Sordelet, D.J., Zhang, L.M.: Mater. Sci. Eng. A **294–296**, 818 (2000)
6. Sordelet, D.J., Besser, M.F., Logsdon, J.L.: Mater. Sci. Eng. A **255**, 54 (1998)
7. Pope, A.L., et al.: Appl. Phys. Lett. **75**, 1854 (1999)
8. Bilušic, A., Pavuna, D., Smontara, A.: Vacuum **61**, 345 (2001)
9. Maciá, E.: Appl. Phys. Lett. **77**, 3045 (2000)
10. Maciá, E.: Phys. Rev. B **64**, 94206 (2001)
11. Yoshimura, M., Tsai, A.P.: J. Alloys Compd. **342**, 451 (2002)
12. Tsai, A.P., et al.: Jpn. J. Appl. Phys. **26**, L1505 (1987)
13. Dolinšek, J., et al.: Phys. Rev. B **76**, 54201 (2007)
14. Benjamin, J.S.: Metall. Trans. **1**, 2943 (1970)
15. Koch, C.C.: Ann. Rev. Mater. Sci. **19**, 121 (1989)
16. Zoz, H.: Mater. Sci. Forum **179–181**, 419 (1995)
17. Basset, D., Matteazzi, P., Mani, F.: Mater. Sci. Eng. A **168**, 149 (1993)
18. Barua, P., Murty, B.S., Mathur, B.K., Srinivas, V.: J. Appl. Phys. **91**, 5353 (2002)
19. Lin, C.-M., Lin, S.-H.: Chinese Journal of Physics **31**, 297 (1993)
20. Cahn, J.W., Shechtman, D., Gratias, D.: J. Mater. Res. **1**, 13 (1986)
21. Biggs, B.D., et al.: Phys. Rev. B **43**, 8747 (1991)
22. Chien, C.L., et al.: Phys. Rev. B **45**, 12793 (1992)
23. Weisbecker, P., Bonhomme, G., Bott, G., Dubois, J.M.: J. Non-Cryst. Solids **351**, 1630 (2005)
24. Cornier-Quiquandon, M., Quivy, A., Lefebvre, S., Elkaim, E., Heger, G., Katz, A., Gratias, D: Phys. Rev. B **44**, 2071 (1991)
25. Yaro, M.: Estudio experimental del cuasicristal i-$Al_{64}Cu_{23}Fe_{13}$ y cálculo de la conductividad electrónica en modelos afines. Licenciatura-Thesis, National University of San Marcos, Lima, Peru (2007)
26. Stadnik, Z.M., Stroink, G., Ma, H., Williams, G.: Phys. Rev. B **39**, 9797 (1989)
27. Stadnik, Z.M., Takeuchi, T., Tanaka, N., Mizutani, U.: J. Phys., Condens. Matter **15**, 6365 (2003)
28. Brand, R.A., Pelloth, J., Hippert, F., Calvayrac, Y.: J. Phys., Condens. Matter **11**, 7523 (1999)
29. Hippert, F., Brand, R.A., Pelloth, J., Calvayrac, Y.: J. Phys., Condens. Matter **6**, 11189 (1994)
30. Cockayne, E., Phillips, R., Kan, X.B., Moss, S.C., Robertson, J.L., Ishimasa, T., Mori, M.: J. Non-Cryst. Solids **153–154**, 140 (1993)

Hyperfine Interact (2010) 195:111–115
DOI 10.1007/s10751-009-0132-4

Mössbauer and EXAFS spectroscopy investigation of iron and arsenic adsorption to lettuce leaves

Igor F. Vasconcelos · Gabriela C. Silva · Regina P. Carvalho · Maria Sylvia S. Dantas · Virgínia S. T. Ciminelli

Published online: 29 September 2009

Abstract The accumulation of iron and arsenic from aqueous solution by lettuce leaves biomass was investigated using Mössbauer and EXAFS spectroscopic techniques. Mössbauer spectroscopy results show that iron is oxidized during sorption while EXAFS results indicate that iron is coordinated by approximately 6 oxygen and 2 carbon atoms while arsenic is coordinated by approximately 4 oxygen atoms with iron as a second neighbor.

Keywords Mössbauer spectroscopy · EXAFS · Arsenic

1 Introduction

Environment contamination by chemical and radioactive wastes has attracted considerable attention as a consequence of ever increasing mining and industrial activities. Arsenic is one of the most important pollutant metalloid and is naturally found in the environment. It is present in gold and other base-metals ores and large amounts are mobilized in the process of extraction of these metals. Mostly all of arsenic complexes are considered toxic and cause several diseases [1]. Investigations focusing on metal accumulation in vegetables grown in mine waste areas found that lettuce leaves tend to accumulate a larger amount of metal than other vegetable species, such as beans and tomatoes [2]. This accumulation pattern suggests that lettuce leaves have potential active sites for arsenic and metalloid complexation. It has been shown [3, 4] that iron plays an important role on arsenic retention by biomass.

I. F. Vasconcelos (✉)
Dep. Eng. Metalúrgica e de Materiais, Universidade Federal do Ceará,
Fortaleza, Ceará, Brazil
e-mail: ifvasco@ufc.br

G. C. Silva · R. P. Carvalho · M. S. S. Dantas · V. S. T. Ciminelli
Dep. Eng. Metalúrgica e de Materiais, Universidade Federal de Minas Gerais,
Belo Horizonte, Minas Gerais, Brazil

However, detailed information about interaction amongst iron, arsenic, and biomass is still lacking. In this contribution we present Mössbauer and X-ray absorption fine structure spectroscopy results and analysis on iron and arsenic adsorption on dried lettuce leaves with the aim to contribute to the investigation of the arsenic retention mechanisms including the important role of iron on arsenic accumulation.

2 Experimental methods

After dried and ground, lettuce leaves were washed with an aqueous solution at pH 4.0, filtered, and dried. Further treatment was performed like in [5, 6]. For the adsorption experiments, the washed biomass was brought together with solutions of Fe(III) as $FeCl_3.6H_2O$ or Fe(II) as $FeSO_4.7H_2O$. For arsenic sorption, Fe(III)-loaded biomass was brought together with a solution of As(V) as $AsHNa_2O_4.7H_2O$. Values of pH were selected so as to avoid iron precipitation and maximize arsenic loadings. Details of the adsorption process will be published elsewhere and show that arsenic is sorbed only when the biomass has been previously loaded with iron.

Extended X-ray Absorption Fine Structure (EXAFS) measurements on biomass samples loaded with iron and arsenic were performed at Laboratório Nacional de Luz Síchroton (LNLS) in Campinas, Brazil. Fe and As K-edges (7112 eV and 11868 eV) EXAFS data were collected at the XAFS1 workstation in transmission geometry with a Si(111) channel-cut monochromator and ion chamber detectors. The data were analyzed using FEFF8 [7] and ATHENA/ARTEMIS [8] software packages. Analysis of the data was carried out as described in [9]. Mössbauer spectra of iron-loaded samples were recorded using a conventional constant acceleration Mössbauer spectrometer in transmission mode with a ^{57}Co (Rh) source. The data were evaluated with discrete subspectra using a least-square fitting routine provided by the Normos software package. All isomer shift values (δ) quoted are relative to α-Fe.

3 Results and discussion

Figure 1 shows the Mössbauer spectra for samples of biomass loaded with (a) Fe(II) and (b) Fe(III)+As(V) (see Table 1 for hyperfine parameters). It can be observed that the oxidation state of iron inside the biomass is mainly +3, even when the biomass is in contact with a Fe(II) solution, indicating that Fe(II) is oxidized to Fe(III). Previous studies have reported that iron oxidation does not occur in solution [10] but do not offer an explanation for the observed oxidation. It is, thus, assumed that a constituent of the biomass is capable of promoting iron oxidation. Values of quadrupole splitting (Q.S.) indicate that Fe(III) is in an octahedral coordination. While Fe(II) is oxidizing to Fe(III), the opposite likely does not occur, as the attempt to fit the data for the Fe(III)+As(V)-loaded biomass with a model that includes an Fe(II) component yields a negligible (within the uncertainty of the measurement) contribution of this Fe(II) subspectrum (see Fig. 1b).

Figure 2a shows the real part of Fourier transformed EXAFS data (and fit) obtained from the Fe(II)-loaded biomass sample. The fitting model is based on the experimental evidence that iron is mainly complexed to carboxyl groups inside the

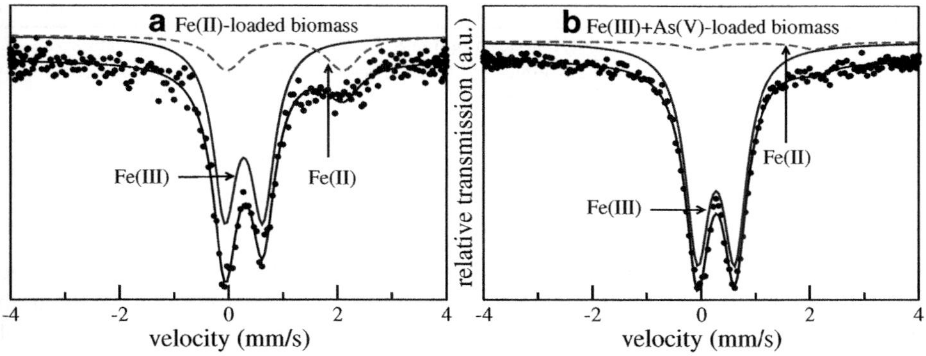

Fig. 1 (color online) Mössbauer spectra of iron-loaded biomass samples: **a** Fe(II)-loaded biomass; **b** Fe(III)+As(V)-loaded biomass

Table 1 Hyperfine parameters resulted from fitting of Mössbauer spectra

Sample	Site	δ (mm/s)	Q.S. (mm/s)	Area
Fe(II)-loaded biomass	Fe(III)	0.39±0.01	0.69±0.01	77%
	Fe(II)	1.15±0.04	2.13±0.08	23%
Fe(III)+As(V)-loaded biomass	Fe(III)	0.37±0.01	0.72±0.01	

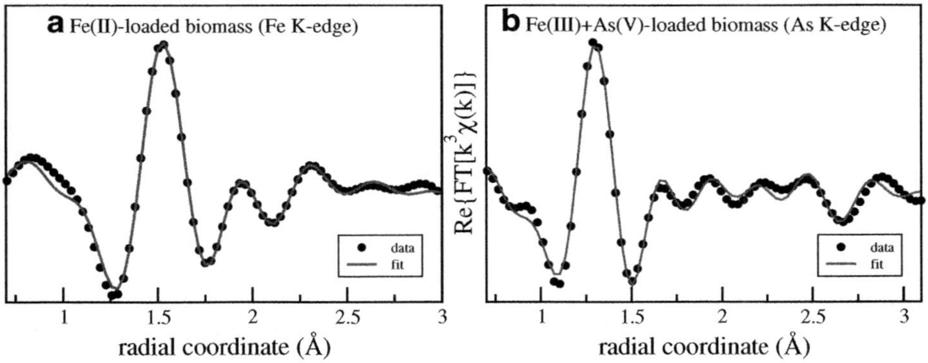

Fig. 2 (color online) Real part of Fourier transformed EXAFS data for ion-loaded biomass samples: **a** Fe(II) at Fe K-edge; **b** Fe(III)+As(V) at As K-edge

biomass. Fe-O and Fe-C scattering paths from a Fe(III) acetate structure were used to fit the data. Table 2 shows the best results for the fitting variables which indicate that iron is bound to 5.4±0.3 oxygen atoms and 3.0±1.0 carbon atoms in the first and second coordination shells. The Fe-O distance of 1.99 Å is consistent with that found in the acetate structure whereas the Fe-C distance of 2.81 Å is shorter than in the acetate group. This can be attributed to Fe bound to carboxyl groups through 2 oxygen atoms, in a bidentate mononuclear geometry, as opposed to the monodentate structure of Fe(III) acetate. The octahedral coordination sphere is completed by two water molecules.

Table 2 Results of fits to EXAFS data

Sample	Shell	N	R (Å)	σ^2 ($\times 10^{-3}$Å)
Fe(II)-loaded biomass	Fe-O	5.4±0.3	1.99±0.02	6±2
Fe K-edge	Fe-C	3.0±1.0	2.81±0.04	10±8
Fe(III)+As(V)-loaded biomass	As-O	4.1±0.7	1.71±0.01	3±1
As K-edge	As-Fe	1 (fixed)	2.93±0.03	6±4

Paths from $FeAsO_4.2H_2O$ (scorodite) structure were used to fit the As K-edge EXAFS spectrum of Fe(III)+As(V)-loaded biomass sample, as can be seen in Fig. 2b. The first coordination shell (see Table 2 for fit results) is composed of 4.1±0.7 oxygen atoms in agreement with the expected tetrahedral geometry of aqueous $H_2AsO_4^-$ present in the original solution [11]. The As-O distance of 1.71 Å found in the fit is consistent with the typical distance for arsenate adsorbed on iron hydroxides [12]. The structure in the Fourier-transformed data between R = 2.5 Å and R = 3.0 Å is properly fitted with an As-Fe scattering path. Freely varying the coordination number of this shell did not yield a stable fit. To circumvent this problem, several fits were attempted with fixed values of N from 1 to 4. The most statistically sound fit was achieved with N = 1 which yielded a As-Fe distance of 2.93±0.03 Å. This result was expected in views of the dependence of arsenic sorption on the presence of iron. The concentration of iron in the biomass samples measured was about twice as large as that of arsenic which explains no evidence of the contribution of arsenic to the second coordination shell of iron. These results suggest that arsenate species are linked to the iron octahedra. The geometry of this complex has yet to be determined.

Acknowledgements IFV acknowledges financial support from FUNCAP. Financial support for this research was provided by CAPES, CNPq, and FAPEMIG. X-ray absorption spectroscopy was performed at the Laboratório Nacional de Luz Síncrotron (LNLS/Campinas/Brazil).

References

1. Jain, C.K., Ali, I.: Arsenic: occurrence, toxicity and speciation techniques. Water Res. **34**, 4304–4312 (2000)
2. Cobb, G.P., Sands, K., Waters, M., Wixson, B.G., Dorward-King, E.: Accumulation of heavy metals by vegetables grown in mine wastes. Environ. Toxicol. Chem. **19**, 600–607 (2000)
3. Jeon, C., Nah, I.W., Hwang, K.-Y.: Adsorption of heavy metals using magnetically modified alginic acid. Hydrometallurgy **86**, 140–146 (2006)
4. Pokhrel, D., Viraraghavan T.: Arsenic removal from an aqueous solution by a modified fungal biomass. Water Res. **40**, 549–553 (2006)
5. Hwang, J.-K., Kim, C.-J., Kim, C.-T.: Extrusion of apple pomace facilitates pectin extraction. J. Food Sci. **63**, 841–844 (1998)
6. Tiemann, K.J., Gardea-Torresdey, J.L., Gamez, G., Dokken, K., Sias, S.: Use of X-ray absorption spectroscopy and esterification to investigate Cr(III) and Ni(II) ligands in alfalfa biomass. Environ. Sci. Technol. **33**, 150–154 (1999)
7. Ankudinov, A.L., Ravel, B., Rehr, J.J.: Real-space multiple-scattering calculation and interpretation of X-ray absorption near-edge structure. Phys. Rev. B **58**, 7565–7576 (1998)
8. Ravel, B., Newville, M.: ATHENA, ARTEMIS, HEPHAESTUS: data analysis for X-ray absorption spectroscopy using IFEFFIT. J. Synchrotron Radiat. **12**, 537–541 (2005)
9. Vasconcelos, I.F., Haack, E.A., Maurice, P.A., Bunker, B.A.: EXAFS analysis of Cd(II) adsorption to kaolinite. Chem. Geol. **249**, 237–249 (2008)

10. Carvalho, R.P., Sousa, A.M.G., Silva, G.C., Krambrock, K., Santos, M.H., Macedo, W.A.A., Ardisson, J.D.: Sorption of iron and copper ions in vegetable biomass. In: Proceedings, 16th IBS, Capetown, South Africa (2005)
11. Bazán, B., Mesa, J.L, Pizarro, J.L, Lezama, L., Arriortua, M.I., Rojo, T.: A new inorganic-organic hybrid iron(III) arsenate: $(C_2H_{10}N_2)[Fe(HAsO_4)_2(H_2AsO_4)](H_2O)$. Hydrothermal synthesis, crystal structure, and spectroscopic and magnetic properties. Inorg. Chem. **39**, 6056–6060 (2000)
12. Sherman, D.M., Randall, S.R.: Surface complexation of arsenic(V) to iron(III) (hydr)oxides: structural mechanism from ab initio molecular geometries and EXAFS spectroscopy. Geochim. Cosmochim. Acta **67**, 4223–4230 (2003)

Hyperfine Interact (2010) 195:117–126
DOI 10.1007/s10751-009-0108-4

Characterization of corrosion products of a carbon steel screw-nut set exposed to mountain weather conditions

**C. P. Ramos · G. Duffó · S. Farina ·
M. Lauretta · C. Saragovi**

Published online: 26 September 2009

Abstract In this work we present the results obtained by means of Mössbauer Spectroscopy to determine and characterize different corrosion products coming from a carbon steel screw-nut set exposed to mountain weather conditions for more than 70 years, in Las Cuevas (Mendoza–Argentina). Measurements at room temperature and 15 K were performed, detecting a great quantity of goethite but also lepidocrocite, hematite, magnetite and maghemite. This study was complemented by material characterization in terms of chemical composition, microscopic observation and X-ray diffraction analysis.

Keywords Corrosion · Carbon steel · Mössbauer spectroscopy

1 Introduction

The present work is part of a project aiming to increase the knowledge of corrosion products of carbon steels in order to obtain information about potential service

C. P. Ramos (✉) · C. Saragovi
Departamento de Física, Centro Atómico Constituyentes,
CNEA Av. Gral. Paz 1499 (1650), Buenos Aires, Argentina
e-mail: ciramos@cnea.gov.ar

G. Duffó · S. Farina
Departamento de Materiales, Centro Atómico Constituyentes,
CNEA Av. Gral. Paz 1499 (1650), Buenos Aires, Argentina

M. Lauretta
Facultad de Ciencias Exactas y Naturales, Universidad de Buenos Aires,
Intendente Güiraldes 2160-Ciudad Universitaria (1428), Cap. Fed.,
Buenos Aires, Argentina

C. P. Ramos · S. Farina
CONICET, Buenos Aires, Argentina

life of steel structures and components. For example, this information is vital with regards to the intermediate level radioactive waste (ILRW) disposal, where the objective is to ensure protection to people and environment for a period of 300 years. Service life of reinforced concrete is limited by the corrosion of reinforcement. The corrosion of iron leads to the formation of various products which occupy a volume much greater than the original iron and may cause internal cracking around the bar resulting in the through cracking of the cover concrete. Nowadays the available experience with modern reinforced concrete dating back only about 100 years old is not enough against the expected life of, for example, the nuclear repository (more than 300 years). This fact requires extrapolation of short term experience to 300 years foreseen life. Accelerated laboratory tests are obviously the straightforward way to assess the problem but doubts arise as whether the extrapolation is correct, particularly regarding the type of corrosion products formed under accelerated conditions. Mechanistic modeling is the other common approach but, in general, the nature and composition of the corrosion products formed are necessary to be used as inputs when modeling the reinforcement corrosion [1].Then, the study of archaeological analogues is necessary to improve the knowledge of long-term corrosion processes of steel components. In this context we present the results obtained by means of Mössbauer Spectroscopy to characterize different corrosion products formed on a carbon steel screw-nut set exposed to mountain weather conditions for more than 70 years, in Las Cuevas (Mendoza–Argentina).

2 Experimental

The analyzed piece, dating back more than 70 years ago, was part of the connecting system of the ex-Trasandino railroad. It was obtained in the neighborhoods of Las Cuevas station (Mendoza–Argentina), whose geographical coordinates are 32°49′ South latitude and 69°55′ West longitude, 2720 m above sea level. The monthly average temperatures oscillate between 0.5°C and 13.4°C, the average annual precipitation is about 300 mm and on average there are 150 frost days per year (data taken from Puente del Inca weather station) [2]. The studied piece consisted on a screw-nut set (Fig. 1) which were observed with both optical (Olympus BX60M) and scanning electron microscope (Philips SEM 500). In some cases, energy dispersion X-ray analysis (EDAX) was conducted at many locations to asses the local chemical composition. Chemical analysis of steel was made by means of combustion followed by emission spectroscopy. Metallographic cross sections were prepared, both in the longitudinal and transversal directions of the screw axis. The mounted sections were mechanically polished with emery papers (grades 220–1500) and then with diamond paste (7–1 μm). Microhardness (H_{V100}) was measured at different depths in the transversal section of the screw.

Mössbauer spectra were obtained in a conventional constant acceleration spectrometer in transmission geometry with a ^{57}Co/Rh source both, at room temperature (RT) and in some cases at 15 K. Spectra were fitted to a hyperfine field distribution, a quadrupole doublet (line width ~0.5 mm/s) and three sextets (line width ~0.3 mm/s) by using the NORMOS least squares fitting program [3]. Compounds assignation was made comparing the hyperfine parameters to those available in literature [4].

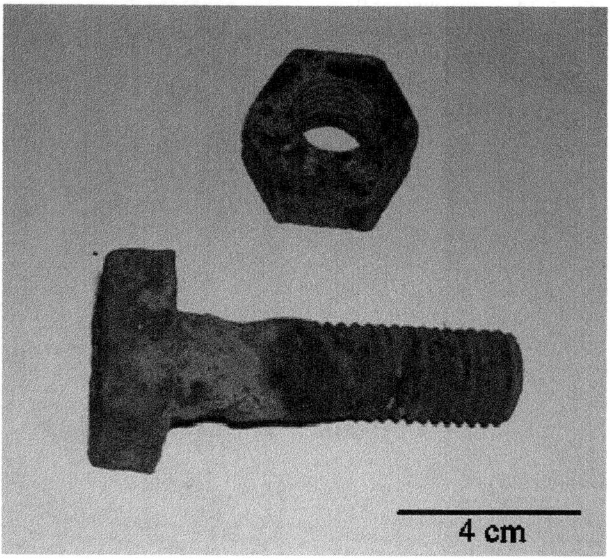

Fig. 1 Picture of the studied screw-nut set

X-ray diffraction (XRD) measurements were performed with a Philips PW 3710 diffractometer with a Cu-anode operating at a generator voltage of 40 KV and a current of 30 mA with the goniometer 2θ values varied from 10° to 70° at a scan rate of 1.0 s/step.

3 Results

3.1 General characterization

The chemical composition of the steel was (in wt.%): C, 0.02; Si, 0.11; Mn, 0.30; Cr, 0.05; Mo, 0.01; S, 0.01; P, 0.007; and Fe balance. These results are consistent with a SAE 1005 steel. Steel microhardness measurements along the screw radius, showed that the value measured at 100 μm from the edge (245 H_{V100}) was significantly higher than those measured between 1 mm from the edge (187 H_{V100}) and 7.5 mm from the edge (170 H_{V100}), at 1 mm steps. Metallographic cross sections showed the existence of numerous and large slag inclusions (Fig. 2). This is in accordance with the fact that ancient steels possess very heterogenous compositions, and they invariably contain entrapped slag inclusions, as the result from the process by which the steel was made. The inclusions in plain carbon and low-alloy steels are MnS, silicates, nitrides, and oxides. MnS inclusions appear as elongated particles (called stringers) and show a medium-gray color in the light microscope. These inclusions are smooth and have rounded edges. Silicate inclusions show a dark gray color [5]. The heterogenous composition of the steel and the presence of Mn and Si in the inclusions were confirmed by EDAX.

The pieces studied were heavily corroded. The surface rust of the nut showed two different color zones, a white zone and a red zone (Fig. 3). These two different

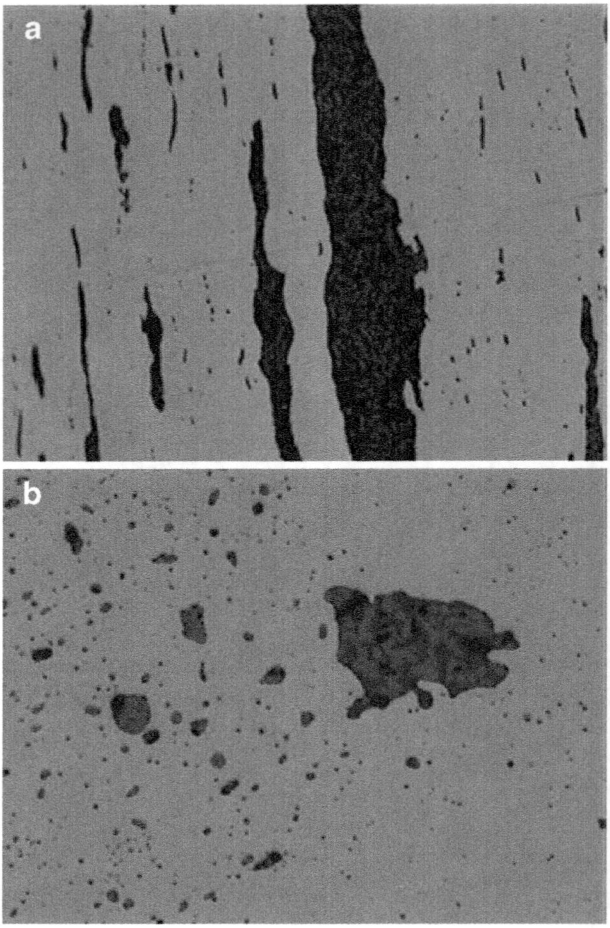

Fig. 2 Microstructure of the steel showing slag stringers in a ferrite matrix. **a** Longitudinal section. **b** Transverse section

corrosion products were analyzed separately, though this differentiation was difficult because white products were deposited over the red ones. SEM observations confirmed the existence of several rust layers (Fig. 4). The surface rust of the screw was mostly red.

3.2 Mössbauer spectroscopy results

3.2.1 The nut

Tables 1 and 2 display the Mössbauer hyperfine parameters of the nut at RT and 15 K for the white and red zones, respectively; and Figs. 5, 6, 7, and 8 show their corresponding spectra.

RT spectra were fitted by using a quadrupole doublet (D), a hyperfine field distribution (d) and three sextets (S1, S2 and S3). In both cases (white and red zones) the greater contributions were from D and d and the sextets contributed to the spectra to a lesser extent. D values were obtained as an average between two

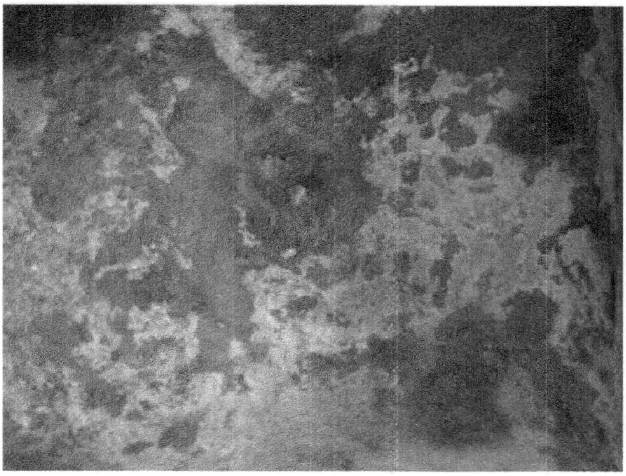

Fig. 3 Surface rust of the nut

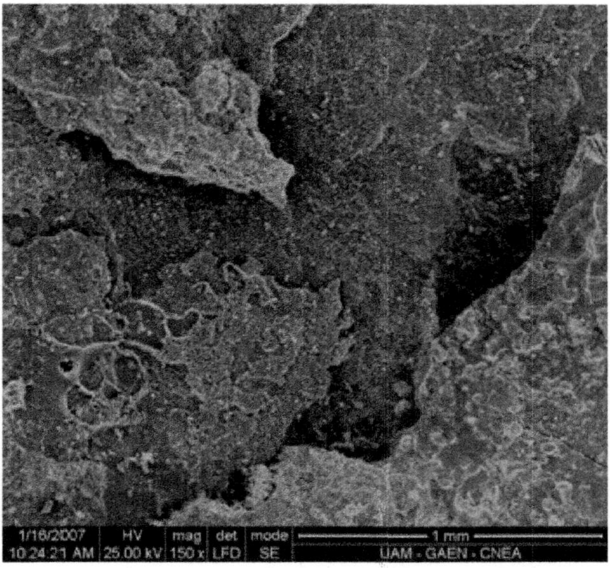

Fig. 4 Detail of surface rust of the nut (SEM)

doublets with noticeably different quadrupole splittings (~0.5 and 0.9 mm/s) but almost equal isomer shifts. As it is difficult to distinguish both contributions at RT, at this stage we can only suppose that this average would take into account the possible presence of superparamagnetic goethite (ultrafine particles) and/or lepidocrocite. The distribution was assigned to goethite with a particle size dispersion. S1 was assigned to hematite and S2 and S3 to non-stoichiometric magnetite ($Fe_{3-x}O_4$). Isomer shift values suggest the possibility to have a mixture of magnetite and maghemite or a partially or totally oxidized magnetite.

Table 1 Hyperfine parameters of the nut at RT and 15 K for the white zone; Bhf [T], QS, IS and 2ε [mm/s]

	RT			15 K	
D	QS	0.59	D′	Q	0.69
	IS	0.37		IS	0.47
d	Bhf	23.9	d′	Bhf	35.06
	2ε	−0.11		2ε	−0.16
	IS	0.41		IS	0.34
S1	Bhf	51.02	S′1	Bhf	53.27
	2ε	−0.28		2ε	−0.16
	IS	0.44		IS	0.48
S2	Bhf	49.1	S′2	Bhf	51.56
	2ε	0.06		2ε	−0.07
	IS	0.34		IS	0.43
S3	Bhf	46.1	S′3	Bhf	50.15
	2ε	0.00		2ε	0.00
	IS	0.57		IS	0.64
–	–	–	S′4	Bhf	47.79
				2ε	0.00
				IS	0.45
–	–	–	S′5	Bhf	44.5
				2ε	−0.45
				IS	0.60

Table 2 Hyperfine parameters of the nut at RT and 15 K for the red zone; Bhf [T], QS, IS and 2ε [mm/s]

	RT			15 K	
D	QS	0.63	D′	Q	0.69
	IS	0.37		IS	0.47
d	Bhf	19.51	d′	Bhf	35.80
	2ε	0.02		2ε	−0.37
	IS	0.40		IS	0.47
S1	Bhf	52.30	S′1	Bhf	54.61
	2ε	0.06		2ε	0.02
	IS	0.33		IS	0.63
S2	Bhf	48.53	S′2	Bhf	52.27
	2ε	−0.09		2ε	0.05
	IS	0.30		IS	0.50
S3	Bhf	46.47	S′3	Bhf	49.99
	2ε	0.00		2ε	0.00
	IS	0.82		IS	0.65
–	–	–	S′4	Bhf	46.99
				2ε	0.00
				IS	0.69
–	–	–	S′5	Bhf	44.35
				2ε	−0.38
				IS	0.46

15 K spectra instead, consisted of a quadrupolar doublet (D′), a hyperfine field distribution (d′) and five sextets (S′1, S′2, S′3, S′4 and S′5). The wide D doublet (still present at 15 K), being also in this case the convolution of two paramagnetic subspectra, would show the presence of poorly crystallized goethite, of ultrafine particle, not allowing to discard the lepidocrocite presence. S′1 would come from

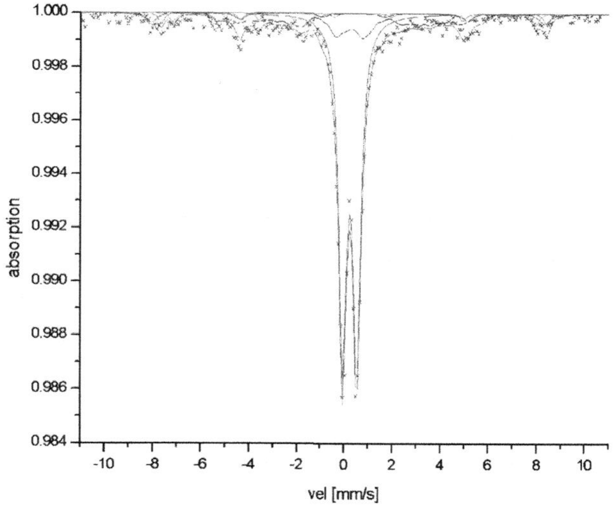

Fig. 5 Fitted Mössbauer spectrum at RT for the white zone of the nut

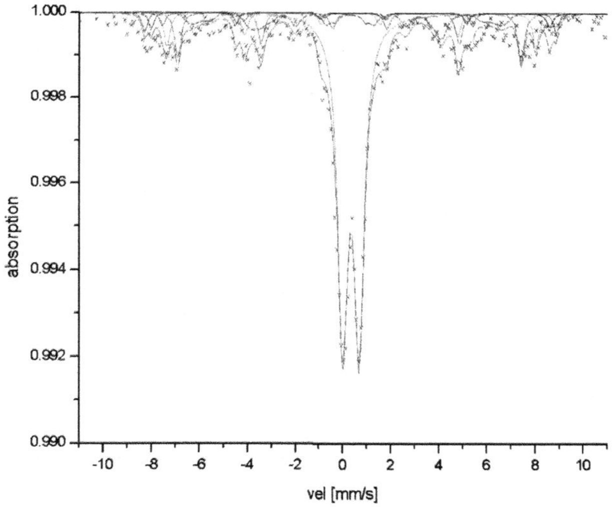

Fig. 6 Fitted Mössbauer spectrum at 15 K for the white zone of the nut

the hematite while S′2 and S′3 would correspond to maghemite, non-stoichiometric magnetite and a mix between them. S′4 was assigned to lepidocrocite, without discarding any contribution of the magnetite to the subspectrum. S′5 and d′ would refer to magnetically ordered goethite with a size particle distribution and with a variable degree of crystallinity.

Taking into account the subspectra areas, the rate between magnetic and superparamagnetic areas increases when lowering temperature. This is a logical result owing to the ordering of the phases which are disordered at RT.

3.2.2 The head of the screw

In this case only a measurement at RT was necessary to observe an appreciable difference with the measurements performed at RT for the nut. The difference

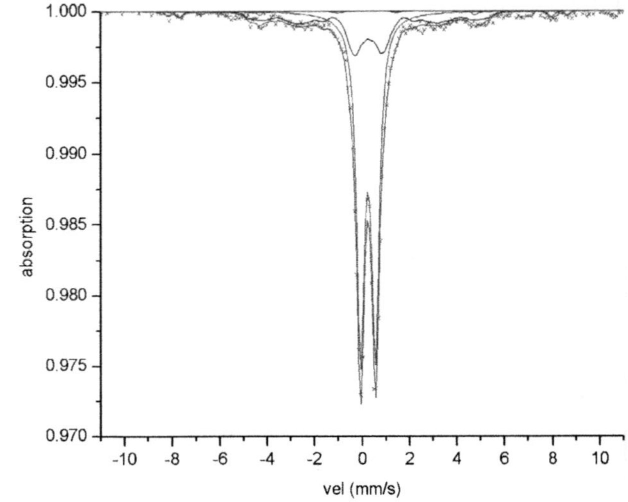

Fig. 7 Fitted Mössbauer spectrum at RT for the red zone of the nut

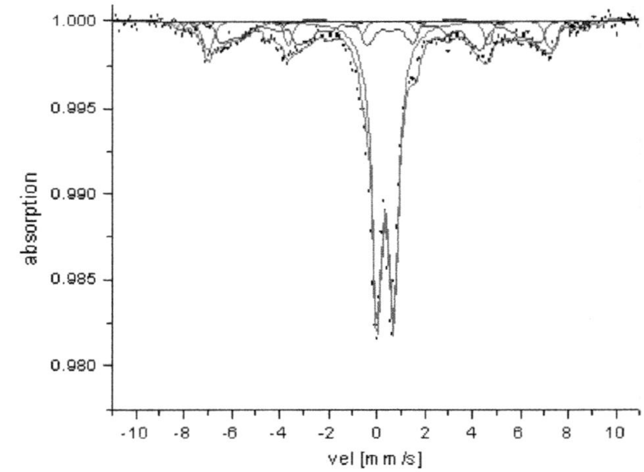

Fig. 8 Fitted Mössbauer spectrum at 15 K for the red zone of the nut

consisted on an increase of the magnetic part, showing a greater quantity of magnetic compounds at RT than those observed in the nut (Fig. 9). The identified compounds are in agreement with those obtained for the nut, but they are present in a different proportion. The magnetic to superparamagnetic area ratios (M/SP) for the nut and the head of the screw, are included in Table 3.

4 Discussion

One of the main differences in the nut between both zones (white and red) is that the hyperfine field distribution denoting goethite presence with a particle size dispersion, has a low field Gaussian component for the red zone which is not visible in the white one. This suggests that the goethite formed in the red zone may be composed by finer particles. Another difference is in the IS value for S′4 which is greater for

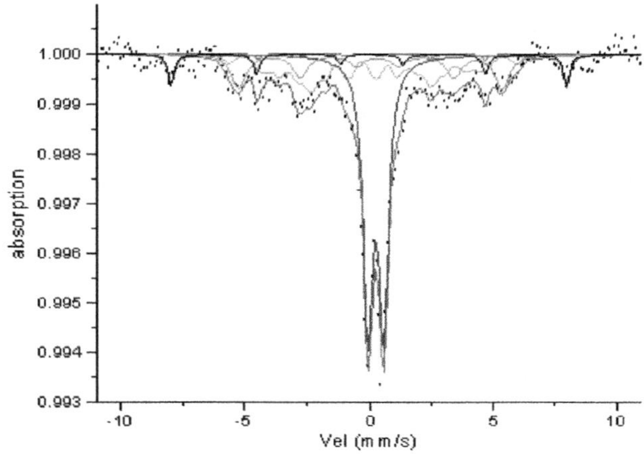

Fig. 9 Fitted Mössbauer spectrum at RT for the head of the screw

Table 3 M/SP area ratios

	M/SP
Nut white zone	0.48
Nut red zone	0.47
Head of the screw	1.23

the red zone as compared to the white zone. Being the ratio between magnetic and superparamagnetic areas almost the same for both zones, the magnetite contribution would be dominant in comparison to lepidocrocite for the red zone. Furthermore, quadrupole shift values for hematite should point out to a spin state coexistence (antiferromagnetic and weakly ferromagnetic). In the red zone antiferromagnetic hematite seems to be predominant, suggesting a bigger hematite particle size than that for the white zone.

Mössbauer spectroscopy results show a coincidence between the iron-bearing components present in both zones of the nut (red and white). However a priori differences are given by the distinct color of the two zones. Our hypothesis was to think about non-iron-bearing compounds present in the sample, so we performed an X-ray analysis to elucidate it. The diffractograms obtained showed abundant calcite (~50%) and quartz (~20%) in the white zone, meanwhile quartz (~25%) and calcite in a significant lower proportion were detected in the red zone. These compounds may come from the soil where the analyzed screw was extracted. Seeking information about Mendoza soils we concluded that this conjecture is highly reasonable, explaining as well the difference between the two colored zones.

Moreover, from the obtained values in Table 3 it can be inferred that the white and the red zones of the nut present the same proportions due to the magnetic contribution (about 50% of the iron-bearing compounds), while the magnetic contribution constitutes a major portion of the head of the screw. A possible explanation consistent with this behavior is that limestone soils have alkaline pH, associated with the presence of carbonates in the soil. Calcite, being the most common form of carbonates in soils, increases the redox potential [6], favoring the Fe-oxyhydroxides formation (superparamagnetic contribution).

5 Conclusions

The SAE 1005 steel screw exposed to mountain weather conditions in Las Cuevas (Mendoza–Argentina) presents abundant goethite, also lepidocrocite and hematite, magnetite and maghemite in a lower proportion. By means of X-ray diffraction, it was possible to detect non-iron-bearing compounds like calcite and quartz. These compounds almost certainly came from the soil where the screw was extracted, explaining the difference between the colored zones (white and red) exhibited in the studied piece.

The present work provides the type of corrosion product that is formed in the environment analyzed. Data concerning their physical properties such as density and molecular weight may be used as inputs for modeling the service life of reinforced concrete exposed to a similar environment.

References

1. Duffó, G.S., Farina, S.B.: A model for time-to-cover cracking of reinforced concrete due to rebars corrosion. Integral Service Life Modelling of Concrete Structures, 277–284 RILEM Publications, Bagneux-France (2007)
2. Secretaría de Minería de la Nación Argentina web page. http://www.mineria.gov.ar/estudios/irn/mendo/tablasclima.asp?donde=m6#t232 (15-09-08)
3. Brand, R.A.: Normos program, Internat. Rep. Angewandte Physic, Univ. Duisburg (1987)
4. Vandenberghe, R.E., De Grave, E., Landuydt, C., Bowen, L.H.: Hyperfine Interact. **53**, 175 (1990)
5. Vander Voort, G.F.: ASM Metals Handbook (9th edition), vol. 9, p. 622. Metallography and Microstructures, ASM, Metals Park, OH (1985)
6. Hernández, J.C., Orihuela, D.L., Pérez-Mohedano, S., Marijuan, L., Furet, N.R.: Estudios de la Zona No Saturada del Suelo Vol. VI. J.Álvarez-Benedí y P. Marinero (2003)

Hyperfine Interact (2010) 195:127–131
DOI 10.1007/s10751-009-0109-3

The role of the spinel phases on the rust activity in corroded steels

F. R. Pérez · C. F. Londoño · C. A. Barrero · K. E. García · J. Tobón

Published online: 26 September 2009

Abstract Carbon, CS, and weathering steels, WS, were totally immersed in a NaCl containing solution. The influence of steel composition and the presence/absence of air flux into the solution on the physical properties of spinel and other iron phases were investigated. Large amounts of defective magnetite were formed only on CS, whereas little amounts of small grain-sized defective maghemite were detected only on WS. The chemical composition of the steels greatly affects the type of spinel phase being formed and their relative abundance. A non-aerated environment favored the formation of magnetite in CS. The protective ability of the rust was unfavored in the presence of large amounts of spinels.

Keywords Spinel phases · Total immersion · Steels

1 Introduction

In general, the main components of the rust layer formed on steels in different environments are α-FeOOH (α), γ-FeOOH (γ), β-FeOOH (β), and non-stoichiometric Fe_3O_4 with a wide distribution of particle sizes [1, 2]. It has been established that the protective ability of the steels specially depends on the characteristics of the rust components and the capability of the rust for adhering to the metal surface. To determine the protective ability index formed on weathering steels, Hara et al. [3] proposed three types of rust layers: (1) protective if $\alpha/(\beta+\gamma+s) > 1$, (2) non-protective and active if $\alpha/(\beta+\gamma+s) < 1$ and $(\beta+s)/(\beta+\gamma+s) > 0.5$ and (3) non-protective and inactive if $\alpha/(\beta+\gamma+s) < 1$ and $(\beta+s)/(\beta+\gamma+s) < 0.5$,

F. R. Pérez (✉)
Grupo de Óptica y Espectroscopía (GOE), Universidad Pontificia Bolivariana,
A. A. 56006, Medellín, Colombia
e-mail: fredy.perez@upb.edu.co, fperez@fisica.udea.edu.co

F. R. Pérez · C. F. Londoño · C. A. Barrero · K. E. García · J. Tobón
Grupo de Estado Sólido, Universidad de Antioquia, A. A. 1226, Medellín, Colombia

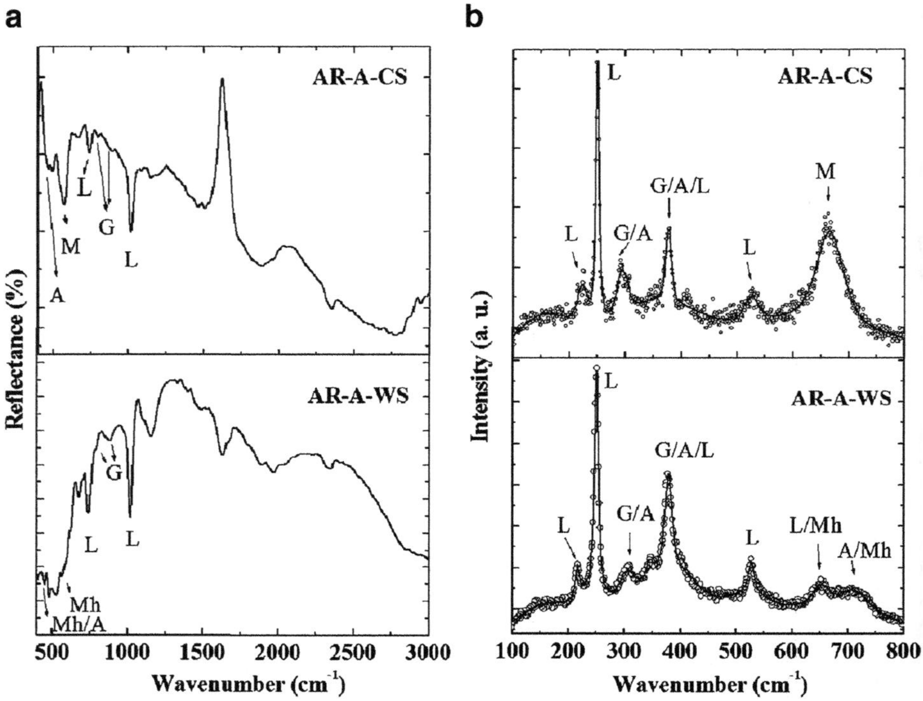

Fig. 1 **a** DRIFT, and **b** Raman spectra corresponding to selected samples. *L* lepidocrocite, *A* akaganeite, *G* goethite and *M* magnetite and *Mh* maghemite

where *s* stands for spinel. Specifically, α, β, γ and *s* represent the respective relative phase abundances. The aim of this work is to identify and fully characterize the spinel and iron phases formed on steels submitted to total immersion tests, and from this information to discuss the protective character of the rust.

2 Experimental procedure and results

The total weight percentages of alloying elements were of 0.63 and of 2.41 for CS and WS, respectively. The composition of WS mainly differs from that of the CS in their higher Ni, Cr, Cu, Si and P content. The immersion process of the coupons into the solutions was similar to that described elsewhere [1]. Aerated and non-aerated 0.6 M NaCl concentrated solutions were studied. After 14 days, the coupons were extracted. The rust scrapped from the steel coupons and the rusts collected from the solutions were called AR and NAR, respectively. The characterization was carried out by using a Nicollet Avatar 330 (of Thermo Electron Co) Fourier transform infrared spectrometer working in diffuse reflectance (DRIFT), an InVia Renishaw Ramanscope system, a D501 Siemens X-ray diffractometer equipped with Cu(Kα) radiation and a conventional transmission Mössbauer spectrometer with a Co-57 source in a rhodium matrix.

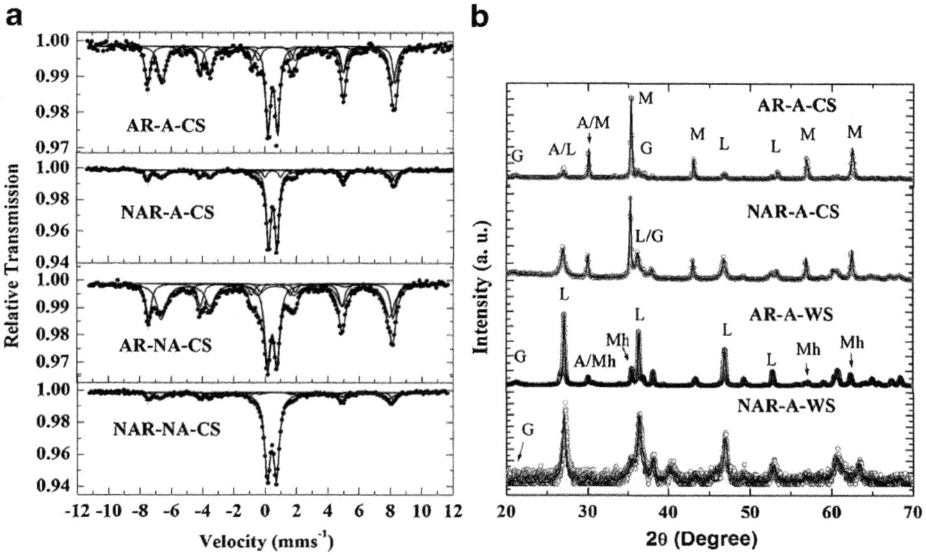

Fig. 2 **a** Room temperature Mössbauer spectra for all AR and NAR samples of CS, and **b** XRD patterns for AR and NAR aerated samples of CS and WS. *L* lepidocrocite, *A* akaganeite, *G* goethite, *M* magnetite and *Mh* maghemite

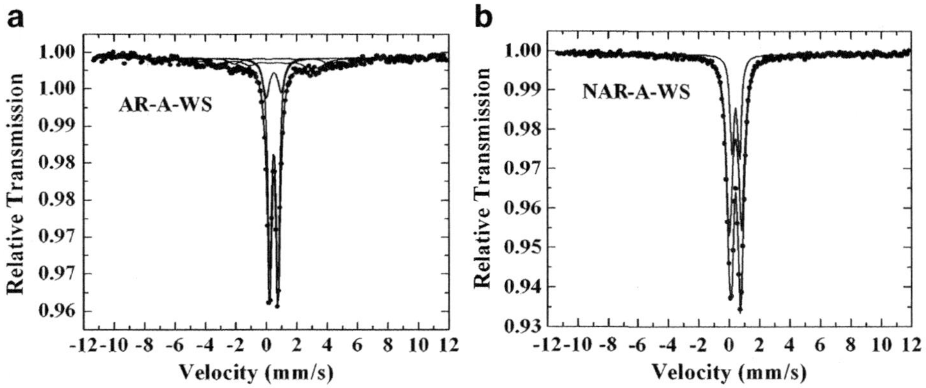

Fig. 3 Room temperature Mössbauer spectra for **a** AR, and **b** NAR aerated samples of WS

The notation in Figs. 1 and 2 has the following meaning: AR and NAR stand for adherent and non-adherent rusts; A and NA stand for aerated and non-aerated samples; and CS and WS stand for carbon and weathering steels, respectively. In all infrared spectra of AR samples, bands of γ-FeOOH, α-FeOOH, and β-FeOOH were found [4] (see Fig. 1a). However, $Fe_{3-x}O_4$ and γ-Fe_2O_3 were detected only in CS and in WS, respectively. Raman spectroscopy supported the DRIFT findings [5] (see Fig. 1b). All Mössbauer spectra of AR and NAR samples from CS were fitted using three lorentzian sextets and one or two lorentzian doublets (see Fig. 2a). The first sextet is due to contributions coming from both Fe^{3+} ions on the magnetite tetrahedral site and unpaired Fe^{3+} ions on the magnetite's octahedral site. The

Table 1 Protective character of the adherent rust as obtained from the relative abundance (vol.%) by XRD data

Sample	$\alpha/(\beta+\gamma+s)$	$(\beta+s)/(\beta+\gamma+s)$	Type of rust
AR-A-CS	0.25	0.64	Non-protective/active
AR-NA-CS	0.06	0.69	Non-protective/active
AR-A-WS	0.03	0.11	Non-protective/inactive

The estimated uncertainty was about ± 0.01

second sextet comes from $Fe^{2.5+}$ ions on the magnetite octahedral site. The third sextet has hyperfine parameters similar to that of magnetic goethite. The doublet(s) has (have) hyperfine parameter(s) associated with akaganeite, lepidocrocite and super-paramagnetic goethite. The magnetite relative sub-spectral areas were of 0.60 and of 0.64 for the AR samples coming from aerated and non-aerated solutions, respectively, whereas the respective values for the NAR were 0.17 and 0.29. Now, the spectrum of the AR sample from WS was fitted using three sextets and two doublets (see Fig. 3a). The hyperfine parameters for one of the sextets were well matched with those of magnetic maghemite. The other two sextets were associated with magnetic goethite exhibiting a wide distribution of particle sizes. The NAR sample from WS was fitted using only two doublets, suggesting the presence of lepidocrocite, akaganeite and probably super-paramagnetic maghemite and/or goethite (see Fig. 3b). Magnetic phases were no detected in the NAR of WS. It can be noticed that more magnetite was formed in the non-aerated solution, and also that there was a higher yield of spinel phases in the CS than in WS. These results can be understood, because poor oxygen environments favor the formation of magnetite, the oxygen diffusion through the corrosion layer forms a gradient with the smallest value at the metal surface, then it oxidizes the Fe^{2+} to Fe^{3+} forming the necessary ingredients for magnetite precipitation. Magnetite in CS seems to be formed mainly in the oxygen diffusion stage through the rust layer either by slow oxidation of $Fe(OH)_2$ via green rust and/or by interaction of Fe^{2+} with the different iron oxy-hydroxides. On the other hand, maghemite in WS seems to be formed by either oxidation of $Fe(OH)_2$ and/or wet oxidation of early precipitated small-particle magnetite. The results suggest that the magnetite formation might be favored in the absence of high abundances of alloying elements such as silicon, phosphorous, chromium and nickel.

Quantitative X-ray diffraction (XRD) analysis was in good agreement with the other techniques (see Fig. 2b). Table 1 shows the classification of the rusts as estimated by XRD analysis using the criterion from Hara et al. [3]. The ratio $\alpha/(\beta+\gamma+s)$ was less than 1 in all cases indicating the formation of a non-protective rust layer. However, the ratio $(\beta+s)/(\beta+\gamma+s)$ was higher than 0.5 for carbon steels (aerated and non-aerated), but less than 0.5 for weathering steels. All these differences were mainly attributed to the steel composition.

3 Conclusions

Different iron oxy-hydroxides phases were detected in all rusts formed on carbon and weathering steels (akaganeite, lepidocrocite and goethite). Large amounts of magnetite were detected both in adherent and non-adherent rust coming from carbon steels. In contrast, small amounts of maghemite were only detected in the adherent rust coming from weathering steels. The relative abundance of magnetite in carbon steels was higher in non-aerated solution in comparison to aerated one. The rust formed on carbon steels can be classified as non-protective and active and that on weathering steels as non-protective and inactive.

Acknowledgements Thanks go to Universidad de Antioquia (project code IN1248CE, Sustainability Programs for GES and GICM) and Universidad Pontificia Bolivariana (project code 889-05/06-27), and the Excellence Center for Novel Materials (CENM) under Contract CENM-Colciencias RC-043-2005 for financial support.

References

1. García, K.E., Morales, A.L., Barrero, C.A., Greneche, J.M.: Corros. Sci. **48**, 2813 (2006)
2. García, K.E., Morales, A.L., Barrero, C.A., Greneche, J.M.: Corros. Sci. **50**, 763 (2008)
3. Hara, S., Kamimura, T., Miyuki, H., Yamashita, M.: Corros. Sci. **49**, 1131–1142 (2007)
4. Weckler, B., Lutz, H.D.: Eur. J. Solid State Inorg. Chem. **35**, 531 (1998)
5. De Faria, D., Silva, V., de Oliveira, M.T.: J. Raman Spectrosc. **28**, 873 (1997)

Hyperfine Interact (2010) 195:133–140
DOI 10.1007/s10751-009-0110-x

On the reaction of iron oxides and oxyhydroxides with tannic and phosphoric acid and their mixtures

J. J. Beltrán · F. J. Novegil · K. E. García · C. A. Barrero

Published online: 10 November 2009

Abstract The actions of tannic acid, phosphoric acid and their mixture on lepidocrocite, goethite, superparamagnetic goethite, akaganeite, magnetite, hematite and maghemite for 1 day and 1 month were explored. It was found that these acids form iron tannates and phosphates. Lepidocrocite and magnetite were the iron phases more easily transformed with the mixture of the acids after 1 month of reaction, whereas hematite was the most resistant phase. In the case of goethite, our results suggest that in order to understand properly the action of these acids, we have to take into account its stoichiometry, surface area and degree of crystallinity.

Keywords Rust converters · Iron tannates and phosphates · Mössbauer spectrometry

1 Introduction

Tannic acid is a natural polyphenolic compound with great significance in food, medical area and other industrial products [1], whereas phosphoric acid is an inorganic acid used to acidify foods and beverages and it has dental and medical utilities, among others applications. The tannic and phosphoric acids are also utilized in the preparation of rust converters, which are environmentally safe, in contrast to toxic inhibitors such as red lead or zinc chromates [2]. These converters can be applied to corroded surfaces to convert the rust and to reduce further corrosion process by forming tannates and phosphates [3–8], and by generating a layer with certain

J. J. Beltrán (✉) · F. J. Novegil · K. E. García · C. A. Barrero
Sede de Investigación Universitaria, Grupo de Estado Sólido, Instituto de Física,
Universidad de Antioquia, A.A. 1226 Medellín, Colombia
e-mail: jjbj08@yahoo.com

J. J. Beltrán · K. E. García
Instituto de Química, Universidad de Antioquia, A.A. 1226 Medellín, Colombia

Table 1 Weight percentage of each component used in the preparation of the converters

Type of converter	Percentage of the components (% w/w)				
	Tannic acid	Phosphoric acid	Isopropyl alcohol	Tert butyl alcohol	Distilled water
TA	5	–	14	21	60
FA	–	35	14	21	30
TFA	5	35	14	21	25

protective value, leaving a suitable roughness for the application of a subsequent system of paintings. There have made many research studies on the application of the rust converters in the iron oxides and oxyhydroxides [3–13]. Although the rust converters have been used with some success for several decades, still exists controversy about their effectiveness and inhibitory capacity against the corrosion. Thus, the reaction mechanisms and specially the transformations taking place at the oxide layer need further research. One of the main drawbacks to understand this phenomenon is that most of the works have been carried out directly on the rust and not on each iron phases individually. For this reason, it is important to understand the effect of each acid on each iron oxide and oxyhydroxide separately, and this is the motivation for the present work. Based upon the results of a previous work [11], we have selected three rust converters and studied their action on each of the most common rust products.

2 Experimental procedure

The different iron phases used in this study were synthesized following hydrothermal methods reported by Schwertmann and Cornell [14]. We prepared poorly crystalline goethite, G-spm, crystalline goethite, G, lepidocrocite, L, and akaganeite, A. The synthetic iron oxides were magnetite, M, and maghemite, Mh. Hematite, H, was obtained commercially. Three types of rust converters were employed, one based in tannic acid, TA, another in phosphoric acid, FA, and the last, one based in a mixture of both acids, TFA. A mixture of alcohol with water was used as solvent. The composition of each converter is shown in Table 1. 500 mg of each sample were mixed with 20.0 mL of each converter and allowed to react for 1 day and 1 month. The products of the reactions were filtrated and dried at 40°C. The obtained samples were characterized by Thermogravimetric Analysis, Fourier Transform Infrared (FTIR), Spectroscopy Mössbauer (MS), Surface BET analysis and X-Ray Diffraction (XRD).

3 Results and discussion

From the thermogravimetric analysis we could calculate the structural OH groups in wt.%, and the stoichiometric formulae for each iron oxyhydroxide. The results are shown in Table 2, together with data obtained by BET analysis. From these values, we can see a great contrast in surface areas between G and G-spm due to the difference in crystallinity of the samples.

By employing FTIR, the products of reaction of the different iron phases with the TA converter for 1 day, showed bands corresponding to iron tannates (TI), whereas with the FA converter, bands due to iron phosphates (PI) were observed. On the other hand, the products of the reaction of each sample with the TFA converter

Table 2 TGA and BET analysis of pures oxyhydroxides

Sample	% H_2O	% OH	% Water total	Stoichiometric formule	$S(m^2/g)$
G	0.3 (3)	10.7 (2)	11.0 (2)	$\alpha\ Fe_{0.32}O_{0.95}OH_{1.05}$	27.3 (2)
G spm	4.4 (3)	13.6 (2)	18.0 (2)	$\alpha\ Fe_{0.24}O_{0.73}OH_{1.27}$	183.1 (2)
L	1.2 (3)	11.8 (2)	14.0 (2)	$\gamma\ Fe_{0.29}O_{0.87}OH_{1.13}$	58.9 (2)
A	1.7 (3)	11.0 (2)	12.7 (2)	$\beta\ Fe_{0.31}O_{0.93}H_{1.07}$	25.0 (2)

Numbers in parenthesis represent the uncertainty in the last digit

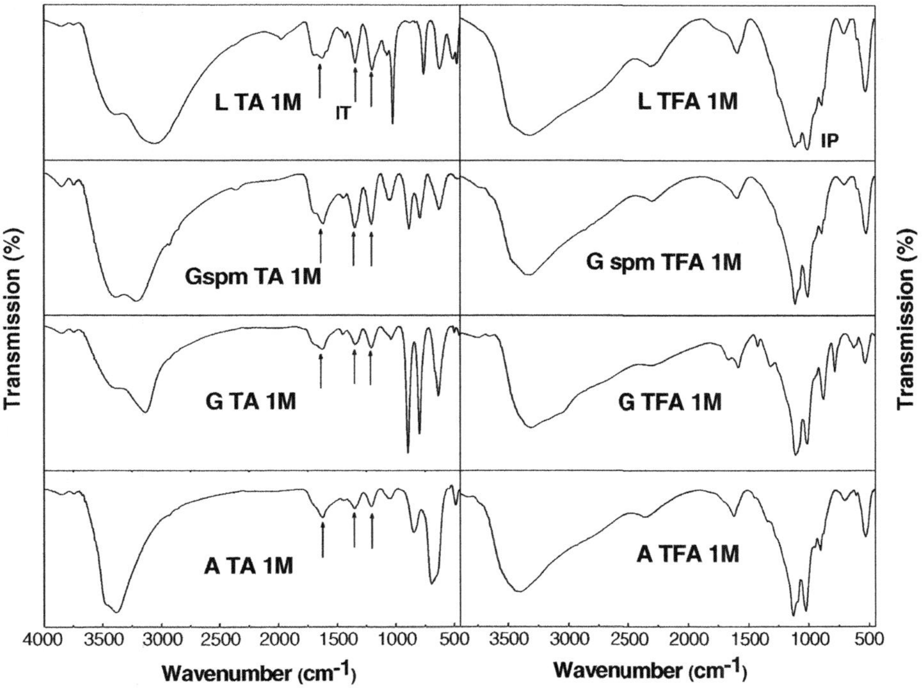

Fig. 1 FTIR spectra of iron oxyhydroxides after the reaction with TA (*left*) and TFA (*right*) converters for 1 month. *Arrows* point to IT bands

for 1 day showed mainly TI and PI bands. The results for 1 month were similar to those obtained for 1 day, but the TI and PI bands were more intense, see Fig. 1. The absorption bands corresponding to TI complexes were much more intense in L, G-spm and M than in the other iron phases, which were presented but with minor intensity.

The XRD patterns of all samples reacting with the TA converter, for both a day and a month, did not clearly show the iron tannates, because these compounds are amorphous [6, 7]. After the reaction with FA and TFA for 1 day, the diffraction patterns of all samples showed peaks assigned to $FeH_3(PO4)_2 \cdot 2.5H_2O$ and to $Fe_3H_9(PO4)_6 \cdot 6H_2O$, this latter compound being presented in minor proportion. The XRD patterns of the oxyhydroxides reacting with the TFA converter for 1 month showed residual peaks of G-spm, A and G, but the characteristic peaks of L were not observed. On the other hand, the XRD patterns of the oxides reacting

Fig. 2 XRD patterns of **a** M (*bottom*), Mh (*middle*) and H (*top*) after the reaction with the TFA converter for 1 month and **b** M after the reaction with FA (*bottom*) and TFA (*top*) for 1 month

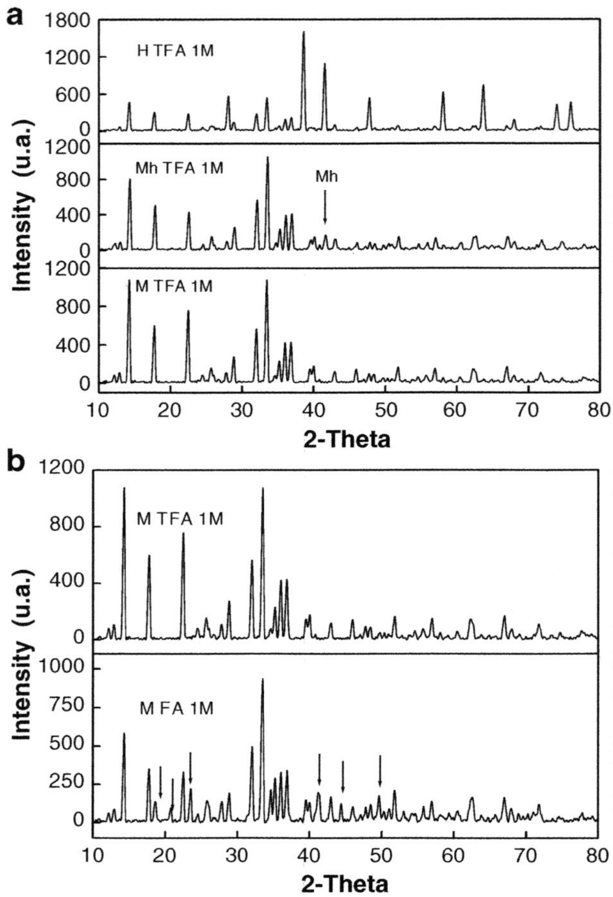

with the TFA converter for 1 month showed all peaks of H, some residual peaks of Mh, but no peaks of M (see Fig. 2a). A special case was the XRD pattern of M reacting with the FA converter, because besides the presence of the phosphate phases identified previously, the pattern also showed the apparition of other iron phosphate, identified as $FePO_4 \cdot 2H_2O$ (strenguite), whose peaks are marked in Fig. 2b. The peaks of vivianite and herraulite could not be detected.

The room temperature Mössbauer spectrum of the reaction of TA with G, M, Mh and H for 1 day gave no indications of the iron tannates complex formation, but with G-spm and L a doublet with quadrupole shift values between 0.97 and 1.12 mm/s and isomer shift values between 0.39 and 0.41 mm/s were observed, suggesting the presence of iron tannates [6, 8]. Figure 3a shows the Mössbauer spectra of products of the reaction of the iron oxyhydroxides with the FA converter for 1 day and for 1 month. The spectra of the reactions for 1 day revealed the presence of one doublet with quadrupole shift values between 0.14 and 0.20 mm/s, and isomer shift values between 0.35 and 0.45 mm/s, indicating the presence of iron phosphates [10]. A similar doublet was observed in the spectra of the iron oxides. In the case of the products of the reaction of iron phases with AF and ATF for 1 month, the spectra

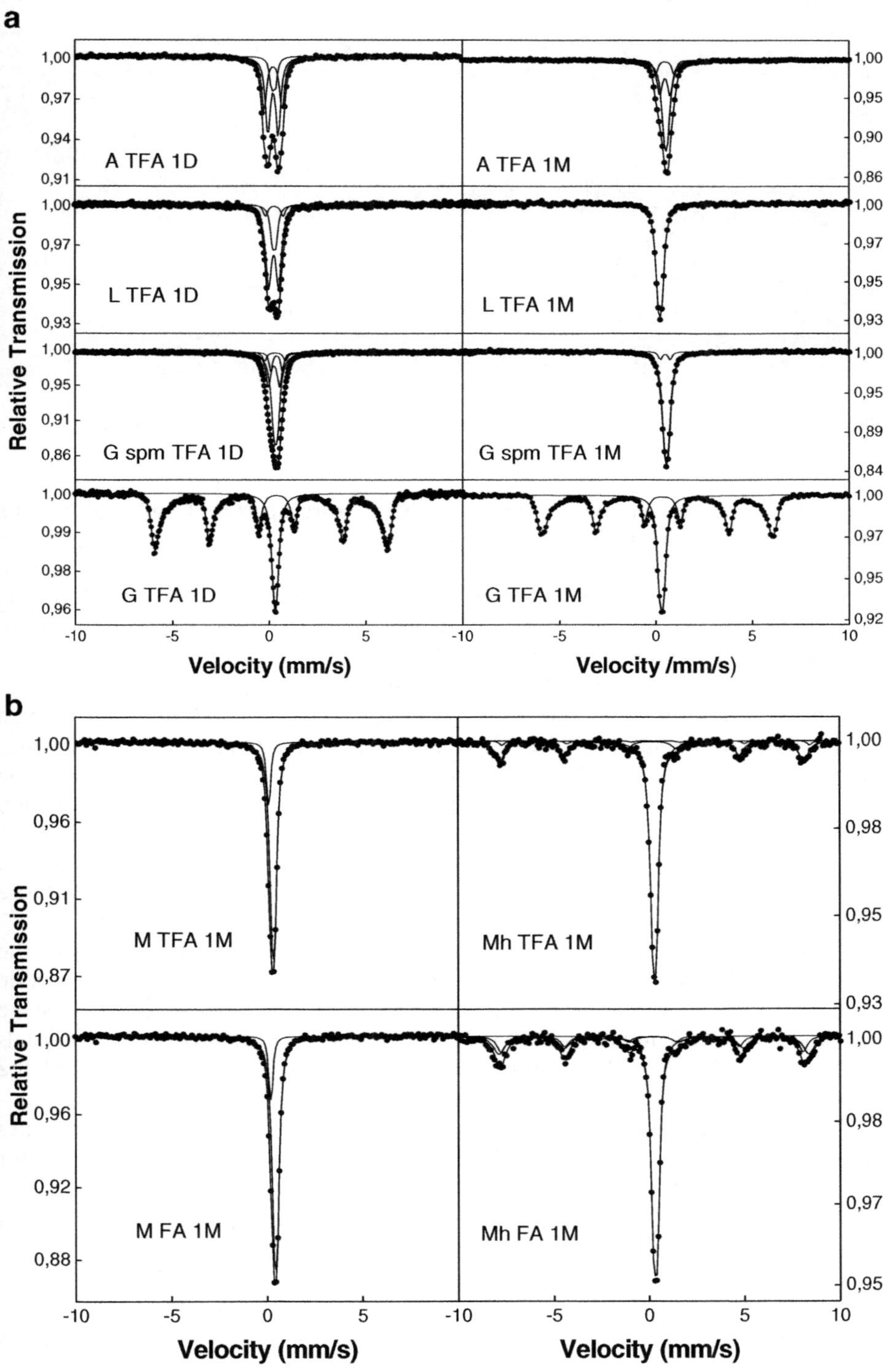

Fig. 3 Room temperature Mössbauer spectra of **a** iron oxyhydroxides after 1 day and 1 month of reactions with the TFA converter and **b** M and Mh after 1 month of reaction with the TFA converter

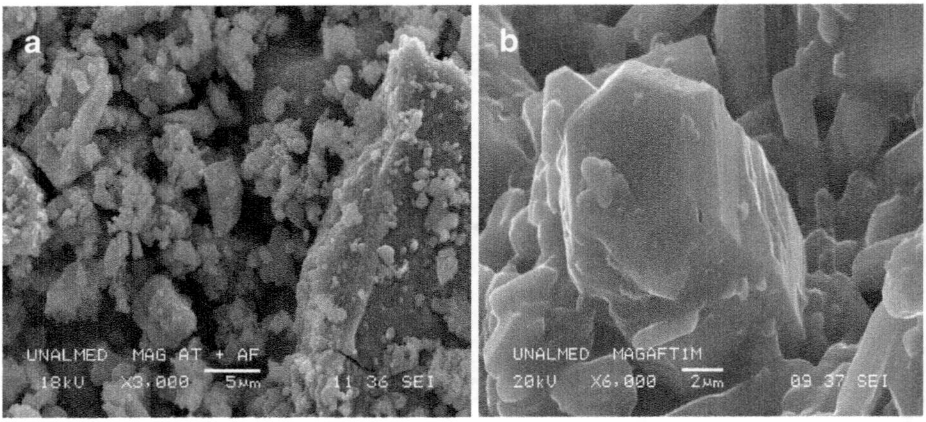

Fig. 4 SEM images of M after the reaction with the TFA converter for **a** 1 day and **b** 1 month

showed an intensive doublet, assigned to the mixture of iron phosphates, but the presence of iron tannates could not be detected. For H, appreciable changes were not observed. The Mössbauer spectrum of Mh showed residual peaks of this iron phase. In the case of the spectrum of M, the best fitting was achieved by introducing two doublets assigned to the mixture iron phosphates without the presence of M, in good agreement with the results of DRX (see Fig. 3b).

The results have shown that the iron phase most resistant to transformation to PI and TI was H, possibly due to its particular crystalline structure and its high thermodynamic stability. On the other hand, the sample that presented the major transformation after 1 day of reaction with the converters was G-spm, because of the poor crystallinity, great surface area and poor stoichiometry. The easier conversion of M in comparison to Mh can possibly be due to that M contains iron in Fe^{2+} oxidation state, which is a direct source of Fe^{3+} for a subsequent transformation to iron phosphates. The more reactive samples after 1 month of interaction with those converters were L and M in agreement with reports by other authors [8, 9].

Figure 4a shows scanning electronic microscopy images of the products of the reaction of M with the TFA converter for 1 day. The fine grains over the oxide surface, suggest the formation of iron phosphate particles. Figure 4b shows the same reaction but after 1 month, where it is clearly observed the formation of crystals with hexagonal habit. These results revealed valuable information about of the crystallographic structure of the iron phosphates.

The phosphate phases can be formed first by a partial dissolution of the samples to produce Fe^{3+} ions, which depends of the pH of the converter. The reactions for the oxyhydroxides and for H and Mh are:

$$FeOOH + 3\,H^+ \rightleftharpoons Fe^{3+} + 2\,H_2O$$

$$Fe_2O_3 + 6\,H^+ \rightleftharpoons 2\,Fe^{3+} + 3\,H_2O$$

These Fe^{3+} ions can react with the predominant specie to this pH to form the main compound.

$$Fe^{3+} + 2\,H_3PO_4 + 2.5\,H_2O \longrightarrow FeH_3\,(PO_4)_2 \times 2.5\,H_2O + 3\,H^+$$

Likely, this compound can be trimerized by a reaction catalyzed by the acid medium to form another hydrated ferric phosphate.

$$3\,FeH_3\,(PO_4)_2 \times 2.5\,H_2O \xrightarrow{H^+} Fe_3H_9\,(PO_4)_6 \times 6\,H_2O + 1.5\,H_2O$$

In the case of M, dissolution reaction produced both, Fe^{2+} and Fe^{3+}.

$$Fe_3O_4 + 8\,H^+ \rightleftarrows 2\,Fe^{3+} + Fe^{2+} + 4\,H_2O$$

These Fe^{3+} ions react with the phosphoric acid to form the compounds identified previously and Fe^{2+} is oxidized to Fe^{3+}, according to:

$$4\,Fe^{2+} + O_2 + 4\,H^+ \rightleftarrows 4\,Fe^{3+} + 2\,H_2O$$

These reactions require an apreciable amount of H^+ ions, which makes that the solution reaches a higher pH. Under these conditionts can be formed estrenguite by reaction between the new source of Fe^{3+} and acid phosphate ions.

$$Fe^{3+} + HPO_4^{2-} + 2\,H_2O \longrightarrow FePO_4 \times 2\,H_2O + H^+$$

Although the reaction of M with ATF for a day can also be oxidized to Fe^{2+}, is not observed the formation of the estreguite possibly due to the contribution in the acidity of AT.

4 Conclusions

The products of the reaction of tannic acid with each iron oxide and oxyhydroxide were iron tannates, with phosphoric acid were iron phosphates and with the mixture of tannic and phosphoric acid the different iron phases were transformed principally into iron phosphates. The produced iron phosphates showed a marked hexagonal structure. The most reactive phases were magnetite and lepidocrocite, whereas the less transformed phase was hematite. In the case of goethite, the sample with highest stoichiometric deviation and the highest surface area showed the major transformation.

Acknowledgements To CODI-Universidad de Antioquia (IN1248CE and Sustainability Program for the Solid State Research Group for the period 2007–2008), Excellence Center for Novel Materials under COLCIENCIAS contract 043-2005, and the doctorate programs of COLCIENCIAS.

References

1. Xu, L., He, N., Du, J., Deng, Y.: Electrochem. Commun. **10**, 1657 (2008)
2. Collazo, A., Novoa, X.R., Pérez, C., Puga, B.: Electrochem. Acta. **53**, 7565 (2008)
3. Barrero, C.A., Ocampo, L.M., Arroyave, C.E.: Corros. Sci. **43**, 1003 (2001)
4. Barrero, C.A., Rios, J.F., Morales, A.L., Bohórquez, A., Pérez-Alcazar, G.: Hyperfine Interact. **148/149**, 211 (2003)
5. Singh, D.D.N., Yadav, S.: Surf. Coat. Technol. **202**, 1526 (2008)
6. Iglesias, J., García de Saldaña, E., Jaen, J.A.: Hyperfine Interact. **134**, 109 (2001)
7. Gust, J., Suwalski, J.: Corrosion **50**, 355 (1994)
8. Jaén, J.A., Arauz, E.Y., Iglesias, J., Delgado, Y.: Hyperfine Interact. **148/149**, 199 (2003)
9. Nasrazadani, S.: Corros. Sci. **39**, 1845 (1997)

10. Nigmam, A.N., Tripathi, R., Dhoot, K.: Corros. Sci. **30**, 799 (1990)
11. Almeida, E., Pereira, D., Figueiredo, M.O., Lobo, V.M.M., Morcillo, M.: Corros. Sci. **39**, 1561 (1997)
12. Bolivar, F., Barrero, C.A., Minotas, J., Morales, A.L., Greneche, J.M.: Hyperfine Interact. **148/149**, 219 (2003)
13. Ocampo, L.M., Margarit, O.R., Mattos, S.I., Cordoba de Torresi Fragata, F.L.: Corros. Sci. **46**, 1515 (2004)
14. Schwertmann, U., Cornell, R.M.: Iron Oxides in the Laboratory. Wiley, Weinheim (2000)

Hyperfine Interact (2010) 195:141–147
DOI 10.1007/s10751-009-0111-9

Design of a conversion electron Mössbauer spectrometer based on an electron multiplier. Evaluation of the mean-escape-depth of the detected signals

Fernando Moutinho · Carlos Rojas · Lisseta D'Onofrio

Published online: 29 September 2009

Abstract A Conversion Electron Mössbauer Spectrometer to be used for the characterization of Fe-containing metal surfaces was designed and installed in an Ultra High Vacuum chamber. The design is based in the use of a Channeltron electron multiplier for the detection of electrons emerging from the sample after an incident γ-Ray is absorbed by resonant nuclear excitation. Using a Monte Carlo simulation for electron trajectories in solids the mean-escape-depth of the detected Mössbauer signal from a metallic iron sample was estimated to be 80 nm, assuming that the main signals being detected correspond to the Fe(M), Fe(L) and Fe(K) conversion electrons as well as the Fe(KLM), Fe(KLL) and Fe(LMM) Auger electrons. The sensitivity to the surface region was also estimated experimentally by acquiring Mössbauer spectra from a series of Fe films of different thickness deposited by magnetron sputtering on 304 stainless steel substrates.

Keywords CEMS · Channeltron · Thin films · Mean-escape-depth

1 Introduction

Conversion Electron Mössbauer Spectroscopy (CEMS) is a well established non destructive technique that has been used in the study of oxidation, corrosion, thin films, surface modification and coatings in iron containing surfaces [1]. The conversion electrons and associated Auger electrons originated from the de-excitation process after γ-Ray resonant nuclear absorption produce signals with a relatively short mean-escape-depth that can be conveniently used for the study of surface phenomena. As there are many surface science laboratories that routinely use conventional ultra high vacuum chambers, what we propose in this work is the use of a simple

F. Moutinho (✉) · C. Rojas · L. D'Onofrio
Centro de Física Experimental del Sólido, Facultad de Ciencias,
Universidad Central de Venezuela, Caracas, Venezuela
e-mail: fmoutinho@fisica.ciens.ucv.ve

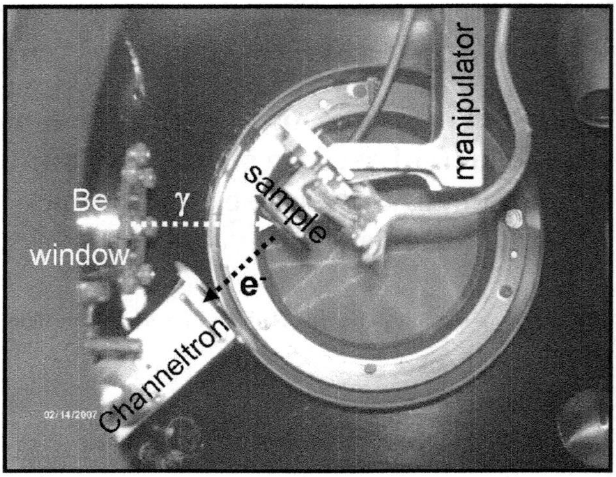

Fig. 1 Experimental arrangement

CEMS spectrometer, easily adaptable to most of those chambers in order to enhance their capabilities with the addition of Mössbauer Spectroscopy. We also present an estimation of the mean-escape-depth of the detected signals based on Monte Carlo simulations [2], to be compared with the results of experimental CEMS spectra of Fe films deposited on stainless steel substrates.

2 Experimental

Figure 1 shows a picture of the experimental assembling used in our CEMS detection system. It is all mounted on an 8-in. flange of an ultra high vacuum chamber normally used for LEED/AES surface analysis. A Channeltron electron multiplier [3], used as the electron detector, was fixed to a stainless steel support rod and tilted 45° upwards in order to face the sample surface, which is oriented 45° downwards. A lead lined stainless steel tube, welded to the flange, ends in a Be/Plexiglass window which lets in the 14.4 keV γ-Rays while filtering out the 6.4 keV Fe(Kα) X-Rays associated with the radioactive source. Three feedthroughs in the same flange allow for the electrical connections necessary to operate the Channeltron: the HV bias, the entrance cone bias and the signal collector. The window-to-sample and Channeltron-to-sample distances are approximately 5–6 cm.

The radioactive source used was a 20 mCi ^{57}Co/Rh mounted on a conventional mechanical driving unit and kept outside the chamber, inside the tube, close to the Be/Plexiglass end window. The γ-Ray footprint on the sample position is uniform over a 25 mm diameter circle. The Channeltron electron multiplier used has an entrance cone of 19 mm diameter and was operated in the pulse counting mode, generating output pulses of about 20 ns width. Its detection efficiency curve as a function of the incoming electron energy is shown in Fig. 2. Mössbauer emission spectra were acquired using a triangle velocity mode and a conventional pulse counting electronics, consisting of the biasing power supplies, preamplifier, amplifier, multichannel analyzer and a personal computer. The spectrometer performance and its surface sensitivity were evaluated by the acquisition of CEMS spectra, at room

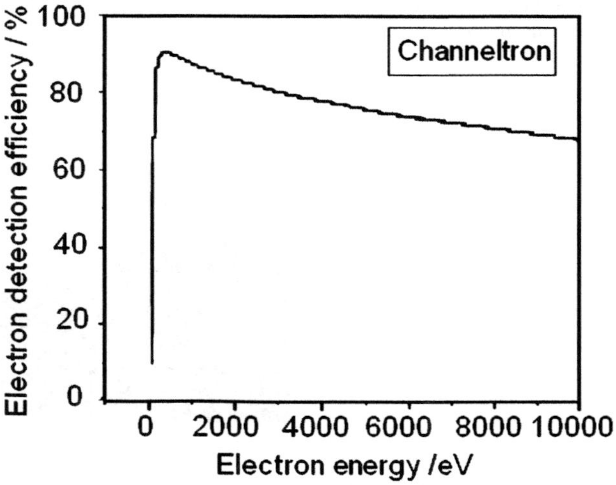

Fig. 2 Electron detector efficiency vs. energy

temperature and under high vacuum conditions, from three Fe thin films samples deposited by magnetron sputtering on 25 mm diameter stainless steel circular plates (SS304). In this way we could compare the relative intensities of the sextet signal coming from the α-Fe films with the singlet signal coming from the underlying SS304 substrate.

3 Results and discussion

The secondary electron image shown in Fig. 3 corresponds to the thicker of the three Fe films; we can appreciate that the thickness is relatively uniform and the structure is columnar. The average thickness of this film corresponds to 200 nm; from this value the thickness of each of the other two films were evaluated according to the deposition times used, resulting in 100 nm and 50 nm.

Due to the high-pass-filter characteristics of the Channeltron, we assume that the main electronic Mössbauer signals being detected in our CEMS experiment are the Fe(M), Fe(L) and Fe(K) conversion electrons as well as the Fe(KLM), Fe(KLL) and Fe(LMM) Auger electrons. Other possible electronic Mössbauer signals like low energy secondary electrons, 47 eV Fe MVV Auger electrons and X-Ray and γ-Ray photoelectrons were not considered. Table 1 lists the electronic Mössbauer signals considered in this work, together with their kinetic energies E_s and their emission probabilities w_s, relative to the number of Fe atoms whose nuclei absorbed resonantly the 14.4 keV γ-Rays. The emission probabilities w_s give the relative proportions in which these signals are originally generated within the sample and were calculated using the values for internal conversion electron coefficients α_i obtained from reference [4] and the values of X-ray fluorescence yields ω_i compiled in reference [5]. The (KLL/KLM) Auger signal ratio was taken as 5 according to the experimental spectra given in reference [6]. The atomic number Z, the density ρ of the material and the electron kinetic energy values E_s are relevant for the transport of these electron signals from the depth d at which they were generated towards the exit surface and then to the electron detector. We performed Monte Carlo electron

Fig. 3 Fe film secondary electrons image

Table 1 Kinetic energy and emission probabilities

Electron signal	Energy/keV	Emission probability
Fe(M)	14.3	1%
Fe(L)	13.6	9%
Fe(K)	7.3	81%
Fe(KLM)	6.3	9%
Fe(KLL)	5.4	43%
Fe(LMM)	0.65	9%

trajectory simulations in order to obtain, for each one of the aforementioned signals, the energy distribution N(E,d) of the electrons leaving the sample after traversing an iron film of thickness d. When the conversion and Auger electrons are generated inside sample, they can suffer straggling and energy losses as they travel through the sample towards the exit surface, resulting in a reduced number of emerging electrons distributed in energy according to N(E,d). As an example Fig. 4 shows the energy distribution $N_{Fe(K)}(E,d)$ of 10^5 Fe(K) conversion electrons escaping through 5, 50 and 100 nm thick iron films.

Upon reaching the Channeltron the energy distribution N(E,d) is modified according to its detector efficiency giving a detected energy distribution D(E,d) whose sum gives the escape probability $P_s(d)$ of each type of detected signal as a function of depth in a metallic Fe sample. By weighting the probabilities $P_s(d)$ with the respective emission probabilities w_s one obtains the relative number of Mössbauer electron signals that could be detected. However, in practice this number does not correspond to the true detected signals because of possible detection coincidence effects. The emission of a given Auger electron is strongly correlated with the emission of its parent conversion electron in such a way that both emissions occur almost simultaneously. Then, there is a high probability that both electrons reach the detector at the same time generating a single pulse if their difference in time of flight is smaller than the Channeltron pulse width. This is the case for the K conversion electron and the KLM and KLL Auger electrons and for the L conversion electron and the LMM Auger electron. If we simply add the weighted $P_s(d)$ probabilities then these coincidence effects would not been taken into account. To include them, each pair of (conversion-Auger) electrons must be combined according to the expression

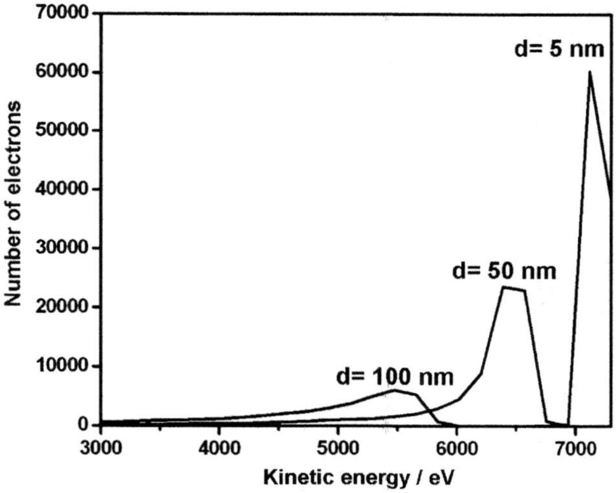

Fig. 4 Fe(K) electron energy distributions

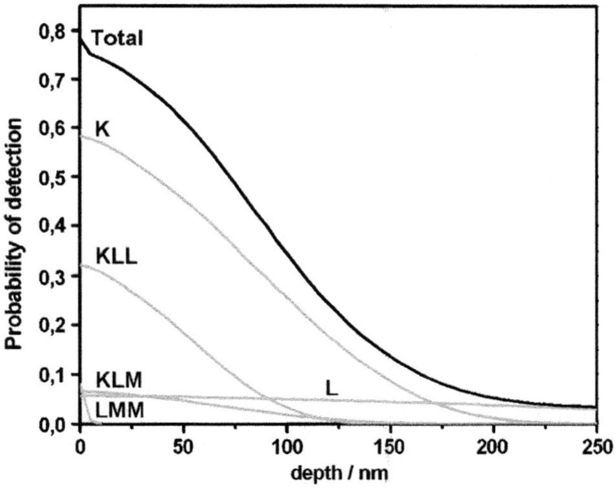

Fig. 5 Probability of electron detection

$w_{conv}P_{conv}(d) + w_{Auger}P_{Auger}(d)\ [1 - P_{conv}(d)]$ [7]. By adding the different escape probability curves according to this expression one obtains the total probability of escape and detection as a function of depth as shown in Fig. 5. The total probability curve is dominated by the contribution from the Fe(K) conversion electrons which provide information from a surface region of about 200 nm thick. Information from deeper regions is provided basically by Fe(L) conversion electrons. Due to its low emission probability the contribution from the Fe(M) conversion electrons is negligible. From the total probability curve we obtained the value of 80 nm as the mean-escape-depth of the detected integrated signals from metal iron samples.

Figure 6 shows the recorded and fitted spectra for the three Fe/SS304 samples, together with the corresponding hyperfine parameters and relative spectral areas for α-Fe and SS304.

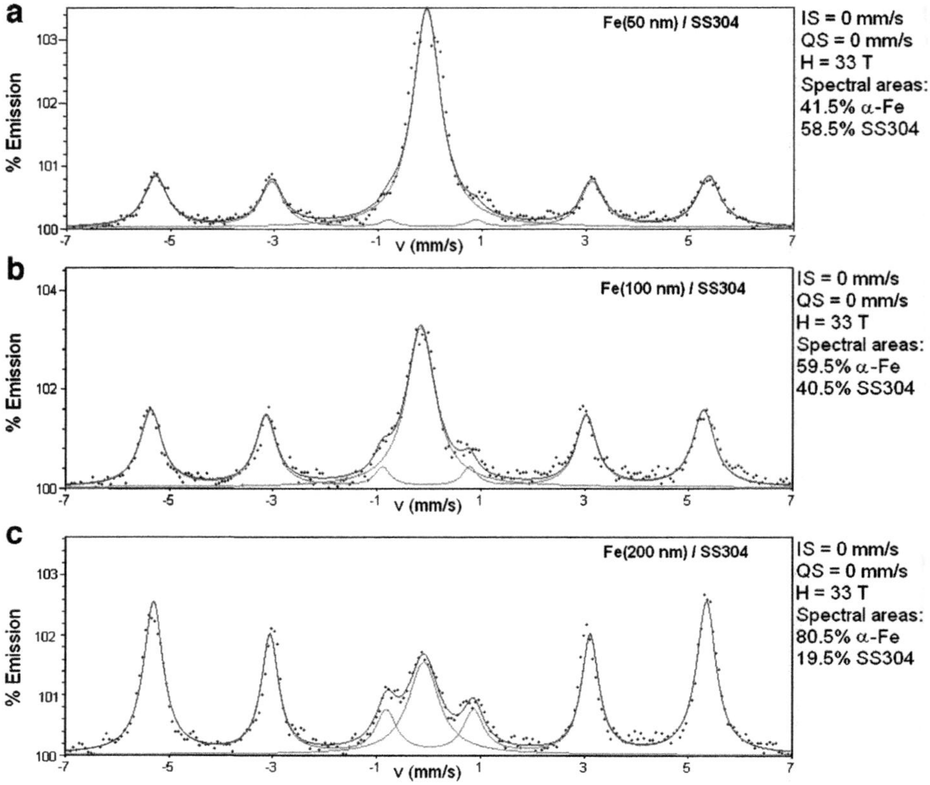

Fig. 6 CEMS spectra of de α-Fe/SS304. **a** 50 nm, **b** 100 nm, **c** 200 nm

As can be noticed, there is still an observable signal from the stainless steel substrate, even after coating it with a 200 nm thick iron film. This can be explained in terms of the columnar structure of the film, typical of the magnetron sputtering deposition method with substrates kept at room temperature [8], which leaves open spaces between the columns, giving for the film an iron density smaller than the bulk one assumed in our Monte Carlo simulations.

4 Conclusion

The simple CEMS spectrometer that we have designed can be easily installed in any conventional Ultra High Vacuum chamber dedicated to surface or thin film analysis, enhancing in this way its analytical capabilities with Mössbauer Spectroscopy. The calculated mean-escape-depth of the Mössbauer electron signals detected by our spectrometer in metallic iron corresponds to 80 nm.

Acknowledgements The advice of Dr. Ramón Gancedo, from the Instituto de Química Física "Rocasolano", Madrid, Spain, and the financial support from CDCH-UCV under Project PI 03-00-5566-2004 and from FONACIT under Project G-2005000449 are gratefully acknowledged.

References

1. Nomura, K., Ujihira, Y., Vértes, A.: J. Radioanal. Nucl. Chem. **202**, 103–109 (1996)
2. Joy, DC: Monte Carlo Modelling for Electron Microscopy and Microanalysis. Oxford University Press, Oxford (1995)
3. Channeltron Electron Multiplier Handbook for Mass Spectrometry Applications. BURLE. www.burle.com/cgi-bin/byteserver.pl/pdf/ChannelBook.pdf (2006). Revised 03 Feb 2006
4. Krakowski, R.A., Miller, R.B.: Nucl. Instrum. Methods **100**, 93–105 (1972)
5. Hubbell, J.H., Treham, P.N., Singh, N., Mehta, D., Garg, M.L., Sing, S., Puri, S.: J. Phys. Chem. Ref. Data **23**, 339–364 (1994)
6. Toriyama, T., Asano, K., Saneyoshi, K., Hisatake, K.: Nucl. Instrum. Methods Phys. Res. **B4**, 170–185 (1984)
7. Salvat, F., Parellada, J.: Nucl. Instrum. Methods Phys. Res. **B1**, 70–84 (1984)
8. Glocker, D.A, Ismat Shah, S. (eds): Handbook of Thin Film Process Technology. IOP Publishing Ltd, Bristol (1995)

Hyperfine Interact (2010) 195:149–154
DOI 10.1007/s10751-009-0112-8

Magnetic properties of γ-Fe_2O_3 nanoparticles encapsulated in surface-treated polymer spheres

J. A. H. Coaquira · C. A. Cardoso · F. Q. Soares · V. K. Garg · A. C. Oliveira · A. F. R. Rodriguez · D. Rabelo · P. C. Morais

Published online: 26 September 2009

Abstract Mössbauer and magnetic characterization of polymer-dispersed γ-Fe_2O_3 nanoparticles treated under different chemical processes are reported in this work. X-ray powder diffraction analysis provides a mean particle size of D $\sim$ 8.0 nm. Whereas Mössbauer spectroscopy data suggest the presence of only Fe^{3+} ions, magnetization measurements indicate the occurrence of a freezing phenomenon in agreement with the thermal evolution of Mössbauer spectra. A core–shell model was used to determine a magnetically disordered layer (shell) of d $\sim$ 1.0 nm covering a region of collinear magnetic moments (core). The chemical treatments with H_2O_2 and $Na_2S_2O_8$ modify notoriously the magnetic response of the polymer-dispersed nanoparticles.

Keywords $\gamma-Fe_2O_3$ nanoparticle · Mössbauer spectroscopy · Magnetization · Core–shell model

J. A. H. Coaquira (✉) · V. K. Garg · A. C. Oliveira · P. C. Morais
Instituto de Física, Núcleo de Física Aplicada, Universidade de Brasília, Brasília, DF 70910-900, Brazil
e-mail: coaquira@unb.br

C. A. Cardoso
Centro de Ciências Exatas e de Tecnologia, Departamento de Física, Universidade Federal de São Carlos, São Carlos, São Paulo 13565-905, Brazil

F. Q. Soares · D. Rabelo
Instituto de Química, Universidade Federal de Goiás, Goiânia, Goiás 74001-970, Brazil

A. F. R. Rodriguez
Departamento de Ciências Exatas e da Natureza, Fundação Universidade Federal de Rondônia, Ji-Paraná, Rondônia 78961-970, Brazil

1 Introduction

Maghemite (γ-Fe_2O_3) nanoparticles are interesting for many areas of science and technology. Since γ-Fe_2O_3 nanocomposites can be useful in biomedical applications, magnetic recording, magnetic refrigeration, magneto-optics and spintronics [1–3], the control of their magnetic properties is of great importance to reach the goals imposed by a specific application. γ-Fe_2O_3 nanoparticles can show characteristic properties such as superparamagnetism, blocking temperature, interparticle interaction, and extra anisotropy contributions, which strongly depend on the history and preparation method and are still not completely understood. For biomedical use, for instance, it is mandatory to have coated nanoparticles to allow biocompatibility and site-specificity and to prevent from aggregation [4]. Those demands have evidenced the interest in γ-Fe_2O_3 nanoparticles with controllable size and size-dispersion in special templates such as polymers, glasses or ceramics [1]. Mossbauer spectroscopy both at room temperature and liquid nitrogen temperature have been used to study the effects of chemical treatment on a set of composite samples containing maghemite nanoparticles prepared by the surfactant-free emulsion method.

2 Experimental

A Poly-Styrene template was used to host co-precipitated maghemite nanoparticles. The polymerization process was conducted in an emulsion of an aqueous phase free from surfactant and using different solvents. The as-produced magnetic nanocomposites were submitted to hydroxylation using two different reagents—hydrogen peroxide (H_2O_2) and sodium persulphate ($Na_2S_2O_8$). Following the hydroxylation step, both sets of samples were treated with stearic acid. In this study we present results regarding the composite samples loaded with 10% of maghemite. Hereafter, the magnetic nanocomposite prepared using the surfactant-free dispersion is named EM13, the hydroxylated sample using H_2O_2 is H27 and using $Na_2S_2O_8$ is H26. The samples treated with stearic acid are tagged EST33 and EST34, corresponding to H27 and H26, respectively.

Transmission electron microscopy (TEM) was used to determine the average particle size and size distribution whereas X-ray powder diffraction (XRD), carried out with CuK_α radiation, was used to determine both the crystalline phase and the particle size of γ-Fe_2O_3 nanoparticles. Mössbauer spectra (MS) at $T = 77$ K and room temperature were obtained using a constant acceleration spectrometer with symmetrical waveform using a $^{57}Co(Rh)$ source. Low-field magnetic measurements were carried out in a vibrating sample (VSM) and SQUID magnetometers in a broad range of temperature (4–300 K).

3 Results and discussion

Powder X-ray diffraction (XRD) patterns of the EM13 (see Fig. 1) show Bragg reflections consistent with the maghemite phase ($\gamma-Fe_2O_3$). The broadened feature of the peaks indicates the formation of small grains of the iron oxide phase. The analysis of XRD data using the Rietveld refinement is consistent with the cubic phase

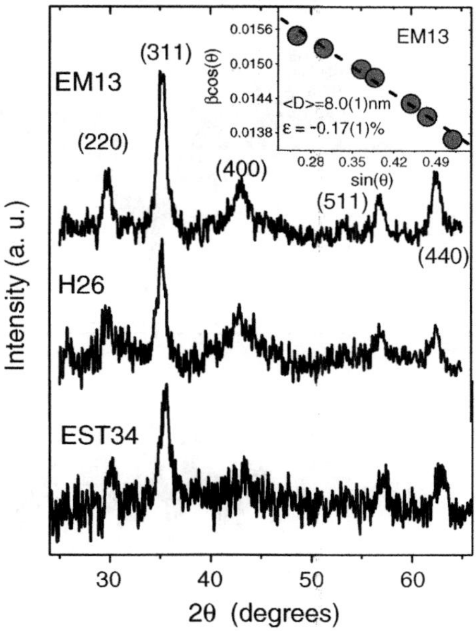

Fig. 1 X-ray powder diffraction pattern of the samples EM13, H26 and EST34. In the *inset* is shown a plot to determine the mean size and the strain of the particles for the sample EM13

(space group *Fd3m*). The line width of the most intense peaks, after correction for instrumental broadening, have been used to determine the mean grain size ($\overline{D}$) and the strain (ε) of the system from the relation [5]: $B\cos(\theta) = \left(0.9\lambda/\overline{D}\right) + 4\varepsilon\sin(\theta)$. In the inset of Fig. 1 is shown the plot of the last relation. An average particle size of $\overline{D} = 8.0 \pm 0.1$ nm and a small but not negligible strain of $\varepsilon = -0.17\%$ has been determined. Since similar particle sizes are determined for the other samples, it seems that the chemical treatments do not affect the composition and morphology of the iron oxide nanoparticles. The analysis of TEM images (not shown here) are well represented by a log-normal distribution function, with an average particle size consistent with the crystalline size determined from XRD data.

Figure 2 shows the Mössbauer spectra (MS) carried out at $T = 77$ K and room temperature for the sample EM13. Similar spectra were obtained for the chemically treated samples but with lower signal-to-noise ratios, in special the samples H26 and EST34. At $T = 77$ K, the spectrum shows magnetic splitting (six lines) with asymmetric lines. However, the room-temperature spectrum suggests the coexistence of a broad distribution of hyperfine components. The room temperature spectrum is well resolved imposing a fit with one broad histogram distribution of magnetic fields and one doublet (see Fig. 2). The onset of the doublet with IS ≈ 0.47 mm/s and QS ≈ 0.6 mm/s suggests the thermal relaxation of the small-diameter part of the size distribution of the iron oxide particles. However, the broad distribution of hyperfine fields indicates a collective collapse from sextets into a doublet at room temperature. The 77 K-spectrum is well-resolved with two Gaussian distributions: one centered at $B_{hf} = 49.7$ T (S_1) with a width ~ 2 T and the second centered at $B_{hf} = 41.4$ T (S_2) with a width $\sim$3 T. The presence of two magnetic subspectra is consistent with iron ions occupying the crystallographic sites of the maghemite structure (tetrahedral and octahedral) [6, 7]. The analysis of spectral areas provides a spectral area ratio

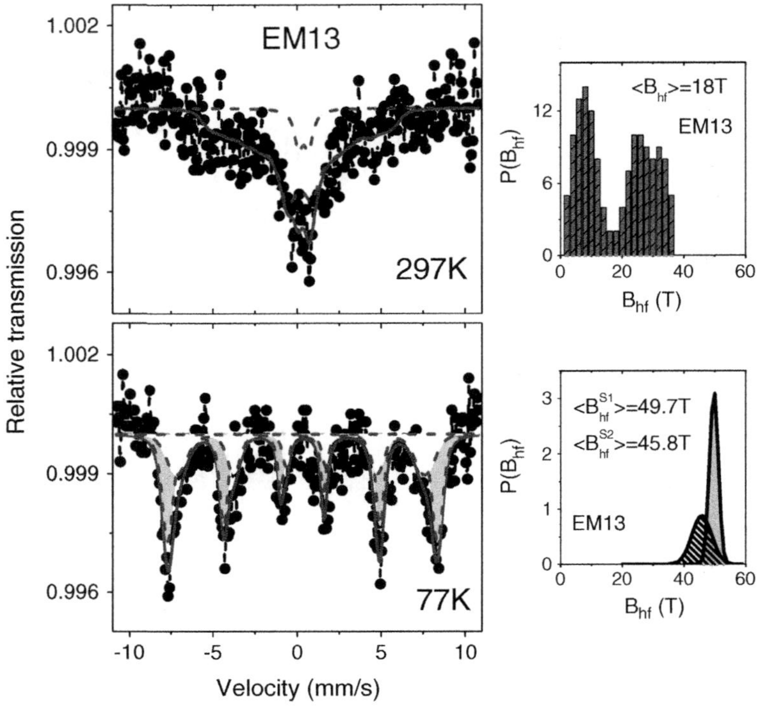

Fig. 2 Typical Mössbauer spectra recorded carried out at $T = 297$ and 77 K. The *solid line* represents the best fit and *dotted lines* represent the subspectra used in the fit. Hyperfine field distributions determined from the fits are presented on the *right-hand side*

of $S_2/S_1 = 0.64 \pm 0.04$, which indicates that cation vacancies are equally distributed in both sites in agreement with our previous report on other family of γ-Fe_2O_3 nanoparticles [8]. An isomer shift of +0.42 mm/s for S_1 and +0.47 mm/s for S_2 strongly suggest the presence of only ferric ions.

The temperature dependence of zero-field-cooled (ZFC) and field-cooled (FC) magnetization curves of the set of samples (not shown here) shows features consistent with superparamagnetic (SPM) behavior. The ZFC curve of the sample EM13 shows a peak centered at $T_{max} = 126 \pm 2$ K. This peak position is shifted to lower temperatures when the polymer-dispersed nanoparticles are hydroxylated. While the hydroxylation with $Na_2S_2O_8$ and H_2O_2 shift the peak position to $T_{max} = 101 \pm 2$ K and 115 ± 2 K, respectively, the final samples' treatment with stearic acid seems not to affect those peak positions. The irreversibility between the ZFC and FC curves remains above the T_{max} and suggests a thermal blocking process of interacting particles. Interacting particles are expected because of the relatively high concentrated dispersion. In Fig. 3 is shown the magnetization (M) as a function of the magnetic field (H). The sample EM13 shows tendency to the saturation at $H \sim 10$ kOe and at higher fields (H > 40 kOe) a slight tendency to decrease is observed. This high-field decreasing behavior has been associated to the diamagnetic contribution of the host template.

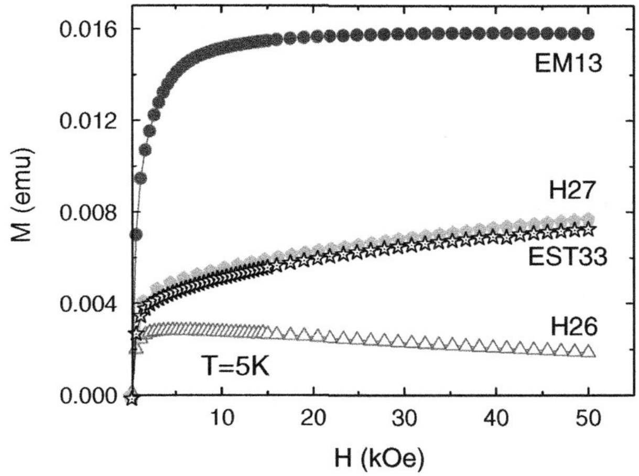

Fig. 3 Isothermal magnetization as a function of the magnetic field obtained at $T = 5$ K for the four different samples

Considering the nominal concentration of γ-Fe_2O_3 in the dispersion (10%), a value of $m_s = 32.3$ emu/g for the saturation magnetization of the EM13 is obtained. This value is largely smaller than the saturation value (87.4 emu/g) expected for the bulk system (defect-free system) at low temperatures [9]. The small value determined for the γ-Fe_2O_3 nanoparticles in our work can be associated to a spin-canting effect which should happen in the bulk and/or near to the particle's surface. Defects related to the cation vacancies originate a non-collinear magnetic structure. Related to the cation vacancy effect, reports suggest a reduction of ~13% of the saturation magnetization of the perfectly collinear system [9]. This reduction cannot explain the low saturation value determined from our measurements. On the other hand, due to the occurrence of symmetry breaking it is commonly accepted the formation of a magnetically disordered layer on the particles surface (core–shell model) [10, 11]. A clear dependence of the saturation magnetization on the particle size has been reported for maghemite nanoparticles [12]. In that model the saturation magnetization is given by: $m_S = M_{0S}\left[(D/2 - d)/D/2\right]^3$, where m_S and M_{0S} is the saturation of the dispersion and bulk system, respectively; D is the particle diameter and d is the thickness of the disordered layer. Using the particle size determined from XRD data and considering the saturation magnetization M_{0S} ~76 emu/g, a disordered layer of $d = 1.0 \pm 0.1$ nm is found. This thickness is in good agreement with values reported for maghemite particles with diameter $D > 8$ nm [12].

The polymer-dispersed nanoparticles hydroxylated with H_2O_2 shows a great reduction in the saturation magnetization (~40%) and a monotonous increasing tendency even at H up to 50 kOe. This finding is related to the formation of either smaller particles or a non-magnetic ferrous phase in agreement with the non-magnetic contribution determined from MS data of samples H27 and EST33. Two clear features are observed after the hydroxylation of the polymer-dispersed nanoparticles with $Na_2S_2O_8$ (sample H26). This process leads to a larger reduction of the saturation (~72%, in relation to the sample EM13) and to the appearance of a diamagnetic contribution. These findings are likely associated to the removal of magnetic material from the dispersion, which enhances the diamagnetic contribution of the polymeric template.

Acknowledgements The authors thank O. F. de Lima and E. M. Guimarães (Institute of Geology—UnB) for the access to their laboratory facilities. The authors also wish to thank S. N. Báo (Institute of Biology—UnB) for TEM experiments. This work was financially supported by the Brazilian agencies FINATEC and MCT/CNPq.

References

1. Millan, A., Palacio, F., Falqui, A., Snoeck, E., Serin, V., Bhattacharjee, A., Ksenofontov, V., Gütlich, P., Gilbert, I.: Acta Mater. **55**, 2201 (2007)
2. Macaroff, P.P., Oliveira, D.M., Ribeiro, K.F., Lacava, Z.G.M., Lima, E.C.D., Morais, P.C., Tedesco, A.C.: IEEE Trans. Magn. **41**, 4105 (2005)
3. Bakuzis, A.F., Neto, K.S., Gravina, P.P., Figueiredo, L.C., Morais, P.C., Silva, L.P., Azevedo, R.B., Silva, O.: Appl. Phys. Lett. **84**, 2355 (2004)
4. Kaufner, L., Cartier, R., Wüstneck, R., Fichtner, I., Pietschmann, S., Bruhn, H., Schütt, D., Thünemann, A.F., Pison, U.: Nanotechnology **18**, 115710 (2007)
5. Dutta, P., Vanimannan, A., Seehra, M.S., Shah, N., Huffman, G.P.: Phys. Rev. B **70**, 174428 (2004)
6. Cohen, R.L.: Applications of Mössbauer Spectroscopy, vol. 1, pp. 92. Academic, New York (1976)
7. Helgason, Ö., Grenèche, J.-M., Berry, F.J., Mørup, S., Mosselmans, F.: J. Phys. Condens. Matter. **13**, 10785 (2001)
8. Coaquira, J.A.H., Rodriguez, A.F.R., Santos, J.G., Silveira, L.B., Oliveira, A.C., Garg, V.K., Soares, F.Q., Rabelo, D., Morais, P.C.: Hyperfine Interact. **176**, 113 (2008)
9. Coey, J.M.D.: Phys. Rev. Lett. **27**, 1140 (1971)
10. Tronc, E., Prené, P., Jolivet, J.P., Dormann, J.L., Grenèche, J.M.: Hyperfine Interact. **112**, 97 (1998)
11. Mørup, S.: J. Magn. Magn. Mater. **266**, 110 (2003)
12. Millan, A., Urtizberea, A., Silva, N.J.O., Palacio, O., Amaral, V.S., Snoeck, E., Serin, V.: J. Magn. Magn. Mater. **312**, L5 (2007)

Hyperfine Interact (2010) 195:155–159
DOI 10.1007/s10751-009-0113-7

Low temperature magnetic ordering in Fe-doped TiO_2 samples

V. Bilovol · A. M. Mudarra Navarro · W. T. Herrera · D. R. Sánchez · E. M. Baggio-Saitovich · C. E. Rodríguez Torres · F. H. Sánchez · A. F. Cabrera

Published online: 30 September 2009

Abstract Fe-doped TiO_2 samples with different Fe content prepared by mechanical alloying have been investigated by means of Mössbauer spectroscopy at 300 and 4.2 K. The results indicate the coexistence of Fe^{2+} and Fe^{3+} ions in paramagnetic states at room temperatures in the rutile structure. All samples present magnetic order at 4.3 K. When the Fe concentration increases the Fe ions in the rutile matrix became closer giving the possibility of strong magnetic interactions between them. The temperature evolution of the magnetic order was followed for the 15 at.% of Fe sample. The Fe-doped oxide formed for this composition orders below 20 K reaching an almost totally magnetic ordered state at 4.3 K.

Keywords Fe-doped TiO_2 · Mössbauer spectroscopy · Rutile matrix

1 Introduction

Oxide diluted magnetic semiconductors (O-DMS), produced by doping nonmagnetic oxide semiconductors with transition metal atoms, are of interest for potential applications in spintronics [1]. There is a continuous search of O-DMSs which present ferromagnetic behavior above room temperature. Much work has been done in the study of this kind of materials, but a real problem is to understand the origin of the observed magnetic behavior [2–5]. This behavior is very dependent on synthesis

V. Bilovol · A. M. Mudarra Navarro · C. E. Rodríguez Torres · F. H. Sánchez ·
A. F. Cabrera (✉)
Departamento de Física, Facultad de Ciencias Exactas,
UNLP–IFLP–CCT-La Plata–CONICET, C.C 67, 1900, La Plata, Argentina
e-mail: cabrera@fisica.unlp.edu.ar

W. T. Herrera · D. R. Sánchez · E. M. Baggio-Saitovich
Rua Dr. Xavier Sigaud, 150—Urca, Río de Janeiro, CEP 22290-180, Brazil

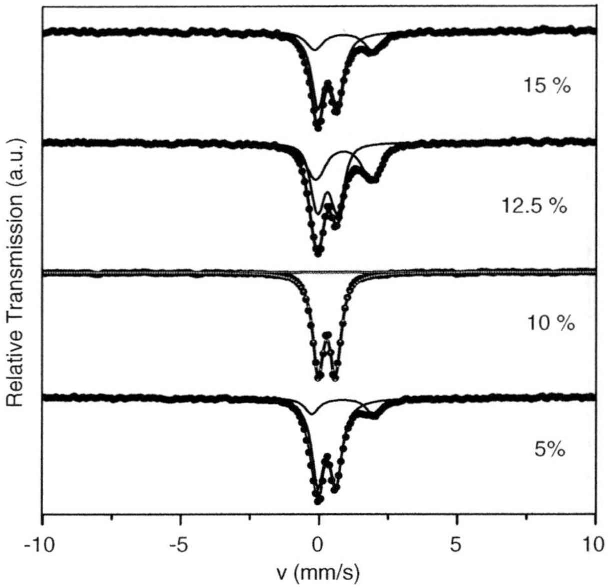

Fig. 1 Mössbauer spectra of $Ti_{1-x}Fe_xO_2$ samples at room temperature

condition of the samples and on their dimensionality (films or powders). In this work we employed Mössbauer spectroscopy in order to characterize Fe-doped TiO_2 powders obtained by mechanical alloying.

2 Experimental

Samples with different concentrations (5, 10, 12.5 and 15 Fe at.%) were mechanically alloyed during 12 h in air at 32 Hz. The former powders (rutile TiO_2 and hematite α-Fe_2O_3, both with 99.9% purity) were milled in a Retsch MM2 horizontal vibratory mill using stainless steel vial and ball. The Mössbauer spectra (MS) were taken between room temperature (RT) and 4.2 K. The isomer shifts are referred to α-Fe. Due to the existence of different iron ion environments, the MS spectra were analyzed using distributed subspectral contributions using a Voigt line shape (Voigt profile is a convolution of a Gaussian profile and a Lorentzian profile).

3 Results and discussion

The RT Mössbauer spectra of Fe 5, 12.5 and 15 at.% samples (Fig. 1) were reproduced using two paramagnetic interactions (quadrupolar electric), one corresponding to Fe^{2+} ($\delta \sim 0.9$ mm s^{-1}) and the other to Fe^{3+} ($\delta \sim 0.38$ mm s^{-1}). The relative fraction of the Fe^{2+} contribution was 20, 44 and 31% for the 5, 12.5 and 15 at.% samples respectively. The Fe 10 at.% sample spectrum was reproduced using one doublet (Fe^{3+}) and one sextet (Fe^{3+} hematite, about 2% of spectral area). The appearance of hematite may be due to failed incorporation of Fe into the rutile TiO_2 lattice. For all the other compositions this incorporation was totally achieved. XRD

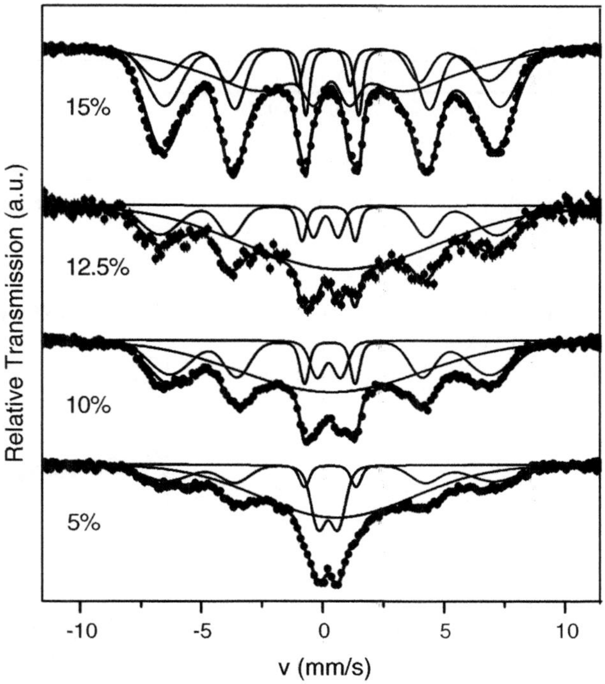

Fig. 2 Mössbauer spectra of $Ti_{1-x}Fe_xO_2$ samples at 4.3 K

results for all samples (not shown here) reveal the presence of the rutile phase, no evidence of other oxides was detected. Therefore both interactions (Fe^{2+} and Fe^{3+}) correspond to iron ions substituting Ti ones in the rutile structure.

The MS taken at 4.3 K are shown in Fig. 2. They reveal a progressive magnetic ordering when the concentration of Fe increases. The spectra of the samples with 5, 10 and 12.5 at.% of Fe were fitted with one distributed doublet, one distributed sextet, both with isomer shifts corresponding to Fe^{3+}, and one distributed singlet. The last contribution is taking into consideration possible relaxation effects. The corresponding Mössbauer parameters are shown in Table 1.

When the concentration of Fe increases the relative area of the paramagnetic doublet attributed to Fe^{3+} decreases progressively and finally for the sample with 15 at.% of Fe disappears. The spectral areas of Fe^{3+} paramagnetic contribution at RT are significantly major than the ones for the same samples at 4.3 K. This means that a part of Fe^{3+} ions are ordered magnetically at low temperature. This fact is confirmed by the appearance of a splitting sextet with an isomer shift value corresponding to Fe ions in an oxidation state 3+. It is difficult to determine if some Fe^{2+} ions remain paramagnetic or become magnetically ordered because of the broad single line used to take into consideration relaxation effects in our fitting procedure. In the cases of 5 and 12.5 at.% Fe-doped samples the central line has an isomer shift value higher than the corresponding to Fe^{3+} which could be associated to the presence of Fe^{2+} ions in the sample. However the central line of the 4.3 K spectrum of the Fe 10 at.% sample has an isomer shift of 0.48 mm/s, corresponding to Fe^{3+} and in good agreement with the results obtained from the RT spectrum (there is no Fe^{2+} contribution).

Table 1 Mössbauer fitting parameters of samples with 5, 10 and 12.5 at.% of Fe at 4.3 K

Sample	Singlet		Doublet			Sextet		
Fe at.%	δ	A	δ	ΔQ	A	δ	B	A
5	0.68_3	60	0.33_1	0.80_2	16	0.41_1	42_1	24
10	0.48_5	56	0.36_2	0.95_2	7	0.40_1	41_1	37
12.5	0.86_7	69	0.24_3	1.02_6	5	0.35_2	43_2	26

δ isomer shift (mm/s), *ΔQ* quadrupole splitting (mm/s), *B* hyperfine field (Tesla), *A* spectral area (%)

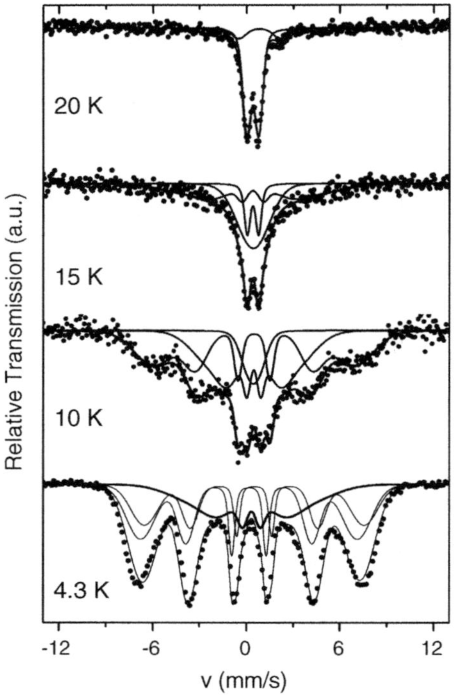

Fig. 3 Mössbauer spectra of $Ti_{0.85}Fe_{0.15}O_2$ measured at 4.3, 10, 15 and 20 K

In order to follow the temperature evolution of the magnetic ordering of the Fe 15 at.% sample, several measurements between 300 and 4.3 K were performed (Fig. 3). The spectra look rather similar from RT down to 20 K. The Fe^{2+} paramagnetic interaction was not observed for temperatures lower than 20 K. Below this temperature the resonant lines are broaden and relaxation effects appear in the Mössbauer spectra. The spectrum for 15 K was well reproduced using a magnetic sextet, the Fe^{3+} doublet and a broad single central line. The last contribution was needed in order to take into account the existence of important relaxation effects. It was not possible to introduce the Fe^{2+} doublet probably due to the superposition of this component with the sextet. At 10 K a second sextet was necessary to reproduce the spectrum. Finally, at 4.3 K the Fe^{3+} quadrupole interaction disappears and all Fe ions are magnetically ordered. The fitted Mössbauer parameters are presented in Table 2.

Table 2 Mössbauer results obtained from the fits of the $Ti_{0.85}Fe_{0.15}O_2$ spectra measured at different temperatures

T (K)	Doublet 1		Doublet 2		Sextet 1		Sextet 2		Sextet 3	
	δ_1	ΔQ_1	δ_2	ΔQ_2	δ'_1	B'_1	δ'_2	$B'_2B'_2$	δ'_3	B'_3
300	0.34_2	0.71_3	0.74_3	2.20_7	–	–	–	–	–	–
92	0.44_2	0.73_3	0.91_3	2.60_7	–	–	–	–	–	–
50	0.44_2	0.74_3	0.89_2	2.54_5	–	–	–	–	–	–
30	0.43_1	0.75_3	0.88_2	2.62_5	–	–	–	–	–	–
20	0.40_5	0.76_8	0.83_4	2.65_9	–	–	–	–	–	–
15	0.41_1	0.78_2	–	–	0.41_5	26.1_3	–	–	–	–
10	0.45_1	0.86_2	–	–	0.47_2	16.8_4	0.49_3	40.0_2	–	–
4.3	–	–	–	–	0.42_5	21.4_2	0.27_2	43.1_3	0.61_3	43.7_4

4 Conclusions

Rutile $Ti_{1-x}Fe_xO_2$ (x = 0.05, 0.1, 0.125 and 0.15) polycrystalline samples were prepared by mechanical alloying. All samples show paramagnetic behavior at room temperature with coexistence of Fe^{3+} and Fe^{2+} ions in the rutile structure. The MS at 4.3 K show magnetic ordering that becomes more evident when the iron content in the sample increases. This evolution of the magnetic ordering with the Fe content could be related to the increasing proximity of the Fe ions in the sample when the Fe concentration increases. Only the sample with the highest Fe content totally orders at the lowest temperature. This ordering begins below 20 K.

Acknowledgement We appreciate financial support of CONICET, Argentina (PIP 6005).

References

1. Janisch, R., Gopal, P., Spaldin, N.A.: J. Phys., Condens. Matter **17**, R657 (2005)
2. Balcells, L., Forntera, C., Sandiumenge, F., Roig, A., Martínez, B., Kouam, J., Monty, C.: Appl. Phys. Lett. **89**, 122501 (2006)
3. Xiaoyan, P., Dongmei, J., Yan, L., Xueming, M.: J. Magn. Magn. Mater. **305**, 388 (2006)
4. Zhu, S., Li, Y., Fan, C., Zhang, D., Liu, W., Sun, Z., Wei, S.: Physica B **364**, 199 (2005)
5. Yamaura, K., Wang, X.H., Li, J.G.F., Ishigaki, T., Takayama-Muromachi, E.: Mat. Res. Bull. **41**, 2080 (2006)

Hyperfine Interact (2010) 195:161–165
DOI 10.1007/s10751-009-0114-6

A quasi-continuous observation of the α-transition of $Fe_{1+x}S$ by Mössbauer line tracking

P. Mendoza Zélis · G. A. Pasquevich · A. Veiga · M. B. Fernández van Raap · F. H. Sánchez

Published online: 3 October 2009

Abstract Mössbauer absorption line tracking methodology, under a constant velocity strategy, is used for a quasi-continuous observation of the α-transition on slightly non stoichiometric $Fe_{1+x}S$ alloy. To this end, two strategies were applied: an intelligent absorption line tracking with a control algorithm that uses the data measured in the previous region to establish the position of the next partial spectral range; and a predetermined line tracking in which temperature evolution of a partial spectral region of interest (ROI) is programmed. The latter uses results from the former, in order to achieve a quasi-continuous partial spectral observation. These experiments clearly demonstrate that line tracking allows a more efficient use of the radioactive source, as the effort is concentrated in a partial region of the spectra from which the desired information can be obtained.

Keywords Mössbauer Line Tracking · FeS alpha-transition · Programmable constant-velocity scaler

1 Introduction

Mössbauer Line Tracking (MLT) [1] is a methodology designed to record the evolution of a spectral region of interest (ROI) while it undergoes changes due to the variation of an external parameter such as temperature, magnetic field, etc. The main advantage of this approach is the possibility of tracking and recording the Mössbauer absorption only of the ROI (which may or not include the center of the spectrum). This features allows to achieved a higher speed at which the external

P. Mendoza Zélis (✉) · G. A. Pasquevich · A. Veiga ·
M. B. Fernández van Raap · F. H. Sánchez
Depto. de Física, Fac. Ciencias Exactas, Instituto de Física La Plata,
Universidad Nacional de La Plata, CONICET, La Plata, Argentina
e-mail: pmendoza@fisica.unlp.edu.ar

physical parameter can be swept. Also, it allows the recording of quasicontinuous experimental response functions and the study of processes which occur too fast to be followed by Mössbauer spectroscopy. In this work we choose the reversible α-transition in $Fe_{1+x}S$ to demonstrate the capabilities of the MLT technique to follow a structural phase transition.

The goal of the configuration described in [1] is to enable the recording of a reduced ROI which can be continuously displaced along the whole velocity range in response to an intentional change in the temperature of the sample. In this way the automatic tracking of a Mössbauer absorption line as its energy position varies can be implemented.

Slightly non stoichiometric $Fe_{1+x}S$ presents a structural transition (α-transition). On heating, the Low Temperature Phase (LTP), the superstructure (*P*62*c*) derived from the NiAs-type structure, transforms at around 400 K into the High Temperature Phase (HTP), a transitional phase which can be described by the MnP (*Pnma*) orthorhombic structure [2–4]. The transition is reversible and of the first order, then both phases coexist within a temperature interval which, depending on stoichiometry, goes from a few degrees to around 200 K. Both phases are antiferromagnetic with iron atoms located at unique crystallographic sites. The probe environments are different enough to allow hyperfine fields from both phases to be distinguished within their coexistence temperature region.

MLT methodology is used to perform a quasi-continuous temperature study of the α-transition. The tracking was performed within a ROI which allows the observation of two absorption lines (one from each phase) and the background simultaneously, while temperature is varied within the coexistence region, both heating and cooling.

2 Experimental procedures

Commercially available FeS powder was treated during 12 h at 870 K and during 72 h at 1,270 K to improve sample homogeneity. Phase identification was achieved through Mössbauer Effect (see Fig. 1a). At RT 92% of the iron atoms are in the LTP, 4% in α-Fe and 4% in a paramagnetic impurity phase barely distinguishable in the spectrum central zone. The presence of the α-Fe signal suggests that FeS stoichiometric is slightly Fe rich, what is confirmed by the fact that the LTP phase spectrum coincides with that of troilite, a phase that is not observed in the case of iron deficiency.

The Mössbauer spectra reflect the α-structural transition. At 430 K it corresponds to the HTP structure plus the already mentioned minor contributions (Fig. 1a). The spectrum recorded at 400 K shows the superposition of LTP and HTP subspectra (Fig. 1a, middle).

The experimental setup for a MLT consists of a conventional pulse height selection branch, a commercial driving system and two recently developed NIM modules serially interfaced with a computer. The first one is a previously introduced [5] programmable constant-velocity scaler that replaces the usual multiscaler. This module allows the independent acquisition of every spectrum channel based on a constant-velocity strategy and consequently the acquisition of partial Mössbauer spectra in selected energy regions. The second module is an analog microprocessor-based input/output interface that records the temperature and commands a linear

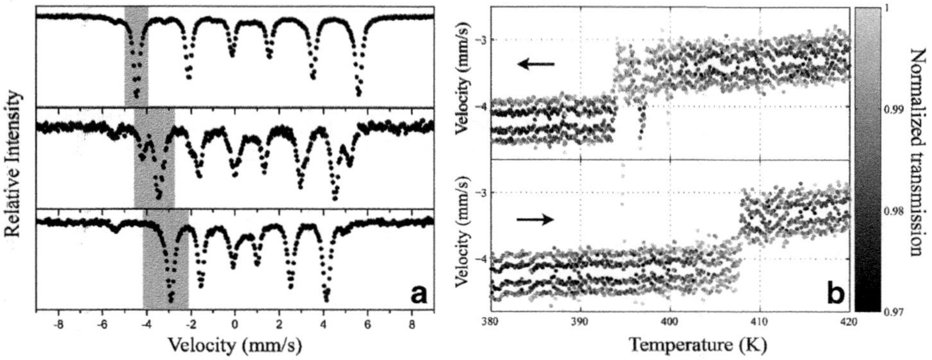

Fig. 1 **a** Mössbauer spectra of $Fe_{1+x}S$ ($x << 1$) at 300, 400 and 430 K from *top* to *bottom*. At high and low temperature a single phase is present, while at 400 K both phases can be observed. The *grey bars* stand to indicate ROI position and extension. **b** Line tracking results over the first line of the magnetically split spectrum, heating (below) and cooling (above)

DC power source that heats the sample. The line tracking and temperature control algorithms are hosted in the computer. These algorithms are written in a high-level technical computing language that interacts directly with hardware, where an environment for algorithm development, data visualization and data analysis was also developed. This configuration enables the programmable scaler and the input/output module to be fully operated using a re-programmable strategy that is closely related with the experiment.

A 25 mCi ^{57}Co*Rh* source was used for all the experiments.

3 Line tracking experiments

As a first approach, we applied an intelligent line tracking methodology where a control algorithm determines the position of the ROI using the measured data. The following result was obtained from a FeS absorber (4% effect at lines 1, 6) while it was warmed up to 435 K from 300 K, and then cooled back down to 300 K. The wall time of the experiment was 13.5 h. A four channel ROI plus background, and a time/channel ratio of 5 s were chosen. The initial ROI was selected in order to match Line 1 of the LTP based on information previously obtained from Mössbauer Spectra. While the temperature was swept the ROI was successfully repositioned by the tracking algorithm, holding the transmission minimum approximately in the center of the ROI (see Fig. 1b). In the temperature transition region, where both phases coexist, the algorithm succeeded in tracking the HTP when the LTP faded. When the temperature was lowered this process occurred in the opposite way.

In a final experiment, a predetermined tracking using a temperature dependent ROI(T) of Lines 1 of both phases was produced, based on the information retrieved from the line tracking experiment discussed above. A 25 point ROI(T) and a time/channel ratio of 5 s were chosen. The background was simultaneously measured at a fixed channel. The sample was cooled to 300 K from 435 K and then warmed back to 435 K. The wall time of the experiment was 70 h. The experimental

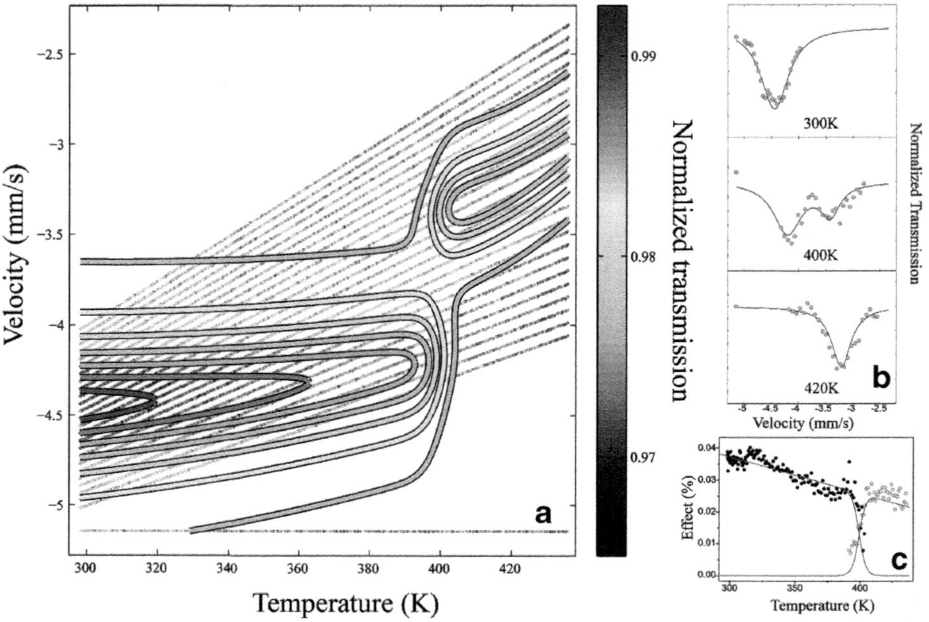

Fig. 2 **a** 3D plot of the Mössbauer transmission as a function of source velocity and sample temperature determined using a temperature dependent ROI chosen to have a good observation of the phase transition. This data correspond to the warming process. In *black* the contour lines of the fitted surface absorption function. **b** Comparison of the function and experimental data at specific temperatures. **c** First absorption line effect obtained from the fit procedure (*solid-line*) compared with those obtained from individual fits at each temperature for both phases (LTP *black points*, HTP *gray point*)

result corresponding to the warming process are shown in the Fig. 2a. The result corresponding to the cooling process shows a similar behavior but a lower transition temperature.

The whole set of data was analyzed as a velocity-temperature absorption surface using an appropriate function f(v, T). This function consists of two Lorentzian lines, each with a fixed width, but with temperature dependent center and intensity. The centers were allowed to vary quadratically with temperature while the intensities were modeled as a product of a linear and a hyperbolic tangent temperature functions. The linear factor accounts for the Mössbauer–Lamb temperature dependence and the hyperbolic tangent for the transformation temperature evolution. To compare the fitted function and the experimental data, level curves of the first one are plotted over the experimental results (Fig. 2a). Comparison of the function and experimental data at 300, 400 and 420 K are shown in Fig. 2b.

Independent fits of each individual ROI with one or two Lorentzian lines (two at phase coexistence temperatures) were also made. In Fig. 2c we show the normalized effects for each phase as a function of the temperature obtained from the individual fits (points) and from the whole surface fit (lines) in the warming process. When warming (cooling) the sample the transition starts at around 390 K (400 K) and the coexistence region size is, in both cases, about 22 K.

4 Conclusions

Mössbauer Line Tracking was applied to record the $Fe_{1-x}S$ α-transition in a quasi-continuous way. Tracking was performed on the lowest energy lines of the low and high temperature magnetically split spectra of the $Fe_{1-x}S$ phases, using a feedback procedure assisted by on-line analysis. The algorithm succeeded in tracking the HTP when the LTP faded and vice versa. With the information retrieved from the line tracking experiment a predetermined temperature dependent ROI(T) experiment was designed from which an absorption vs energy–temperature surface was reconstructed through the fitting of the experimental data with an appropriate function. In this way a detailed quasi continuous observation of the α-transition vs. temperature in terms of absorption spectral line intensities and positions of both phases was accomplished.

Acknowledgements This research was supported by CONICET through PIP 6011 and ANCYPT through grant 12-14526 of República Argentina.

References

1. Veiga, A., Pasquevich, G.A., Mendoza Zélis, P., Sánchez, F.H., Fernández van Raap, M.B., Martínez, N.: Experimental design and methodology for a new Mössbauer scan experiment: absorption line tracking. Hyperfine Interact. **188**, 137–142 (2009). doi:10.1007/s10751-008-9899-y
2. Baek, K.S., et al.: Effects of sulfur vacancies on the crystallographic and spin rotation transition of iron sulfide. Phys. Rev. B **41**, 9024 (1990)
3. Li, F., Franze, H.F.: Phase transition in near stoichiometric iron sulfide. J. Alloys Compd. **238**, 73 (1996)
4. Töpel-Schadt, J., Müller, W.F.: Transmission on meteoritic troilite. Phys. Chem. Miner. **8**, 175–179 (1982)
5. Veiga, A., Martínez, N., Mendoza Zélis, P., Pasquevich, G.A., Sánchez, F.H.: Advances in constant-velocity Mössbauer instrumentation. Hyperfine Interact. **167**, 905–909 (2006)

Hyperfine Interact (2010) 195:167–171
DOI 10.1007/s10751-009-0115-5

Mössbauer study of Fe–Co alloys with Cr additions synthesized by mechanical alloying

L. D'Onofrio · G. González · D. Oleszak · A. Sagarzazu · R. Villalba

Published online: 25 September 2009

Abstract Cr has been added to FeCo substituting 10 at.% of Co or Fe in the alloy. The alloys $Fe_{50}Co_{40}Cr_{10}$, $Fe_{40}Co_{50}Cr_{10}$ and $Fe_{50}Co_{50}$ were prepared by mechanical alloying for 2, 5, 10, 20, 40 and 60 h. The formation of the alloy and the incorporation of the elements have been followed by X-Ray Diffraction (XRD) and Mössbauer Spectroscopy. The kinetics of mixing occurs by incorporation of Co and Cr into the Fe structure. After prolonged milling it seems that Cr incorporates itself into both α-Fe and α-FeCo structures and a mixture of FeCoCr rich in Cr and FeCoCr rich in Co solid solutions is obtained.

Keywords Nanostructured alloys · Mechanical alloying · Mössbauer spectroscopy

1 Introduction

Mechanical alloying is widely used to produce nanophase alloys with interesting structural and magnetic properties, due to size effects and to the disorder created by the high density of defects. Among Fe-alloys, the Fe–Co system shows a high permeability and the highest magnetization saturation. In particular the $Fe_{50}Co_{50}$ alloy has a considerably large permeability, low magnetocrystalline anisotropy and

L. D'Onofrio (✉)
Centro de Física Experimental del Sólido, Facultad de Ciencias,
Universidad Central de Venezuela, AP 20513, Caracas 1020-A, Venezuela
e-mail: lonofrio@fisica.ciens.ucv.ve

G. González · A. Sagarzazu · R. Villalba
Lab. Ciencia e Ing. Materiales, Dpto. Ing., Instituto Venezolano
de Investigaciones Científicas, Caracas, Venezuela

D. Oleszak
Faculty of Materials Science and Eng., Warsaw University of Technology,
Woloska 141, 02-507, Warsaw, Poland

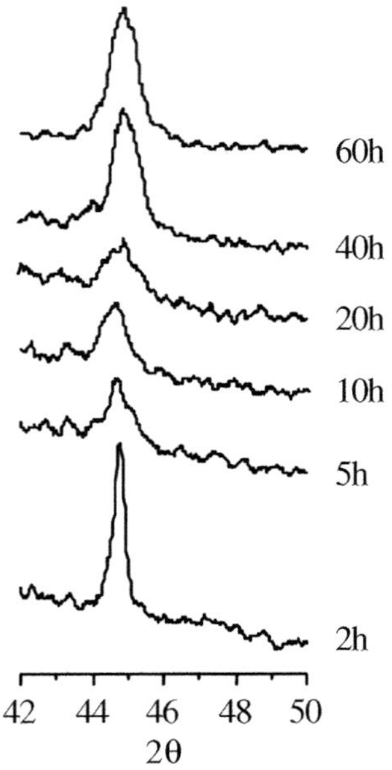

Fig. 1 XRD for different milling times of $Fe_{50}Co_{40}Cr_{10}$ alloys

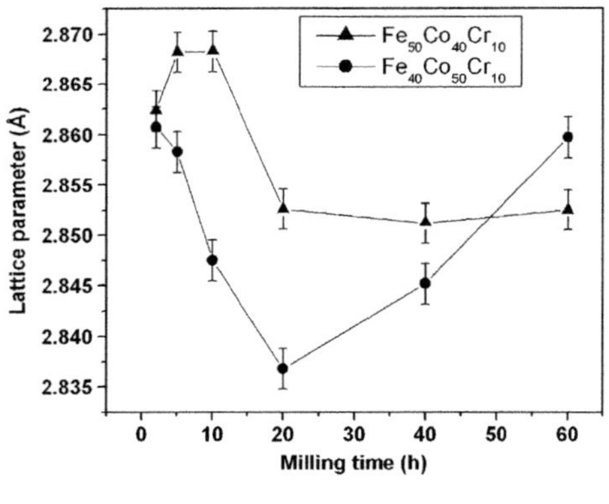

Fig. 2 Lattice parameter vs. milling time for $Fe_{50}Co_{40}Cr_{10}$ and $Fe_{40}Co_{50}Cr_{10}$ alloys

low electrical resistivity. Near equiatomic Fe–Co alloys with additions of a third element can change significantly their structural, electrical and magnetic properties [1]. The addition of Cr increases the resistivity and improves their dynamic properties, slightly decreasing the saturation magnetization by dilution of the magnetic atoms [2]. Recent work on Fe–Co–Cr has been done by Bentayeb et al. [3]; however, to our knowledge, no reports on the near equiatomic $Fe_{50}Co_{50}$ with additions of Cr, obtained by mechanical alloying have been found in the literature. In this work we present a Mössbauer Spectroscopy study of $Fe_{50}Co_{40}Cr_{10}$ and $Fe_{40}Co_{50}Cr_{10}$ alloys prepared by mechanical alloying to observe the effect of the incorporation of Cr into the Fe–Co structure. For comparison $Fe_{50}Co_{50}$ alloys were also prepared.

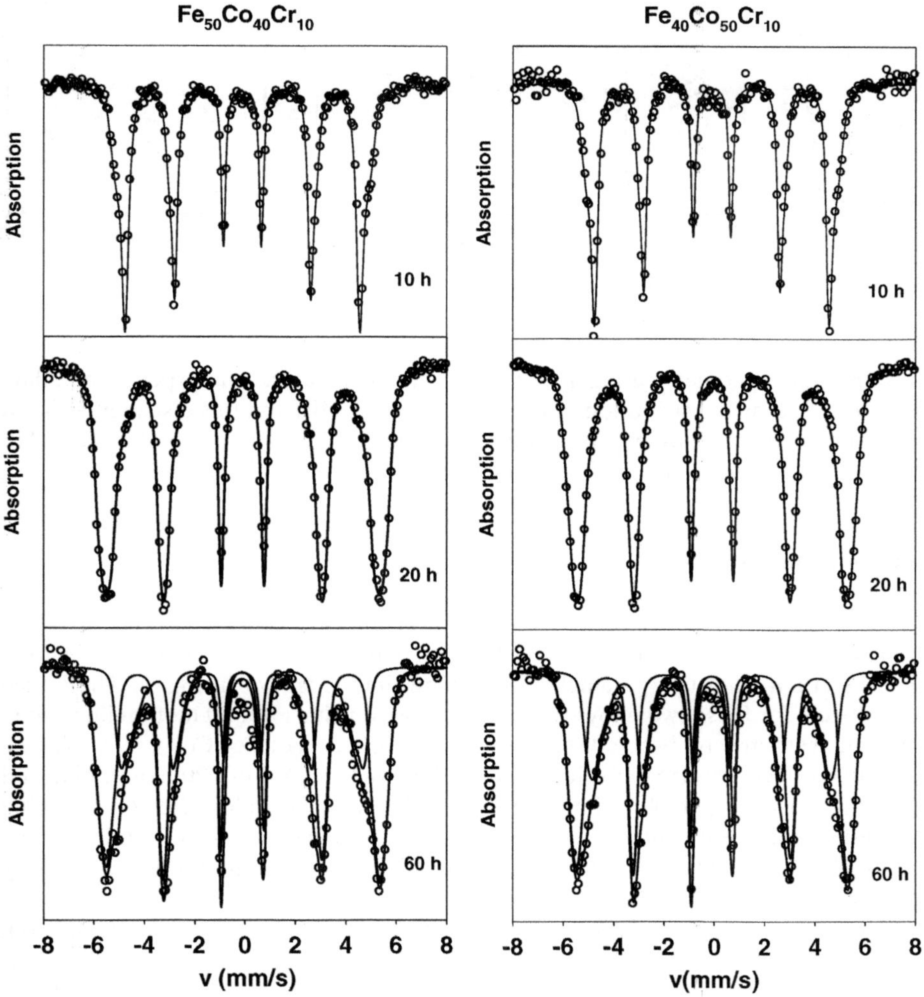

Fig. 3 Mössbauer spectra at room temperature for different periods of milling

2 Experimental

Powders of Fe, Co and Cr (ABCR Germany) with a purity of 99.8% and particle size below 50 μm were used. For each sample 10 g of powder were mixed in a Fritsch P5 planetary ball mill, equipped with hardened steel vials and balls of 10 mm diameter, with a rotational speed of 250 rpm and a ball-to-powder weight ratio of 10:1. The milling experiments and all powder handling were performed under protective argon atmosphere and the periods of milling were 2, 5, 10, 20, 40 and 60 h. Small amounts of powder were withdrawn after selected milling times for structural examinations. XRD was performed on a Siemens 5005 X-ray diffractometer, using Cu-Kα (Ni filter) operating at 40 keV and 20 mA. ^{57}Fe Mössbauer spectra were obtained with a spectrometer running in the triangular symmetric mode for the velocity in

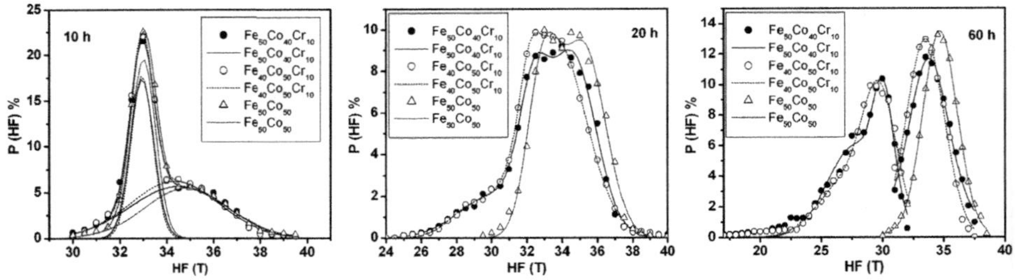

Fig. 4 Hyperfine field distribution for 10, 20 and 60 h of milling times

a transmission geometry using a ^{57}Co source in a Rh matrix. The experimental data were fitted using a distribution of hyperfine fields (DHF) and this distribution fitted with Gaussian functions in order to obtain the mean values of the hyperfine fields HF for each alloy. The isomer shifts IS are given relative to metallic iron.

3 Results and discussion

In Fig. 1 are shown the XRD for the main diffraction peak of the $Fe_{50}Co_{40}Cr_{10}$ alloys milled for different periods. Similar diffraction patterns are obtained for other alloys. The variation of lattice parameter vs. milling time is shown in Fig. 2. The lattice parameters for these two alloys reached the lowest value after 20 h of milling, when maximum strain was observed (results to be published). For $Fe_{50}Co_{40}Cr_{10}$ the lattice parameter decreases from a value near to 2.863 to 2.852 Å, after 20 h of milling and remains constant up to 60 h of milling. For the $Fe_{40}Co_{50}Cr_{10}$ alloy after 2 h of milling the lattice parameter value is 2.862 Å and progressively decreases with milling time to 2.837 Å after 20 h, and then increases to 2.860 Å after 60 h of milling, probably due to an increase in the degree of order with prolonged milling.

The decrease in the lattice parameter may be due to the allotropic transformation, of hcp Co to fcc Co, induced by deformation. For all the alloys the peak broadens with milling, suggesting a decrease in the grain size. After 60 h of milling all the alloys reached a grain size between 2 and 5 nm.

The Mössbauer spectra at room temperature of $Fe_{50}Co_{40}Cr_{10}$ and $Fe_{40}Co_{50}Cr_{10}$ alloys for milling periods of 10, 20 and 60 h are shown in Fig. 3. The HFD of both alloys were plotted together with that of the FeCo alloy for comparison and can be observed in Fig. 3 for the same milling times. For both alloys the HFD for 10 h of milling have been fitted with two Gaussian functions, Fig. 4, giving for both alloys a $HF_1 = 33.0$ T, corresponding to Fe-bcc that has not reacted yet and a $HF_2 = 34.5$ T for $Fe_{50}Co_{40}Cr_{10}$, and a $HF_2 = 34.3$ T for $Fe_{40}Co_{50}Cr_{10}$ which correspond to an FeCo alloy [4], suggesting that diffusion of Cr is slower than that of Co.

The HFD after 20 h of milling could indicate that Cr starts to diffuse into the crystalline structure in both the Fe-bcc and the Fe–Co bcc. In this case three Gaussian functions fit best, Fig. 3, giving for $Fe_{50}Co_{40}Cr_{10}$ $HF_1 = 30.2$ T, $HF_2 = 32.5$ T, $HF_3 = 34.6$ T and IS = 0.03 mm/s, and for $Fe_{40}Co_{50}Cr_{10}$ $HF_1 = 28.9$ T, $HF_2 = 32.1$ T, $HF_3 = 33.6$ T and IS = 0.03 mm/s.

After 60 h of milling, the spectra for both alloys can be fitted with two HFD since it seems that there are two separated compounds with different values of isomer shift.

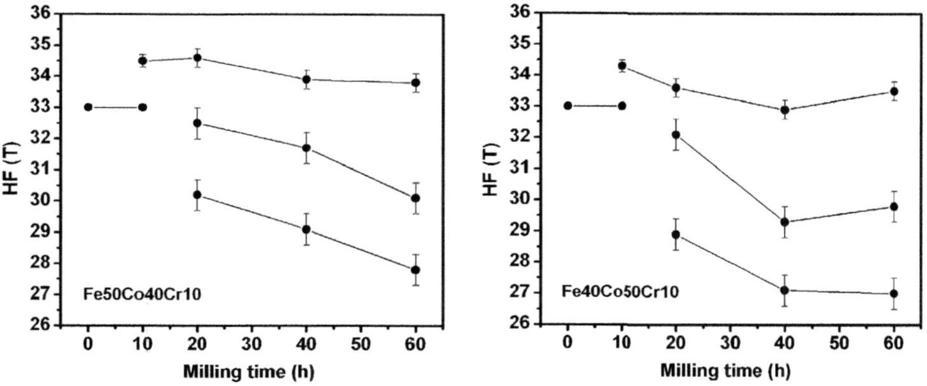

Fig. 5 Hyperfine field vs. milling time for $Fe_{50}Co_{40}Cr_{10}$ and $Fe_{40}Co_{50}Cr_{10}$ alloys

For $Fe_{50}Co_{40}Cr_{10}$ we obtained $HF_1 = 27.8$ T and $HF_2 = 30.1$ T with an IS = 0.00 mm/s and $HF_3 = 33.8$ T with an IS = 0.026 mm/s. For $Fe_{40}Co_{50}Cr_{10}$ we got $HF_1 = 27.0$ T and $HF_2 = 29.8$ T with an IS = 0.00 mm/s and $HF_3 = 33.5$ T with an IS = 0.026 mm/s. For comparison the $Fe_{50}Co_{50}$ alloy gave us a HF = 34.7 T and IS = 0.042 mm/s. The first two values of HF correspond to a FeCr alloy [5], with Co additions, and the third value corresponds to a FeCoCr alloy, since the additions of Cr in FeCo decreases the HF.

In Fig. 5 are displayed the HF vs. Milling time obtained from Mössbauer spectra. It is interesting to note that the alloys with more amount of iron have a HF higher than the alloys with lesser amount of iron, and that the incorporation of Cr into the structure seems to be slower and decreases the HF. The diffusion coefficient of Co in α-Fe at high temperatures is around 6.91×10^{-4} m^2/g and the diffusion coefficient of Cr in α-Fe at high temperatures is 2.53×10^{-4} m^2/g what agrees with these results.

4 Conclusion

The mixing of Fe, Co and Cr powders by mechanical alloying suggests that Cr and Co diffuse into the α-Fe, but Cr diffuses more slowly than Co. For both alloys after 10 h of milling α-Fe remains unreacted and a FeCo alloy is obtained. After 60 h Cr incorporates into both α-Fe and α-FeCo structures and a mixture of FeCoCr rich in Cr and FeCoCr rich in Co alloys seems to be obtained.

Acknowledgements This work was supported by the projects PI 03-00-5561-2007 CDCH-UCV and G-2005000449-FONACIT.

References

1. Sourmail, T.: Prog. Mater. Sci. **50**, 816 (2005)
2. Couderchon, G., Tiers, J.F.: J. Magn. Magn. Mater. **26**, 196 (1982)
3. Bentayeb, F.Z., Alleg, S., Bouzabata, B., Greneche, J.M.: J. Magn. Magn. Mater. **288**, 282 (2005)
4. Moumeni, H., Alleg, S., Greneche, J.M.: J. Alloys Compd. **386**, 12 (2005)
5. Lemoine, L., Fnidiki, A., Lemarchand, D., Teillet, J.: J. Phys. Condens. Matter **11**, 8341 (1999)

Hyperfine Interact (2010) 195:173–177
DOI 10.1007/s10751-009-0116-4

Structure and soft magnetic properties of FINEMET type alloys: $Fe_{73.5}Si_{13.5}Nb_{3-x}Mo_xB_9Cu_1$ ($x = 1.5, 2$)

Josefina M. Silveyra · Javier A. Moya · Victoria J. Cremaschi · Dušan Janičkovič · Peter Švec

Published online: 26 September 2009

Abstract FINEMET type ribbons ($Fe_{73.5}Si_{13.5}Nb_{3-x}Mo_xB_9Cu_1$, x = 1.5, 2 at.%) were produced by the planar flow casting technique and subsequently heat treated at 823 K to induce nanocrystallization and to optimize its soft magnetic properties. The coercivity, measured by conventional fluxmetric method, resulted in $H_C = 0.53 \pm 0.10$ and 0.41 ± 0.05 A/m for $x = 1.5$ and 2 respectively. A correlation between magnetic properties and the amorphous and nanocrystalline phases when Nb was partially substituted with Mo was studied by means of Mössbauer spectroscopy and X-ray diffraction.

Keywords Nanocrystalline alloys · Substitution · Structure · Magnetic properties

1 Introduction

There has been a great effort in the past decades on the development of new soft magnetic materials for their technological application such as transformer cores, inductive devices, magnetic shielding, sensors, etc. [1]. In 1988 Yoshizawa et al. [2] reported FINEMET ($Fe_{73.5}Si_{13.5}Nb_3B_9Cu_1$), a nanocrystalline ferromagnetic material, i.e. FeSi nanocrystals (grain size d = 10–20 nm) embedded in a ferromagnetic

J. M. Silveyra (✉) · J. A. Moya · V. J. Cremaschi
Laboratorio de Sólidos Amorfos, INTECIN, Facultad de Ingeniería,
UBA-CONICET, Paseo Colón 850, (C1063ACV), Buenos Aires, Argentina
e-mail: jsilveyra@fi.uba.ar

J. A. Moya · V. J. Cremaschi
Member of Carrera del Investigador, CONICET, Buenos Aires, Argentina

D. Janičkovič · P. Švec
Institute of Physics, Slovak Academy of Sciences,
Dúbravská cesta 9, 845 11 Bratislava, Slovakia

amorphous matrix. Since then, numerous works on this type of alloys have been published in order to improve and understand their magnetic properties. The amorphous structure is achieved through a rapid solidification technique. Then, the alloy is annealed and the nanocrystallization takes place: Cu acts as nucleation centres and Nb controls grain growth.

Good soft magnetic properties can also be obtained by replacing Nb by another refractory metal such as Mo, Ta and W [2]. However, it was said [3] that Nb is the most preferable element to improve the soft magnetic properties and it seems to be related to the fact that it is the most effective element for decreasing d: for small grains ($d < 40$ nm) coercivity, H_C, is proportional to d^6 [4].

Numerous researchers have investigated the effects on the magnetic properties and structure of the total [3] or partial [4–8] substitution of Nb and/or Fe by other refractory metals on FINEMET-like alloys. These are some of the achieved conclusions from the latter:

1. When 2% of the Nb content was replaced by another refractory element:
   - Crystalline volume fraction: Nb = Ta < Mo < V [5, 6].
   - Nanocrystals Si content: V < Nb < Mo [6].
   - Crystallization temperature: Nb = Ta < W < Mo < V [5].
   - d: Nb = Ta < W < Mo < V [5, 6].
   - H_C: the results depended not only on the refractory metal substitution, but also on Fe/Si/B ratios and annealing temperatures [5, 6].
2. When Nb was gradually replaced by Mo:
   - Nanocrystals Si content: There was an increase of Si content while partially replacing Nb with Mo [7].
   - Crystallization temperature: decreased when increasing Mo content [8].
   - d: with the exception of the 3 at.% Mo alloy, which exhibited a ~30% larger d, the partial substitution of Mo by Nb slightly increased d [7, 8].
   - H_C: alloys containing Nb generally exhibited much lower coercivities than the 3 at.% Mo alloy. The lowest H_C was developed in the alloy with mixed refractory metals [7]. On the other hand, a different result was obtained in a FINEMET-like composition [8] where H_C increased with Mo concentration.

In sum, the former statement that the best magnetic properties are achieved in FINEMETs using Nb as refractory element is not so straight forward: for some still unknown reason and although there are some discrepancies in the results, better properties can be attained with a partial substitution of Nb, i.e. mixed refractory elements. In this work, we looked for a further comprehension of the influence of Nb partial substitution by Mo, complementing X-ray diffraction (XRD) and magnetic measurements with Mössbauer spectroscopy (MS). Results were compared to typical FINEMET ($Fe_{73.5}Si_{13.5}B_9Nb_3Cu_1$, called from now on Mo0) previously reported [9].

2 Experimental

FINEMET type ribbons 10 mm wide and 20 μm thick of compositions $x = 1.5$ and 2 (Mo1.5 and Mo2 respectively) were produced by the planar flow casting technique in

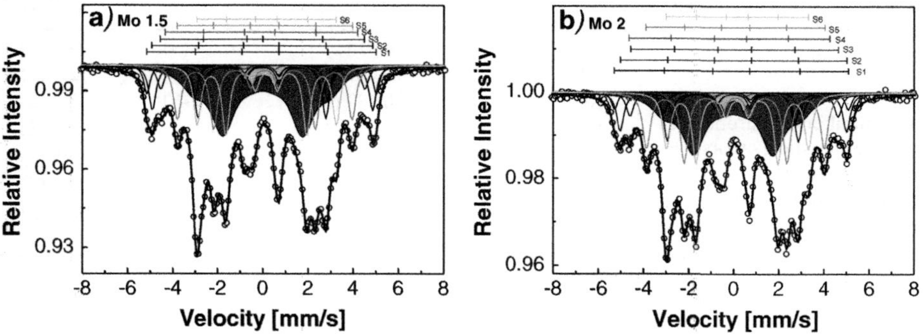

Fig. 1 Mössbauer spectra and fitting for annealed samples: **a** Mo1.5 and **b** Mo2. *Shaded* subspectra represent amorphous phases (*dark grey* Am1, *light grey* Am2), while the *lines* (S1–S6) correspond to crystalline sites

air. The amorphous structure was checked on both sides by XRD and the chemical composition by inductively coupled plasma spectroscopy. Nanocrystallization was induced through annealing in vacuum at 823 K for 1 h.

The structure of the annealed samples was analysed by XRD and MS at room temperature. XRD was performed using a Rigaku Geiger Flex D-Max II TC diffractometer. All scans were run at 40 kV/20 mA with monochromatic radiation Cu K_{α} (1.5418 Å) and Ni filter. MS was carried out in transmission geometry using a constant acceleration drive and a ^{57}Co in Rh source. The calibration was done with an α-Fe foil and isomer shifts (IS) are given relative to α-Fe. The spectra were fitted using the NORMOS programme [10] considering six sextets for the different crystalline sites of Fe atoms in a Fe_3Si non-stoichiometric DO_3 structure (S1 to S6) and two wide sextets for fitting the subspectra corresponding to the amorphous phase (Am1 and Am2) [9].

Hysteresis loops were obtained using a quasistatic fluxmetric method by applying a longitudinal magnetic field on the sample ($H_{max} = 740$ A/m).

3 Results and discussion

Mo1.5 and Mo2 Mössbauer spectra were very similar to each other (Fig. 1) and even to that of Mo0 alloy. As an example, some hyperfine parameters obtained from the fitting of Mo2 sample spectrum are shown in Table 1. The greatest difference found was between the areas corresponding to crystalline and amorphous sites: the resonant area relative to the amorphous phase decreased with increasing Mo content. Another difference was related to the chemical composition of the nanocrystals when compared Mo1.5 and Mo2 to the Mo0 alloy: the Si content of the crystalline phase was obtained from the probability of occurrence of each kind of Fe environment (S1 to S4, S5 and S6) using a binomial distribution [11]. Our results indicated a Si % of about 20.5 ± 0.3 for Mo1.5 and Mo2 (see Table 2), showing a higher Si concentration than Mo0 (∼19%). This trend was in agreement with the decrease in the lattice parameter on Mo containing alloys obtained in [7]. Likewise, we also obtained from XRD a nanograins lattice parameter of 2.834 ± 0.001 Å for both

Table 1 Parameters obtained from the Mössbauer spectrum fitting for annealed Mo2

Subspectrum	IS (mm/s)	BHF (T)	Area (%)
S1	0.01 (1)	32.14 (9)	4.8 (2)
S2	0.081 (5)	31.09 (4)	12.4 (2)
S3	0.137 (9)	28.74 (8)	7.4 (2)
S4	−0.01 (2)	27.4 (2)	3.7 (2)
S5	0.187 (4)	24.39 (3)	19.5 (2)
S6	0.264 (4)	19.51 (3)	19.0 (2)
Am1	0.05 (1)	18.7 (1)	30.6 (3)
Am2	0.11 (1)	4.5 (1)	2.6 (1)

The numbers in parentheses are the statistical errors in the last digit
IS isomer shift, *BHF* hyperfine magnetic field, *Area* relative resonant Fe area

Table 2 Chemical composition of the crystallites and of the amorphous matrix of Mo0, Mo1.5 and Mo2 (in at.%)

	Mo 0	Mo 1.5	Mo 2
Crystalline fraction	61 (1)	64 (1)	67 (1)
Grains composition			
Fe	81.0 (3)	79.7 (3)	79.5 (3)
Si	19.0 (3)	20.3 (3)	20.5 (3)
Amorphous fraction	39 (2)	36 (2)	33 (2)
Matrix composition			
Fe	62 (2)	63 (2)	62 (2)
Si	5 (2)	2 (1)	0 (1)
Cu	2.4 (1)	2.6 (1)	2.8 (1)
Nb	7.3 (3)	4.0 (1)	2.8 (1)
Mo	–	4.0 (1)	5.7 (2)
B	22.1 (8)	24.0 (9)	26.0 (9)

The numbers in parentheses are the statistical errors in the last digit. Percentages may not add up to 100% because of rounding

alloys, which was in accordance with the Si % values obtained by Mössbauer [12]. By fitting FeSi 110 peak with a Voigt distribution and applying Scherrer formula [13], a mean grain size of 18 ± 1 nm for Mo1.5 and Mo2 was estimated, in agreement with [8].

Knowing the relative fraction of ^{57}Fe in the amorphous and crystalline phases (which could be considered approximately proportional to the resonant area of each phase), the chemical composition of the crystalline phase and the nominal chemical composition of the as-quenched sample, it was possible to estimate the chemical composition of the matrix. These results are enlisted in Table 2 and compared with Mo0. Although there was an increment in the crystalline fraction, the Fe content in the matrix did not vary significantly. This is an important fact in order to maintain a good ferromagnetic intergranular amorphous matrix to ensure the soft magnetic properties. However, on the other hand, there was an increase on the content of B and Nb + Mo in the matrix that could detriment the soft magnetic properties.

Nevertheless, a decrease in H_C with Nb partial replacement was measured: Mo0 (0.58 ± 0.50 A/m) < Mo1.5 (0.53 ± 0.10 A/m) < Mo2 (0.41 ± 0.05 A/m). We associated it to: (1) a decrease of magnetocrystalline anisotropy due to the increase of Si % in the grains and, (2) a reduction on the magnetostriction constant of the alloy when Mo was added [5]. Both factors reduce the effective anisotropy constant ⟨K⟩ and consequently increment the soft magnetic properties of the nanocrystalline alloy.

4 Conclusions

A new contribution was made to the comprehension of the FINEMET system with the partial replacement of Nb by Mo. By means of the Mössbauer technique we were able to obtain the crystalline and amorphous mass fractions of nanocrystallized alloys and their chemical composition, as well as to make a correlation with magnetic properties. We noticed an increment on crystalline fraction when Mo content ascended from 1.5% to 2%, possibly related to the decrease of the thermal stability [8]. Nanocrystals Si content was estimated on ~20.5 for both alloys (error: 0.3%), which represented ~1% more than in our FINEMET composition alloy (Mo0). The interesting thing was that the increase of Si was attained by the balance of refractory elements whilst keeping Fe, Si and B contents in the alloy constant, as well as the heat treatment.

With respect to the intergranular amorphous matrix, because of the increase on the crystalline fraction, the content of B and Nb + Mo—that could deteriorate the soft magnetic properties of the material—rose, but on the contrary, the Fe content remained nearly the same as in Mo0. Moreover, such adverse effect could be counteracted. Our results and the ones reported in [7] agreed on the decrease of H_C with the partial substitution of Nb by Mo. We explain the good soft magnetic properties in terms of a reduction on the effective anisotropy constant via the decreasing on both, the magnetocrystalline anisotropy of the nanocrystals due to the higher Si content and on the magnetostriction constant. Further work will be made in order to clarify these preliminaries results.

References

1. Herzer, G., Vazquez, M., Knobel, M., Zhukov, A., Reininger, T., Davies, H.A., Grössinger, R., Sanchez Ll, J.L.: Round table discussion: present and future applications of nanocrystalline magnetic materials. J. Magn. Magn. Mater. **294**, 252–266 (2005)
2. Yoshizawa, Y., Oguma, S., Yamauchi, K.: New Fe-based soft magnetic alloys composed of ultrafine grain structure. J. Appl. Phys. **64**, 6044–6046 (1988)
3. Yoshizawa, Y., Yamauchi, K.: Magnetic properties of Fe–Cu–M–Si–B (M = Cr, V, Mo, Nb, Ta, W) alloys. Mater. Sci. Eng. **A133**, 176–179 (1991)
4. Herzer, G.: Grain size dependence of coercivity and permeability in nanocrystalline ferromagnets. IEEE Trans. Magn. **26**, 1397–1402 (1990)
5. Müller, M., Mattern, N., Illgen, L., Hilzinger, H.R., Herzer, G.: The influence of partial substitution of Nb by refractory elements on the structure and on the magnetic properties in nanocrystalline soft magnetic FeBSi–CuNb alloys. Key Eng. Mater. **81–83**, 221–228 (1993)
6. Borrego, J.M., Conde, C.F., Millan, M., Conde, A., Capitan, M.J., Joulaud, J.L.: Nanocrystallization in $Fe_{73.5}Si_{13.5}B_9Cu_1Nb_1X_2$ (X = Nb, Mo and V) alloys studied by X-ray synchrotron radiation. Nanostruct. Mater. **10**, 575–583 (1998)
7. Frost, M., Todd, I., Davies, H.A., Gibbs, M.R.J., Major, R.V.: Evolution of structure and magnetic properties with annealing temperature in novel Al-containing alloys based on Finemet. J. Magn. Magn. Mater. **203**, 85–87 (1993)
8. Liang, G., Huang, Y., Friedman, G.: Effects of Mo substituted Nb on magnetic properties of Finemet alloy. Trans. Nonferr. Met. Soc. China **12**, 189–192 (2002)
9. Moya, J., Cremaschi, V., Sirkin, H.: From Fe_3Si towards Fe_3Ge in Finemet-like nanocrystalline alloys: Mössbauer spectroscopy. Phys. B. **389**, 159–162 (2007)
10. Brand, R.: Normos program. Internal Report, Angewandte Physik, Universität Duisburg (1987)
11. Rixecker, G., Schaaf, P., Gonser, U.: On the interpretation of Mössbauer spectra of ordered Fe–Si alloys. Phys. Status. Solidi. (a) **139**, 309–320 (1993)
12. Bozorth, M.: Ferromagnetism. Van Nostrand, New York (1951)
13. Klug, H., Alexander, L.: X-Ray Diffraction Procedures. Wiley, New York (1954)

Hyperfine Interact (2010) 195:179–184
DOI 10.1007/s10751-009-0117-3

Influence of silicon and atomic order on the magnetic properties of $(Fe_{80}Al_{20})_{100-x}Si_x$ nanostructured system

G. Y. Vélez · G. A. Pérez Alcázar · Ligia E. Zamora · J. J. Romero · A. Martínez

Published online: 7 November 2009

Abstract Mechanically alloyed $(Fe_{80}Al_{20})_{100-x}Si_x$ alloys (with $x = 0$, 10, 15 and 20) were prepared by using a high energy planetary ball mill, with milling times of 12, 24 and 36 h. The structural and magnetic study was conducted by X-rays diffraction and Mössbauer spectrometry. The system is nanostructured and presents only the BCC disordered phase, whose lattice parameter remains constant with milling time, and decreases when the Si content increases. We found that lattice contraction is influenced 39% by the iron substitution and 61% by the aluminum substitution, by silicon atoms. The Mössbauer spectra and their respective hyperfine magnetic field distributions show that for every milling time used here, the ferromagnetism decreases when x increases. For samples with $x \geq 15$ a paramagnetic component appears. From the shape of the magnetic field distributions we stated that the larger ferromagnetic phase observed in the samples alloyed during 24 and 36 h is a consequence of the structural disorder induced by mechanical alloying.

Keywords Alloys FeAl · Alloys FeSi · Mechanical alloying · X-rays diffraction · Mössbauer spectrometry

1 Introduction

The FeSi and FeAl binary alloys have been of major interest for several decades, due to their wide range of both magnetic and structural properties, as well as their diversity of applications. For example, the FeSi alloys are economics and show a

G. Y. Vélez (✉) · G. A. Pérez Alcázar · L. E. Zamora
Departamento de Física, Universidad del Valle, A.A. 25360, Cali, Colombia
e-mail: gyovelca@univalle.edu.co

J. J. Romero · A. Martínez
Instituto de Magnetismo Aplicado IMA, P. O. Box 155 28230,
Las Rozas de Madrid, Spain

good soft magnetic behavior, the principal reason for their commercial use in the elaboration of transformers cores [1], whereas the FeAl alloys are characterized for being resistant to corrosion.

The magnetic properties of the FeAl alloys are very much linked with their structural order. It has been observed on alloys prepared by techniques such as sputtering [2], implantation [3] and more recently by mechanical alloying [4], that the increase of their ferromagnetism is associated to the increase of their structural disorder. Recent studies have shown that when silicon is added to the $Fe_{70}Al_{30}$ system prepared by mechanical alloying, the milling time plays an important role on the magnetic properties of the material [5]. For this reason we realized by X-rays diffraction (XRD) and Mössbauer spectrometry (MS) the structural and magnetic study of the $(Fe_{80}Al_{20})_{100-x}Si_x$ system (with $x = 0$, 10, 15 and 20) mechanically alloyed during 12, 24 and 36 h.

2 Experimental procedure

Using highly pure fine powder ($\geq$99.90%) of Fe, Al and Si, samples of the system $(Fe_{80}Al_{20})_{100-x}Si_x$ (with $x = 0$, 10, 15 and 20) where prepared by mechanical alloying, during milling times of 12, 24 and 36 h. For the alloying, a high energy planetary ball mill *Fritsch-Pulverisette 5* was used. The millings of the powders were performed in stainless steel jars with balls of the same material inside. The vacuum inside the jars was of 4.5 10^{-2} torr and the ratio between balls and powder masses was 15:1. The XRD patterns were realized in a *Siemens* powder diffractometer with Cu-Kα radiation and they were refined by using the GSAS program [6]. Mössbauer spectra where taken on a conventional spectrometer with a ^{57}Co/Rh source of 25 mCi. All the spectra were fitted with the MOSFIT program [7], using a hyperfine magnetic field distribution (HMFD) and in some cases with the addition of a paramagnetic site. All the studies were realized at room temperature.

3 Results and discussion

Figure 1a shows the XRD patterns for all the samples alloyed during 12 h. All the patterns show the corresponding (110), (200), (211) and (220) of a BCC disordered phase. It can be noted in this figure that the peaks present a small shift toward greater angular positions when the Si concentration increases, indicating that the lattice parameter decreases with the x increase. The patterns for the samples alloyed during 24 and 36 h show a similar behavior to that shown in Fig. 1a.

The behavior of the lattice parameter vs. Si concentration for the different milling times is shown in Fig. 1b. This figure also shows that the lattice parameter presents decreasing values when Fe and Al atoms are substituted by Si atoms (near 2%) for all the milling times. This compression is due to the small atomic size of the Si as compared to those of Fe and Al atoms. It can be noted also that the lattice parameters are not appreciably affected by the milling time (the largest range of variation is of 0.008 Å, near 0.3%, corresponding to $x = 0$).

Experimental results obtained by Abdellaoui, in the $Fe_{1-x}Si_x$ system, showed that the lattice parameter decreases when x increases [8, 9]. A comparison between

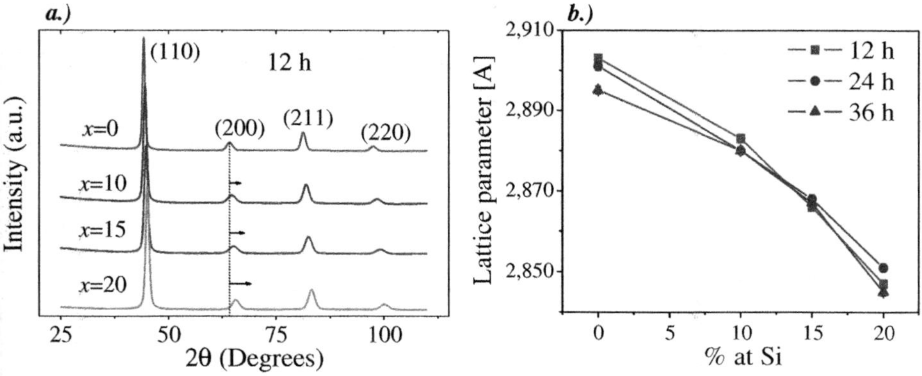

Fig. 1 **a** XRD patterns for samples alloyed during 12 h. **b** Lattice parameter vs. at.% Si for different milling times

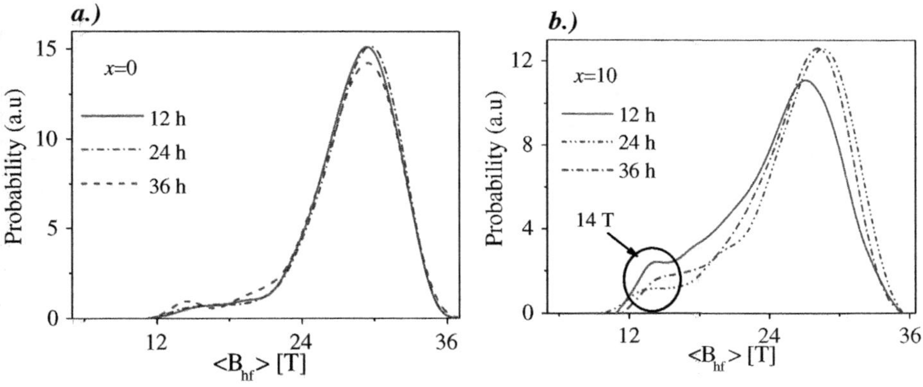

Fig. 2 Hyperfine magnetic field distributions for the samples with **a** $x = 0$, and **b** $x = 10$

our results and the Abdellaoui's results, shows that when the Si concentration changes from 0 up to 20 at.%, the lattice parameter decreases in 1.79% and 0.70%, respectively. Then these results permit us to conclude that nearly a 61% of the compression is due to the substitution of Al by Si atoms and that a 39% of this is due to the substitution of Fe by Si atoms. The Al atom presents the biggest atomic size in this ternary system.

For all the milling times the crystallite sizes are of nanometric order (between 18 and 33 nm) and exists a decreasing tendency in the mean crystallite size as x increases, this is a consequence of the effect of silicon on the system which increases the internal micro tensions. These previous structural results show that inclusive for 36 h milling the properties are still changing and the stationary state has not been reached.

The Mössbauer spectra (MS) show that for all the milling times and compositions the samples are ferromagnetic and this behavior decreases as silicon concentration increases. For $x = 15$ it appears a broad paramagnetic site which is larger for the $x = 20$ sample. This behavior is principally due to the substitution of Fe atoms, which

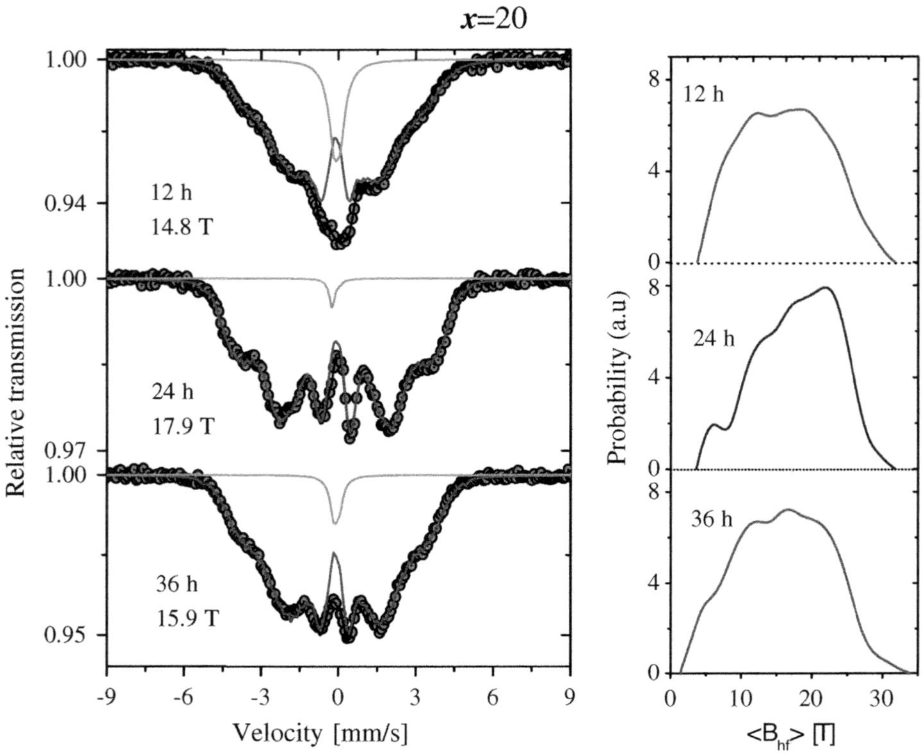

Fig. 3 Mössbauer spectra and their respective hyperfine magnetic field distributions for the samples with $x = 20$

present magnetic moment, by Si atoms which do not present a magnetic moment and behave as diluters.

Figure 2a shows the HMFDs obtained for the $x = 0$ samples. It can be noted that this sample presents basically the same HMFD for all the milling times. The obtained mean hyperfine magnetic field value of 28 T matches well with those values found for disordered samples of the same composition prepared by techniques such as melt spinning [10] and melting followed by rapid quenching [11], respectively. The nature of these preparation techniques guarantees a high level of disorder, permitting us to conclude that in our sample the aluminum diffuses quickly reaching a maximum grade of disorder before 12 milling hours.

From the HMFDs for $x = 10$ (Fig. 2b), three major facts can be observed (also valid for $x = 15$ and 20). (a) The HMFDs are displaced to lower fields compared to those HMFDs of samples without Si (Fig. 2a). The displacement of the 12 h sample is larger, followed by that of 36 and then that of the 24 h sample. (b) The HMFD of the sample milled during 12 h presents a major asymmetry with respect to the peak of maximum probability (around 29 T) and this asymmetry decreases for the 36 h sample and then for the 24 h sample. (c) A peak near 14 T is observed which is more intense for 12 h, decreases for 36 h and increases a little for 24 h milling. This appreciation suggests that in alloys with Si (ternary ones) the high grade of disorder,

Table 1 Mössbauer parameters obtained from the fit of the different spectra

X	12 h		24 h		36 h	
	$<B_{hf}> \pm 0.2$ [T]	IS $\pm$ 0.003 [mm/s]	$<B_{hf}> \pm 0.2$ [T]	IS $\pm$ 0.003 [mm/s]	$<B_{hf}> \pm 0.2$ [T]	IS $\pm$ 0.003 [mm/s]
0	28.2	0.055	28.3	0.059	28.0	0.053
10	24.8	0.065	26.5	0.066	25.9	0.066
15	17.2	0.123	19.2	0.120	19.1	0.120
20	14.8	0.119	17.9	0.117	15.9	0.122

showed by the binary system, was not reached. This is so due that the presence of the additional peak suggests that some local site (field) is now appearing with bigger probability and this is an evidence of local order.

The MS and their corresponding HMFDs for samples with $x = 20$ are shown in Fig. 3. In this case all the HMFDs are displaced for lower field values due to the bigger Si content. From these HMFDs it can be noted that the previous discussion, for samples with $x = 10$, can be extended for the present samples. However, for the $x = 20$ samples the peak around 14 T appears with bigger probability (especially for $t = 12$ and 36 h) and new peaks appear for lower fields.

All the previous results indicate that the ferromagnetism of current alloys is not only dependent on the Si content but also strongly dependent on the milling time. When milling time increases from 12 to 24 h it appears an increase of the ferromagnetic spectral area and of the mean hyperfine magnetic field, which can be associated to the increase of the disorder of the sample, in the same way as was previously established for Fe–Al alloys [4]. In our case the ferromagnetic behaviour present a small decrease for samples milled during 36 h, compared with that of samples milled during 24 h due the increase of the order which is proved by the increase of the new peaks intensity.

The Mössbauer parameters obtained during the fit of the spectra for the different samples are reported on Table 1. It can be noted that the isomer shift values increase with the Si content increase, especially between $x = 10$ and $x = 15$. Then the s electron density in the Fe nucleus decrease and the reason for this increase is the same given for binary Fe–Si alloys [12]. As Si atoms increase in the sample, more electrons pass from Si to the $3d$ shell of the Fe atoms decreasing not only the magnetic moment but also the shielding of its s electrons on the Fe nucleus

4 Conclusions

In this work it was shown that all the samples of the $(Fe_{80}Al_{20})_{100-x}Si_x$ system at room temperature, present the non stable structural BCC disordered phase, with a lattice parameter which remains nearly constant with the milling time and decreases with the Si concentration. The obtained results permit us to conclude that nearly a 61% of the compression is due to the substitution of Al by Si atoms and that a 39% of this is due to the substitution of Fe by Si atoms. It was determined that the ferromagnetism decreases with x due primarily to the non-magnetic nature of silicon. By means of the HMFDs it was established that the increase on the ferromagnetism with the milling time is due to the structural disorder caused by the mechanical alloying, giving a reason why the samples milled during 12 h are less ferromagnetic.

Acknowledgements The authors would like to thank Colciencias, Colombian Agency, and Universidad del Valle for financial support for the realization of this work.

References

1. Barrett, W.F., Brown, W., Handfield, R.A.: Sci. Trans. R. Dublin Soc. **7**, 67 (1900)
2. Hsu, J.H., Chien, C.L.: Hyperfine Interact. **69**, 451 (1991)
3. Fnidiki, A., Eymery, J.P., Delafond, J.: J. Magn. Magn. Matter. **40**, 130 (1983)
4. Zamora, L.E., Pérez Alcázar, G.A., Vélez, G.Y., Betencur, J.D., Marco, J.F., Romero, J.J., Martínez, A., Palomares, F.J., González, J.M.: Phys. Rev. B **79**, 094418 (2009)
5. Pérez Alcázar, G.A., Zamora, L.E., Betancour-Ríos, J.D., Tabares, J.A., Greneche, J.M., González, J.M.: Physica B **384**, 313–315, (2006)
6. Larson, A.C., Von Dreele, R.B.: General Structure Analysis System (GSAS). Los Alamos National Laboratory Report LAUR 86-748 (2004)
7. Varret, F., Teillet, J.: Unpublished MOSFIT Program
8. Abdellaoui, M., Barradi, T., Gaffet, E.: J. Alloys Compd. **198**, 155–164 (1993)
9. Abdellaoui, M., Djega-Mariadassou, C., Gaffet, E.: J. Alloys Compd. **259**, 241–248 (1997)
10. Schmool, D.S., Araujo, E., Amado, M.M., Alegría Feio, M., Martín Rodríguez, D., Garitaonandia, J.S., Plazaola, F.: J. Magn. Magn. Mater. **272–276**, 1342–1344 (2004)
11. Pérez Alcázar, G.A., Galvão da Silva, E.: J. Phys. F Met. Phys. **17**, 2323–2335 (1987)
12. Stearns, M.B.: Phys. Rev. **129**(3), 1136 (1963)

Hyperfine Interact (2010) 195:185–189
DOI 10.1007/s10751-009-0118-2

Fe-doped SnO_2 nanopowders obtained by sol–gel and mechanochemical alloying with and without thermal treatment

L. C. Sánchez · J. J. Beltran · J. Osorio · A. M. Calle · C. A. Barrero

Published online: 25 September 2009

Abstract The present work is aimed to compare the physical properties of $Sn_{1-x}Fe_xO_{2-\delta}$ ($x = 0$, and 0.05) nanopowders obtained by sol–gel method, mechanochemical alloying, and mechanochemical alloying followed by thermal treatment. The X-ray diffraction of $Sn_{1-x}Fe_xO_{2-\delta}$ samples prepared by sol–gel showed peaks due to the cassiterite phase of SnO_2 and their Mössbauer spectra showed ferromagnetic and paramagnetic signals. The samples obtained by the milling process of SnO_2 mixed with α-Fe showed Bragg peaks due to SnO_2 (rutile) with a line broadening caused by the reduction of grain sizes and the presence of microstrains. Mössbauer spectra for these samples revealed the presence of Fe^{3+} as well as unreacted α-Fe. In the case of mechanochemical alloying with thermal treatment, the incorporation of Fe^{3+} in the SnO_2 structure with the presence of impurities was observed.

Keywords Fe and SnO_2 · Sol–gel method · Ball milling · Thermal treatment · Mössbauer spectrometry

1 Introduction

The progress in the field of spin-based electronics critically depends on the development of new semiconductors. Oxide diluted magnetic semiconductors, ODMS, in which nonmagnetic oxide semiconductors are doped with just a low percent of transition metal ions are expected to play an important role in the development of semiconductor spintronics [1, 2]. Transition metal doped SnO_2 belong to these classes of materials [3–5]. Several synthetic methods to produce them have been

L. C. Sánchez · J. J. Beltran · J. Osorio · A. M. Calle · C. A. Barrero (✉)
Grupo de Estado Sólido, Sede de Investigación Universitaria,
Universidad de Antioquia, Medellín, Colombia
e-mail: cbarrero@pegasus.udea.edu.co

explored, but many methods have the drawback of producing impurity phases and/or magnetic ion clustering. Therefore, the search for proper experimental conditions to successfully prepare these materials has become an important part of this research field. In this work we compare some physical properties of $Sn_{1-x}Fe_xO_{2-\delta}$ ($x = 0$, and 0.05) powders obtained by three different methods.

2 Experimental

By using sol–gel method, solutions of Sn^{2+} and Fe^{2+} precursors were employed. The solutions were mixed in the appropriate amounts with citric acid and ethyleneglycol, evaporated between 80–90°C, calcinated at 200°C for 4 h, and annealed in air at 500°C for 2 h. Mechanochemical alloying was performed in a planetary ball mill Fritsch Pulverisette 5. The starting materials were rutile SnO_2 (Merck 99%) and α-Fe (Merck 99%). Samples were milled for 12 h in air at atmospheric pressure, using Cr-based stainless steel jars and balls of 12 mm in diameter. Fe nominal concentration was of 4 at.%. The rotation velocity of the disc was of 390 rpm, and the ball to powder ratio was of 20:1. In the mechanochemical processing with thermal treatment, the starting reactants were $SnCl_2$, $FeCl_3$ and Na_2CO_3, with NaCl added as diluent under inert atmosphere. The precursor concentrations were adjusted to give a final Fe nominal concentration of 5 at.% in Fe-doped SnO_2. Samples were milled for 3 h in argon atmosphere. The rotation velocity of the disc was 250 rpm, and the ball to powder ratio were of 20:1. The as-milled mixtures were subsequently heat treated at 600°C during 1 h, in air atmosphere, and finally washed in double deionized water to remove the NaCl. The overall reaction is accounted for by: $SnCl_2 + Na_2CO_3 \rightarrow SnO + 2NaCl + CO_2$, and the heat treatment of the milled-product in oxygen is followed in order to oxidize SnO to SnO_2 [6]. In a previous work we have studied proper conditions to obtain Fe-doped SnO_2 by mechanochemical alloying and mechanochemical alloying followed by thermal treatment [7]. The crystalline structure was investigated by X-ray diffraction (XRD) using Rietveld method, and Mössbauer spectroscopy studies were performed using a conventional constant acceleration drive and a ^{57}Co:*Rh* source. The velocity calibration was done with a room temperature α-Fe absorber and the isomer shifts values are relative to that of iron.

3 Results and discussion

Figure 1a shows the XRD pattern of the sample prepared by mechanochemical milling with thermal treatment. Only Bragg peaks due to the presence of SnO_2 rutile phase were observed. Figure 1b shows the XRD patterns for the high energy milled SnO_2 mixed with metallic iron. Besides the presence of the SnO_2 rutile phase, also peaks due to metallic iron were observed. The Bragg lines are broad and this effect is attributed to the diminution of the crystallite size and to the increment of microstrains due to the milling processes. Additionally, the XRD patterns of SnO_2 and iron-doped SnO_2 obtained by sol gel method are shown (Fig. 1c and d, respectively). Both samples evidence the formation of only the rutile phase of SnO_2. By using Rietveld method average crystallite sizes of approximately 51, 37 and 22 nm

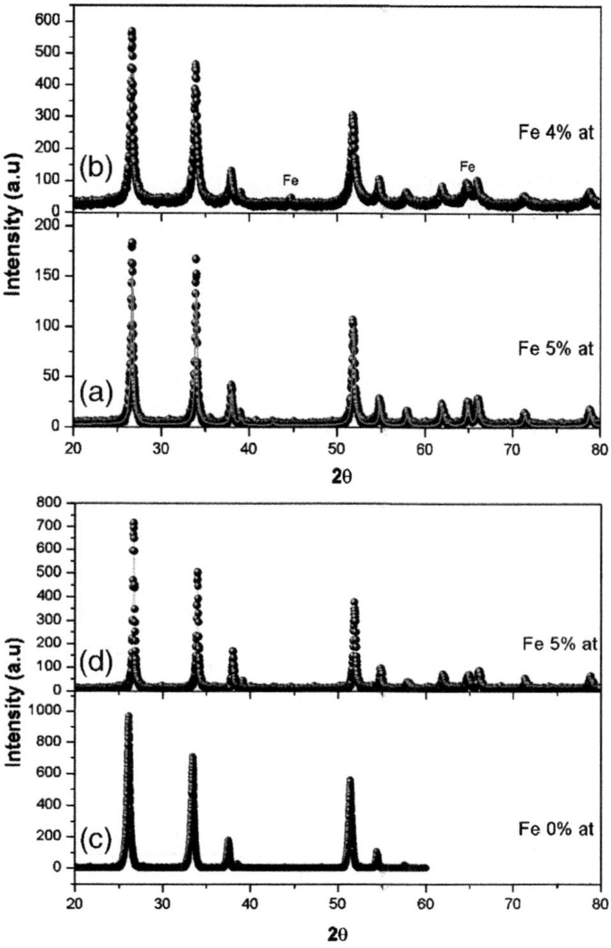

Fig. 1 XRD patterns for: **a** iron-doped SnO_2 obtained by mechanochemical milling and thermal treatment, **b** iron-doped SnO_2 obtained by mechanical alloying, **c** pure SnO_2 obtained by sol–gel method, and **d** iron-doped SnO_2 obtained by sol gel method

for the samples obtained by sol gel method, mechanochemical alloying and thermal treatment, and mechanochemical alloying, respectively, were estimated.

Room temperature Mössbauer spectra for the samples are shown in Fig. 2. The parameters derived from the fitting are listed in Table 1. Mössbauer spectra were fitted with two components: one sextet (S1) and one doublet (D1). The hyperfine parameters of the sextet components point to the presence of metallic iron for the case of the sample prepared from mechanochemical alloying, and a magnetic phase, probably hematite, for the samples obtained from mechanochemical alloying followed by thermal treatment and sol–gel method. The α-Fe phase corresponds to the starting iron materials that do not react with SnO_2. The parameters of the D1 point to the presence of Fe^{3+} ions substituting for Sn^{4+} in the tetragonal SnO_2 structure. This interpretation is in agreement with the reports by Cabrera et al. [8] and Sorescu et al. [9]. In the case of the sample obtained by sol–gel, the Mössbauer spectra shows a well defined doublet D1, indicating a major incorporation of Fe^{3+} in the SnO_2 structure, and the sextet S1, probably assigned to hematite or Sn-doped hematite, in agreement with Berry et al. [10].

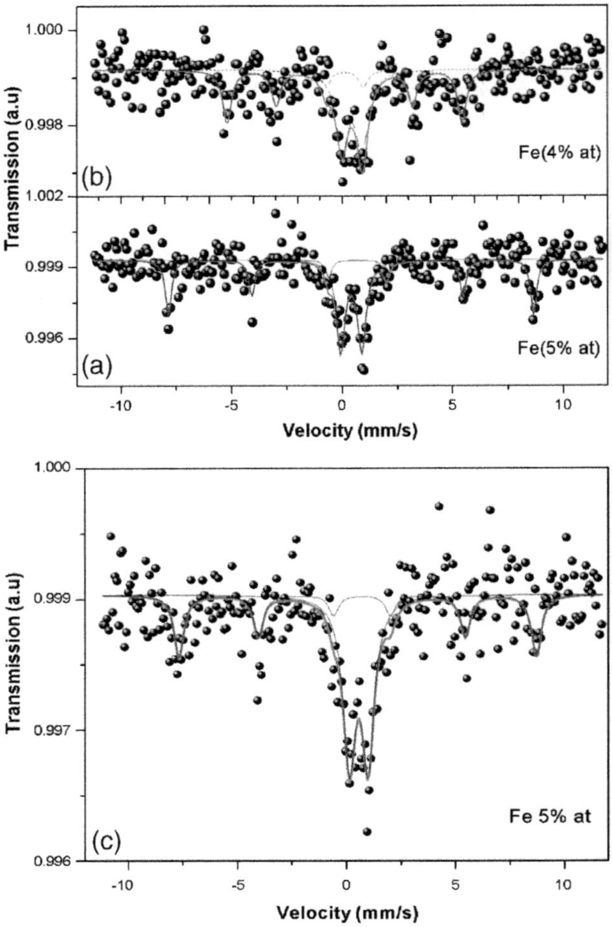

Fig. 2 Room temperature Mössbauer spectra for: **a** iron-doped SnO_2 obtained by mechanochemical milling and thermal treatment, **b** iron-doped SnO_2 obtained by mechanical alloying, and **c** iron-doped SnO_2 obtained by sol gel method

Table 1 Parameters derived from the fit of the Mössbauer spectra for samples obtained by mechanochemical milling with and without thermal treatment, and sol gel method

S1			D1	
B_{hf} (T)	δ (mm/s)	Δ (mm/s)	δ (mm/s)	Δ (mm/s)
Ball milling with thermal treatment				
51.2	0.55	−0.29	0.41	0.96
Ball milling				
33.0	0	0	0.45	0.90
Sol gel method				
50.8	0.50	−0.20	0.45	0.86

4 Conclusions

The mechanochemical alloying of the mixture of rutile SnO_2 and α-Fe powders produces Fe-doped SnO_2 and α-Fe. On the other hand, the mechanochemical alloying of $SnCl_2$, $FeCl_3$, Na_2CO_3 and NaCl followed by thermal treatment, produce Fe-doped SnO_2 and hematite. The sample produced by sol–gel method contains

Fe^{3+} in a paramagnetic phase, which is attributed to iron incorporated in the SnO_2 structure, and probably hematite or Sn-doped hematite.

Acknowledgements This work was supported by CODI-Universidad de Antioquia (Sustainability program for the solid state research group 2007–2008), CENM, and COLCIENCIAS through the doctorate programs (cod. 1983).

References

1. Coey, J.M.D.: Curr. Opin. Solid State Mater. Sci. **10**, 83 (2006)
2. Janisch, R., Gopal, P., Spaldin, N.A.: J. Phys. Condens. Matter **17**, R657 (2005)
3. Rykov, A.I., Nomura, K., Sakuma, J., Barrero, C.A., Yoda, Y., Mitsui, T.: Phys. Rev. B **77**, 014302 (2008)
4. Nomura, K., Barrero, C.A., Sakuma, J., Takeda, M.: Phys. Rev. B **75**, 184411 (2007)
5. Sakuma, J., Nomura, K., Barrero, C.A., Takeda, M.: Thin Solid Films **515**, 8653 (2007)
6. Yang, H.M., Hu, Y.H., Tang, A.D., Jin, S.M., Qiu, G.Z.: J. Alloys Comp. **363**, 271 (2004)
7. Sánchez, L.C., Calle, A.M., Arboleda, J.D., Osorio, J., Nomura, K., Barrero, C.A.: Microelectron. J. **39**, 1320 (2008)
8. Cabrera, A.F., Mudarra Navarro, A.M., Rodríguez Torres, C.E., Sanchez, F.H.: Physica B **398**, 215 (2007)
9. Sorescu, M., Diamandescu, L., Tarabasanu-Mihaila, D., Teodorescu, V.S., Howard, B.H.: J. Phys. Chem. Solids **65**, 1021 (2004)
10. Berry, F.J., et al.: J. Solid State Chem. **151**, 157 (2000)

Hyperfine Interact (2010) 195:191–197
DOI 10.1007/s10751-009-0119-1

Crystallographic and ^{119}Sn and ^{155}Gd Mössbauer analyses of $Gd_5Ge_2(Si_{1-x}Sn_x)_2$ (x = 0.23 and x = 0.40)

J. C. P. Campoy · A. O. dos Santos · L. P. Cardoso · A. Paesano Jr · M. T. Raposo · J. D. Fabris

Published online: 2 October 2009

Abstract We report the structural characterization of $Gd_5Ge_2(Si_{1-x}Sn_x)_2$ (x = 0.23 and x = 0.40) compounds by means of 100 and 298 K-X-ray diffractometry (XRD) and 4 K-^{155}Gd and 298 K-^{119}Sn Mössbauer spectroscopy. These compounds order ferromagnetically at 218.4 and 172.7 K, respectively. At ~100 K, it was identified the Gd_5Si_4-orthorhombic phase (type I) for both samples. At ~298 K, it was identified a $Gd_5Si_2Ge_2$-monoclinic phase, for x = 0.23 and a Sm_5Sn_4-orthorhombic phase (type II), for x = 0.40. The Rietveld analysis of XRD data suggests a first order magneto-structural transition at Curie temperature for both compositions.

J. C. P. Campoy · L. P. Cardoso
Instituto de Física "Gleb Wataghin",
Universidade Estadual de Campinas, 13083-970,
Campinas, SP, Brazil

A. O. dos Santos
CCSST, Universidade Federal do Maranhão,
65900-410, Imperatriz, MA, Brazil

A. Paesano Jr
Departamento de Física, Centro de Ciências Exatas,
Universidade Estadual de Maringá,
87020-900, Maringá, Paraná, Brazil

M. T. Raposo (✉)
Departamento de Ciências Naturais,
Universidade Federal de São João Del Rei,
C.P. 110, 36301-160, São João Del Rei, MG, Brazil
e-mail: mtraposo@ufsj.edu.br

J. D. Fabris
Departamento de Química—ICEx,
Universidade Federal de Minas Gerais,
Campus-Pampulha, 31270-901,
Belo Horizonte, MG, Brazil

Mössbauer results are well consistent with the proposed crystallographic models for these systems.

Keywords X-ray diffraction · Magnetocaloric effect · Mössbauer

1 Introduction

The giant magnetocaloric effect (GMCE) was discovered for the $Gd_5Si_2Ge_2$ system by Pecharsky and Gschneidner Jr. [1], suggesting potential technological application in refrigeration area. Since then variations in the stoichiometry and/or different preparation methods and alloying [2–4] are being studied looking at increasing temperatures in the more intense region of MCE to, consequently, enrich the structural and magnetic knowledge of this family of compounds.

In this context chemical substitution is an interesting way to modify magneto-elastic properties and at the same time to understand their electronically properties. In particular, complete replacement of germanium by tin tends to diminish the MCE peak, changing the magnetic transition to a second order behavior [5]. In contrast, small amount of Sn-doping $Gd_5Si_{1.95}Ge_{2.05}$ changes its magneto-elastic properties, increasing its MCE peak retaining the first-order character of the magnetic-structural transition [6]. Replacing tin by silicon has a similar effect but a clear dependence in magnetically ordered transition temperature with tin content is observed [7].

In an earlier work [7] we studied $Gd_5Ge_2(Si_{1-x}Sn_x)_2$ ferromagnetic compounds with $x = 0.23$ and 0.40, by characterizing them with XRD and ^{119}Sn Mössbauer spectroscopy at room temperature. $Gd_5Si_2Ge_2$-monoclinic and Sm_5Sn_4 orthorhombic-type II phases were identified for $x = 0.23$ and $x = 0.40$, respectively. Also, at room temperature, ^{119}Sn Mössbauer results showed the occurrence of two non-equivalent sites for Sn in these solid matrices.

In this paper, we studied these same samples focusing on their structural and magnetic features, using powder X-ray diffractometry (XRD), ^{119}Sn, and ^{115}Gd Mössbauer spectroscopy, below and above the magnetic transition temperature.

2 Experimental details

Primary components of $Gd_5Ge_2(Si_{1-x}Sn_x)_2$ compounds, with $x = 0.23$ and 0.40, were sequentially three times melted in a water cooled copper heart arc furnace under Ar atmosphere, at ambient pressure, with negligible (<0.2 mass%) mass losses as it is described in [7].

X-Ray powder diffraction (XRD) measurements were performed in a Phillips diffractometer, Cu-Kα radiation, graphite monochromator, between 10° and 60°, at 100 and 298 K, scanning steps of 0.03° and 10 s counting time per step; the collected data were refined by the Rietveld method. The crystal structure refinement of each crystallographic phase was obtained by applying the Rietveld method of the GSAS package. We considered three types of structures: Gd_5Si_4-(orthorhombic type I), $Gd_5Si_2Ge_2$-(monoclinic) and Sm_5Sn_4-(orthorhombic type II), with Pnma, $P112_1/a$, and Pnma space groups, respectively.

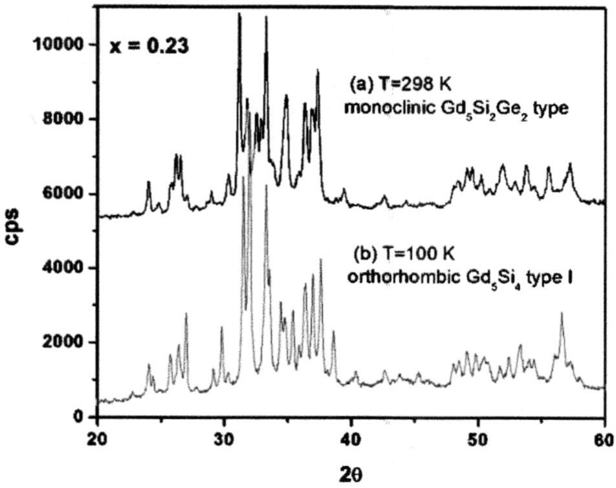

Fig. 1 X-ray diffraction patterns for the x = 0.23 sample taken at **a** 298 K and **b** 100 K. Patterns show a Gd_5Si_4 orthorhombic type I structure, in the magnetically ordered region, and a $Gd_5Si_2Ge_2$ monoclinic phase, in the paramagnetic state

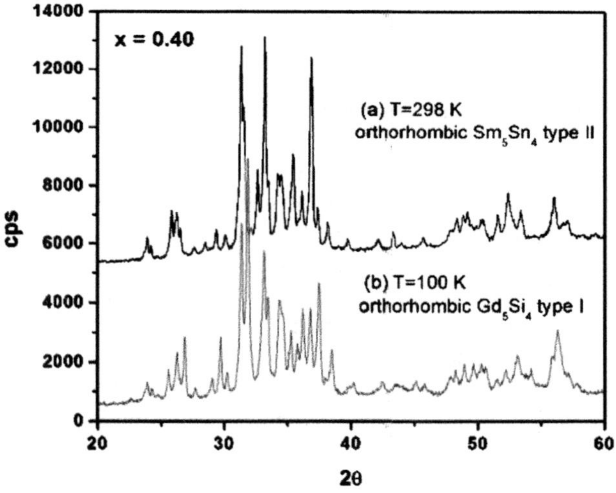

Fig. 2 X-ray diffraction patterns for x = 0.40 sample taken at **a** 298 K and **b** 100 K. Patterns show a Gd_5Si_4 orthorhombic type I structure, in the magnetically ordered region, and a Sm_5Sn_4 orthorhombic type II phase, in the paramagnetic state

4.2 K-^{155}Gd and 298 K-^{119}Sn Mössbauer transmission spectra were collected using a $SmPd_3$- and a $CaSnO_3$-embedded ^{155}Eu and ^{121}Sb, respectively, matrix sources, at the same temperature as the sample. The isomer shift values are quoted relative to the gamma-sources.

3 Experimental results and discussion

3.1 X-ray diffraction

The XRD patterns for x = 0.23 taken at 100 and 298 K, shown in Fig. 1, indicate that orthorhombic Gd_5Si_4 type I structure at the lower temperature changes to monoclinic $Gd_5Si_2Ge_2$ structure at the higher temperature. On the other hand, XRD patterns for x = 0.40, taken at 100 and 298 K (Fig. 2), indicate that the

Table 1 Structural parameters obtained from the Rietveld analysis of powder X-ray diffraction patterns at 100 and 298 K for sample x = 0.23

x = 0.23 space group	100 K Pnma—Gd_5Si_4 orthorhombic type I	298 K P112$_1$/a $Gd_5Si_2Ge_2$ monoclinic
a/Å	7.5377 (8)	7.6047 (9)
b/Å	14.842 (2)	14.873 (2)
c/Å	7.8239 (7)	7.8307 (9)
V/Å^3	875.3 (4)	884.7 (5)
Γ		92.72 (8)°
$\frac{R_p}{R_{wp}}$/%	7.5/10	6.6/8.2

Table 2 Structural parameters obtained from the Rietveld analysis of powder X-ray diffraction patterns at 100 K and 298 K for sample x = 0.40

x = 0.40 space group	100 K Pnma—Gd_5Si_4 orthorhombic type I	298 K Pnma—Sm_5Sn_4 orthorhombic type II
a(Å)	7.5670 (8)	7.6997 (9)
b(Å)	14.903 (2)	14.922 (3)
c(Å)	7.866 (7)	7.840 (8)
V(Å)^3	887.1 (4)	900.7 (6)
R_p/R_{wp}(%)	7.5/10	7.0/10

orthorhombic Gd_5Si_4 type I structure of this material at the lower temperature changes to orthorhombic Sm_5Sn_4 type II structure at the higher temperature.

From the present Rietveld refinement of XRD data for x = 0.23 (Table 1), one can infer an increase of ~1.0% of the unit cell volume. Only a small difference ($R_{wp} \leq 10\%$) was found between the experimental and refined profiles for the sample at either temperature. It indicates that the unit cell dimensions are properly and accurately determined. Rietveld refined parameters for x = 0.40, as obtained from XRD data, are presented in Table 2. Similarly to the x = 0.23 compound, an expansion of ~1.0% of the unit cell volume accompanies the structural change. A small difference ($R_{wp} \leq 10\%$) is found between experimental and refined profiles for both temperatures.

Magnetization measurements in both samples show that saturation magnetization moment at 4 K is around 7.3 μ_B/Gd^{3+}, in well agreement with the expected saturated moment of 7 μ_B/Gd^{3+}. The excess of 0.3 μ_B/Gd^{3+} is probably due to an itinerant contribution. In a previous magnetic characterization of these materials [7] it was found a first order ferromagnetic transition at 218.4 K, for x = 0.23, and at 172.7 K, for x = 0.40. Those results suggest that magnetic transition in both compounds is likely triggered by a magneto-structural change.

In short, the XRD data for the samples at 100 K evidence that the magnetically ordered phase for both cases is related to the occurrence of the Gd_5Si_4-orthorhombic type I crystallographic structure [8], reinforcing the hypothesis that a first order structural transition somehow accompanies the magnetic change to the paramagnetic state at T_C = 218.4 K, for x = 0.23 (P112$_1$/a monoclinic), and at T_C = 172.7 K, for x = 0.40 (Pnma orthorhombic type II) [7].

3.2 Mössbauer spectroscopy

4.2 K-^{155}Gd Mössbauer spectra are showed in Fig. 3 and corresponding fitted parameters with NORMOS® computer program are presented in Table 3. These correspond to relative subspectral areas ratios of 87:13, for sample with x = 0.23, and

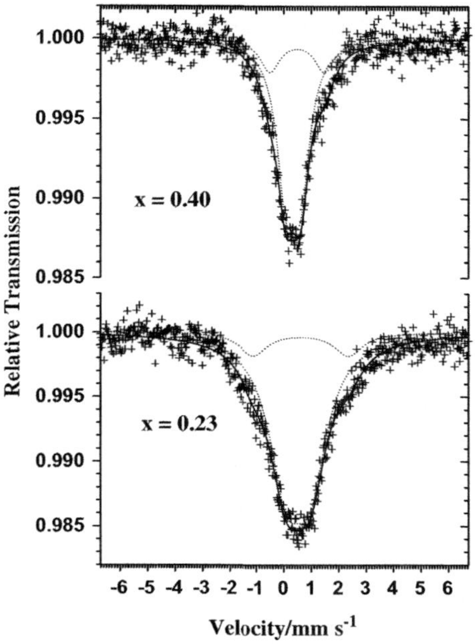

Fig. 3 4.2 K-^{155}Gd Mössbauer spectra for $Gd_5Ge_2(Si_{1-x}Sn_x)_2$, with x = 0.23 and x = 0.40

Table 3 ^{155}Gd Mössbauer parameters Mössbauer parameters for $Gd_5Ge_2(Si_{1-x}Sn_x)_2$ (x = 0.23 and x = 0.40) at 4.2 K

x	δ/mm s^{-1}	Δ/mm s^{-1}	Γ/mm s^{-1}	RA/%	
0.23	0.51	0.87	1.49	87	~7 : 1
	0.61	3.47	1.23	13	
0.40	0.34	0.53	0.81	79	~4 : 1
	0.48	1.99	0.88	21	

δ isomer shift relative to the $SmPd_3$ gamma-source at 4.2 K, Δ quadrupole splitting, Γ resonance line-width, *RA* relative spectral area

79:21, for x = 0.40, meaning the co-existence of non-equivalent Gd1, in 4(c); Gd2, in 8(d) and Gd3, in 8(d) sites of the orthorhombic Gd_5Si_4-type structure [9]. Also, all 8(d) sites are presumably equivalent and present higher ionic occupation than 4(c) sites.

Earlier reported ^{119}Sn room temperature parameters (Table 4) evidenced the co-existence of two non-equivalent hyperfine sites for the Sn^{4+} ions, consistently with what is expected for both $Gd_5Ge_2Si_2$ (monoclinic) and Sm_5Sn_4 (type II) crystallographic structures. For the orthorhombic x = 0.40, one ^{119}Sn Mössbauer doublet is associated to the $4c_2$ and 8d equivalent sites and the other with the $4c_1$ site [7, 10]. This is the same for the monoclinic x = 0.23 sample, where the 8d site is split into two symmetrically fourfold sites. When the temperature is lowered below the Curie point $T_C = 218.4$ K, for x = 0.23, and 172.7 K, for x = 0.40, a structural transition is observed.

In spite of the magnetically ordered character of these compounds, the hyperfine structure for the nuclear 86.5 keV transition in ^{155}Gd is such that it may be much less evident separating corresponding Zeeman resonance lines. Even for a moderately

Table 4 ^{119}Sn Mössbauer parameters for $Gd_5Ge_2(Si_{1-x}Sn_x)_2$ (x = 0.23 and x = 0.40) at 298 K (from [7])

x	Site	δ/mm s^{-1}	Δ/mm s^{-1}	Γ/mm s^{-1}	RA/%
0.23	B	2.139 (6)	0.74	0.80	22 (4)
	A	1.912 (1)	1.13	0.78	78 (5)
0.40	B	2.051 (2)	0.70	0.77	30 (4)
	A	1.892 (2)	1.07	0.77	70 (4)

δ isomer shift relative to the $Ca^{119}SbO_3$ matrix, Δ quadrupole splitting, Γ resonance line-width, *RA* relative spectral area

strong hyperfine field, values for energy transitions due to coupling with the nuclear dipole moment may be close to those resulting from a sole electric field gradient (EFG)–quadrupole moment interaction. This circumstance is greatly different from that expected, for instance, for equivalent magnitudes of magnetic field and EFG couplings on the ^{57}Fe nuclear probe. This happens for at least three fundamental reasons: (1) the intrinsic atomic magnetic moment are larger for the lanthanide Gd^{3+}, with $[Xe]4f^7$, than for Fe^{3+}, with $[Ar]3d^5$ electronic configuration; (2) absolute values of nuclear electric quadrupole and dipole magnetic moments (actual values are reported e.g. in [11–13]) for the 86.5 keV ($I_{ex} = \frac{5}{2}+$; mean life-time, $\tau = 9.1$ ns) and ground $\left(I_{ex} = \frac{3}{2}-\right)$ states of ^{155}Gd are roughly thrice those for the 14.4 keV $\left(I_{ex} = \frac{3}{2}-; \tau = 141.5\,\text{ns}\right)$ and ground $\left(I_{ex} = \frac{1}{2}+\right)$ states of ^{57}Fe and (3) the natural resonance line (two times the natural line-width of the Mössbauer energy level) for ^{155}Gd is 0.50 mm s^{-1}, comparatively to 0.19 mm s^{-1}, for ^{57}Fe.

4 Conclusions

100 K-XRD data evidence that $Gd_5Ge_2(Si_{1-x}Sn_x)_2$ (x = 0.23 and x = 0.40) systems do contain a Gd_5Si_4-orthorhombic, crystallographic structure type I, suggesting a first order magneto-structural transition to paramagnetic state at $T_C = 218.4$ K for x = 0.23 ($P112_1/a$ monoclinic) and $T_C = 172.7$ K for x = 0.40 (Pnma orthorhombic type II). ^{155}Gd-Mössbauer spectra for the samples at 4.2 K, despite the magnetic structure, are numerically resolved as two quadrupolar patterns, for both samples, with relative areas ratios 87:13, for x = 0.23, and 79:21, for x = 0.40, corresponding to the proportion of non-equivalent distribution of Gd1 in 4(c), Gd2 in 8(d), and Gd3 in 8(d) sites of the orthorhombic Gd_5Si_4-type structure. Here, all 8(d) sites are presumably equivalent and present higher ionic occupation than 4(c) sites.

Room temperature ^{119}Sn Mössbauer spectroscopy for compounds with x = 0.23 and 0.40 indicate the co-existence of two non-equivalent hyperfine sites for the Sn^{4+} ions, as expected for both $Gd_5Ge_2Si_2$ (monoclinic) and Sm_5Sn_4 (type II) crystallographic structures.

Acknowledgements This work was supported by CNPq, FAPEMIG and FAPESP (Brazil). CNPq/Prosul (grant no. 490132/2006-5) also supported the international mission by JDF to attend the Latin American Conference on the Applications of the Mössbauer Effect, LACAME'2008, in La Plata, Argentina.

References

1. Pecharsky, K., Gschneidner, K.A. Jr.: Phys. Rev. Lett. **78**, 4494 (1997)
2. Pecharsky, V.K., Gschneidner, K.A. Jr.: J. Magn. Magn. Mater. **167**, L179–L184 (1997)
3. Pecharsky, A.O., Gschneidner, K.A. Jr., Pecharsky, V.K.: J. Appl. Phys. **93**, 4722 (2003)
4. Pecharsky, A.O., Gschneidner, K.A. Jr., Pecharsky, V.K.: J. Magn. Magn. Mater. **267**, 60 (2003)
5. Campoy, J.C.P., Plaza, E.J.R., Carvalho, A.M.G., Coelho, A.A., Gama, S., Von Ranke, P.J.: J. Magn. Magn. Mater. **272**, 2375 (2004)
6. Li, J.Q., Sun, W.A., Jian, Y.X., Zhuang, Y.H., Huang, W.D., Liang, J.K.: J. Appl. Phys. **100**, 073904 (2006)
7. Campoy, J.C.P., Plaza, E.J.R., Nascimento, F.C., Coelho, A.A., Pereira, M.C., Fabris, J.D., Raposo, M.T., Cardoso, L.P., Persiano, A.I.C., Gama, S.: J. Magn. Magn. Mater. **316**, 368 (2007)
8. Levin, E.M., Pecharsky, V.K., Gschneidner, K.A. Jr., Miller, G.J.: Phys. Rev. B **64**, 235103 (2001)
9. Pecharsky, V.K., Gschneidner, K.A. Jr.: J. Alloys Compd. **260**, 98 (1997)
10. Wang, H.B., Altounian, Z., Ryan, D.H.: Phys. Rev. B **66**, 214413 (2002)
11. Raghavan, P.: Table of nuclear moments. At. Data Nucl. Data Tables **42**(2), 189–291 (1989)
12. Stevens, J.G., Dunlap, B.D.: J. Phys. Chem. Ref. Data **5**, 1093–1121 (1976)
13. Czjzek, G.: Mössbauer spectroscopy of new materials containing gadolinium. In: Long, G.J., Grandjean, F. (eds.) Mössbauer spectroscopy applied to magnetism and materials science, vol. 1. pp. 373–429. Plenum, New York (1993)

Hyperfine Interact (2010) 195:199–204
DOI 10.1007/s10751-009-0120-8

A Mössbauer effect study of the $Fe_{2+x}Mn_{1-x}Al$ Heusler alloys

C. Paduani · C. A. Samudio Pérez · J. Schaf · J. D. Ardisson · A. Y. Takeuchi · M. I. Yoshida

Published online: 1 October 2009

Abstract In this work the Mössbauer spectroscopy has been used to study the magnetic properties of $Fe_{2+x}Mn_{1-x}Al$ alloys with small deviations of composition from the stoichiometric 2:1:1. The Mössbauer parameters obtained for the $L2_1$ phase indicate H_{hf} fields of about 25 T and 30 T at 80 K for Fe atoms at X sites in the ordered X_2YZ structure of the $L2_1$ full Heusler alloys.

Keywords Heusler alloys · Mössbauer spectroscopy · Ferromagnetism

PACS 71.20.Lp · 71.15.Nc · 75.50.Ss · 75.50.Bb

1 Introduction

The disordered FeAl and FeMnAl alloys have been intensively investigated in the last past years, and have attracted attention from both experimentalists and

C. Paduani (✉)
DF-UFSC, Florianópolis, CEP 88040-900, Santa Catarina, Brazil
e-mail: paduani@fisica.ufsc.br

C. A. Samudio Pérez
ICE-UPF, Passo Fundo, CEP 99001-970, Rio Grande do Sul, Brazil

J. Schaf
IF-UFRGS, Porto Alegre, CEP 91501-970, Rio Grande do Sul, Brazil

J. D. Ardisson
CDTN, Belo Horizonte, CEP 30123-970, Minas Gerais, Brazil

A. Y. Takeuchi
CBPF, Rio de Janeiro, CEP 22290-180, Brazil

M. I. Yoshida
DQ-ICEX-UFMG, Belo Horizonte, CEP 31270-901, Minas Gerais, Brazil

theoreticians [1–6]. The disordered ternary alloys are interesting magnetic materials considering its promising technological applications in view of their similarities with conventional stainless steel as well as in view of the different types of magnetic phases that can be detected. In the austenitic phase the high workhardening rate makes it a good candidate in wear resistant application. In engineering applications these are offered as cryogenic materials, presenting high strength with low density, with good oxidation resistance, good toughness and strength and as cheap substitutes for some conventional stainless steels. This has encouraged the experimental research to improve its performance for a large variety of applications. This system provides a plentiful field of interesting subjects of investigation, like order-disorder structural transitions, wide variety of magnetic phases and transitions, critical behavior, competing interactions and dilution.

Heusler alloys otherwise are ternary intermetallic compounds with the stoichiometry X_2YZ, i.e., the $L2_1$ structure. The large majority of Heusler alloys order ferromagnetically and saturate in applied magnetic fields of less then 5 kOe at low temperatures [7–16]. Some indeed do order antiferromagnetically, particularly those compounds containing a group 3B element in which the magnetic moment is confined to the Y sites, usually occupied by Mn. Fe_2MnAl alloys can be obtained in the X_2YZ composition as full Heusler alloys (s.g. Fm3m). Theoretical predictions for Fe_2MnAl in the $L2_1$ structure yield moments of −0.311 and 2.633 for Fe and Mn atoms, respectively [7–9]. Although being metals, these intermetallic compounds have localized magnetic properties and are ideal model systems for studying the effects of both atomic disorder and changes in the electron concentration on magnetic properties. Extensive studies have been carried out using Heusler alloys by employing hyperfine methods to understand the coupling mechanism responsible for the magnetic hyperfine fields at non magnetic sites. In this work we investigate the magnetic properties of FeMnAl alloys with Mössbauer spectroscopy. Our aim is to determine the Mössbauer parameters of the $L2_1$ phase besides analyze the effect of changes in the composition on the relative occupancy of Fe sites as well as on these parameters. This work is a further step from previous investigation [13].

2 Experimental

Polycrystalline ingots of $Fe_{2+x}Mn_{1-x}Al$ alloys with nominal compositions $x = 0.0$ and ± 0.1 were prepared by arc-melting of high-purity initial elements in argon atmosphere. Each sample was remelted several times to promote homogeneity and then further encapsulated in quartz tubes filled with argon and annealed for 3 days at 1273 K, then 673 K for more 3 days, and cooled in the furnace by itself. The XRD patterns were taken by using the Cu $K\alpha$ radiation. Magnetization measurements above room temperature (RT = 293 K) were also performed with a magnetobalance in an arbitrary field in order to verify magnetic transitions at higher temperatures. Powder absorbers for the Mössbauer spectroscopy (MS) were obtained from the central part of each sample. The MS spectra were collected at RT and at 80 K, in the transmission geometry with a conventional spectrometer, operating with constant acceleration mode (driven in the triangle mode), using a 50 mCi $^{57}Co/Rh$ source. The standard is natural α-iron, for which all quoted isomer shifts are reduced to its centroid (δ(Fe)).

3 Results and discussion

The XRD measurements showed that all the studied alloys unambiguously exhibit a $L2_1$ structure of the full Heusler alloy as the main phase. A minority phase therefore was also detected and identified as a face centered tetragonal (fct) phase (s.g. P4/mmm). The lattice parameters of the $L2_1$ phase is constant with a value 5.816 Å, exactly the same indicated in the literature [10]. The cell volume remains unchanged despite the variations from the stoichiometric composition. The fct phase also presents practically the same lattice parameters in all the samples. As in a ideal ordered $L2_1$-type structure the lattice consists of four interpenetrating face centered cubic sublattices, with four atoms as basis: X at (000) and $(\frac{1}{2}\frac{1}{2}\frac{1}{2})$, Y at $(\frac{1}{4}\frac{1}{4}\frac{1}{4})$, and Z at $(\frac{3}{4}\frac{3}{4}\frac{3}{4})$ in Wyckoff coordinates. The X sub-lattice's are occupied by Fe atoms and Y and Z are occupied by manganese and aluminum.

As observed in previous work [13] the main ordered $L2_1$ phase in these alloys order ferromagnetically with $T_C \approx 300$ K for $x = 0.0$ and -0.1, and $T_C = 380$ K for $x = 0.1$. On the other hand, experimental evidence was found that the fct phase otherwise shows both antiferromagnetic (AF) and ferromagnetic (FM) exchanges. The Mn atoms carry the largest moments in the ordered $L2_1$ phase (2.633 μ_B), whereas the Fe atoms have much smaller moments (-0.311 μ_B) [7–9], and a competition mechanism between Fe-Fe, Fe-Mn and Mn-Mn FM exchanges and Fe-Mn and Mn-Mn AF exchanges therefore is expected. The experimental data show a broad transition at the disordering temperature T_C, which has been interpreted as being due to a large distribution of the FM interaction energies which arises from the loss of FM couplings with the random substitution of Fe for Mn. In Fig. 1 are shown results of measurements with a magnetobalance. The transition temperatures T_C are indicated. As can be seen, the transition indeed takes place gradually, and some residual magnetic contribution from a ferromagnetic entity still remains in the alloys up to higher temperatures until the collapse of the total magnetization.

The fitted Mössbauer spectra collected at 80 K are shown in Fig. 2. Only the spectra at low temperature are shown, since at RT these spectra are very similar. In the previous work the M(T) curves show a plateau up to T_C, from which a monotonic decrease is observed, as one can also see in Fig. 1. Hence, the similarities of the Mössbauer spectra at RT and 80 K can be understood as follows: for $x = 0.1$ T_C and T_N are well above RT (380 K and 350 K, respectively [13]); for $x = 0$ the transition temperature T_N of the AF entity also is higher than RT, whereas T_C for the ordered phase is near RT, and thereby thermal fluctuations in the environmental conditions can affect the results. For $x = -0.1$, the disordered phase has a significant contribution to the total magnetization and exhibits both FM and AF exchanges; the AF component gives a lesser contribution and the FM component is thus majority, whose T_C is comparable to that one of the ordered phase, which is also near RT. In Fig. 1 a transition is clearly seen at 350 K, which might be ascribed either to the $L2_1$ phase or to the fct phase. If one considers that the strongest FM couplings belong to the ordered phase, where the Mn atoms possess large moments, this T_C is associated with the former. Thus, this results makes a correction in the previously reported value of 300 K for T_C of the $L2_1$ phase in these alloys at this composition.

From the discussion above, the Mössbauer spectra of these alloys at 80 K therefore are considered as a superposition of two contributions, one associate do the $L2_1$ ordered phase and another due to the fct disordered phase. Hence, a distribution

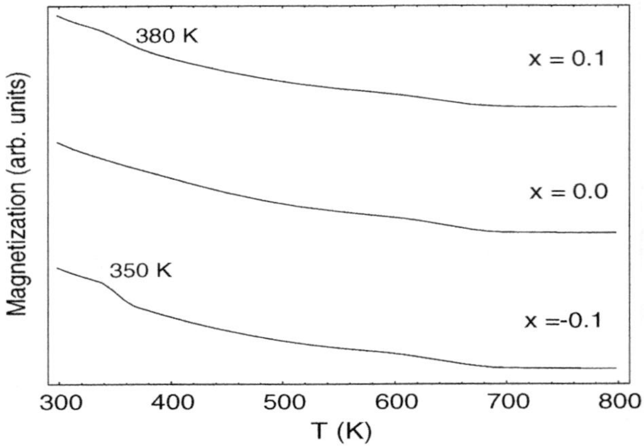

Fig. 1 M(T) curves above room temperature registered with a magnetobalance

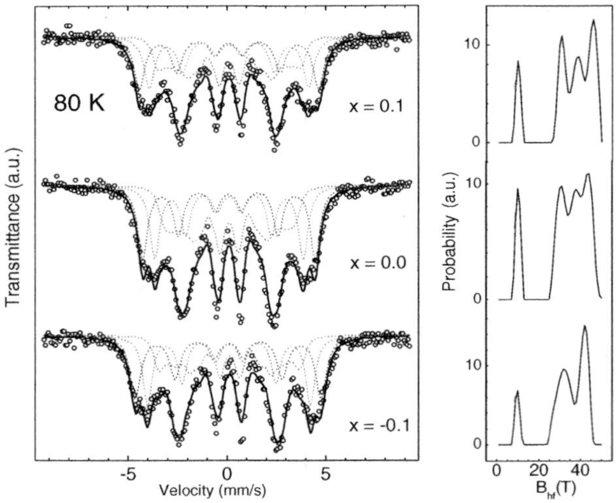

Fig. 2 Mössbauer spectra at 80 K of $Fe_{2+x}Mn_{1-x}Al$ alloys. On the *right* is shown the histogram of each H_{hf} distribution profile, which is plotted in the spectrum envelope by *crosses*

of magnetic hyperfine field (H_{hf}) was used in the fitting procedure to account for the contribution of the fct phase to the total envelope of each spectrum, which was performed by using 75 sextets. The linewidths were constrained to have widths corresponding to bcc iron. As partial atomic disorder in the structure cause changes in the Fe-sites neighborhood, in order to fit the contribution of the $L2_1$ phase to the spectra we adopted two sextets, representing the X sublattices in the X_2YZ structure. The results of the fitting procedure are listed in Table 1 and the corresponding distribution profiles are shown on the right in Fig. 2. In the H_{hf} histograms one sees a block of several peaks, where a prominent peak is seen at both in the region of high fields and low fields. The peaky structure in the intermediate region accounts for the different contributions which arises from the several possible configurations of the Fe sites in the alloys. The higher field values might be associated with those Fe sites which are strong coupled to Mn atoms, whose ordering is broken off at about 700 K, as can be seen in Fig. 1. In the central part of each spectrum a paramagnetic contribution can be identified, which is associated with Fe sites surrounded by Al

Table 1 Alloy composition, isomer shift (δ(Fe)), quadrupole splitting (2ϵ), hyperfine magnetic field (H_{hf}), line width (Γ) and subspectral area of $Fe_{2+x}Mn_{1-x}Al$ alloys (in the $L2_1$ structure 2 sextets were used in the fitting procedure; for the fct phase H_{hf} is the average field value of the distribution)

x	Structure	δ(Fe) (mm/s)	2ϵ(mm/s)	H_{hf}(T)	Γ(mm/s)	A (%)
0.1	$L2_1$	0.07	0.14	29.01	0.69	27
		−0.02	−0.04	24.93	0.56	27
	fct	−0.09	0.17	18.10	0.26	46
0.0	$L2_1$	0.05	0.16	28.38	0.68	20
		−0.03	0.02	24.40	0.59	26
	fct	−0.09	0.17	17.60	0.26	54
−0.1	$L2_1$	0.05	0.02	30.00	0.63	26
		0.03	−0.02	25.80	0.56	29
	fct	−0.10	0.18	19.10	0.26	45

atoms and with smaller moments. A range of values for the Fe moments thus is expected.

The prominent sextets in each spectrum are tentatively ascribed to the principal Fe sites in the ordered $L2_1$ phase. One may argue that this might be an artefact, but it represents a first attempt to obtain a reasonable description for the observed results. In Table 1 are listed the corresponding parameters. It is noteworthy that at an equivalent composition in the disordered bcc phase of ternary $Fe_{50}Mn_{25}Al_{25}$ alloys a $H_{hf} = 13.2$ T has been reported at room temperature [1]. Therefore, the profiles of the spectra shown in Fig. 2, as well as the distribution profiles (on the right) are completely different from that one corresponding to a disordered $Fe_{50}Mn_{25}Al_{25}$ alloy. This reflects the contribution of the ordered structure in the alloys, for whom a large mean magnetic moment has been calculated [7–9, 16]. By fitting the subspectra (sextets), although leaving free all parameters for each one of them, it is somewhat surprising that similar results are obtained in all the cases. From this one may infer that these quantities yield a reasonable description for the magnetic contribution of the ordered phase in building the envelope of each spectrum. At 80 K the alloy with $x = -0.01$ is below T_N of the fct phase, which results in a higher average H_{hf} value, as seen in Table 1. Furthermore, as at 80 K the Mössbauer spectrum of the alloy with $x = -0.10$ is similar to that one at RT, it may be concluded that most of the contribution of the fct phase to the spectral area is due to the magnetic entity which persists up to higher temperatures, as seen in Fig. 1. Thus both FM and AF exchanges are ascribed to the fct phase in this alloy. This study with the Mössbauer spectroscopy complements results of low temperature magnetization measurements in the interpretation of the observed trends.

Summarizing, from the results above one may conclude that the fct phase exhibits both FM and AF exchanges in the studied alloys. However, it is not possible distinguish between these contributions for the average magnetic hyperfine field. The ordered phase, which is responsible for the main FM entity in these alloys, has a well defined magnetic contribution to the total envelope of the Mössbauer spectra, as can be promptly verified in Table 1, and at 80 K show H_{hf} values lower than α-Fe. This is consistent with results of first-principles calculations in Fe_2MnAl Heusler alloys, which give 1.98 μ_B for the total magnetic moment [7–9]. Moreover, since at 80 K the spectra of the alloy with x = −0.10 is similar to that one at RT, and $T_N = 205$ K, one may infer that the AF component gives a small contribution for the spectral area as well as for H_{hf}, which is then preponderantly due to the FM entity, which exists in both phases. Particularly in this case (x = −0.10), with the Mn atoms carrying the largest moments, only a small fraction of Fe atoms is thus expected to be weakly coupled to the nearest neighbors and the large majority is strongly coupled to the surrounding magnetic neighbors.

References

1. Pérez Alcázar, G.A., Plascak, J.A., da Silva, E.G.: Phys. Rev. B **38**(4), 2816 (1988)
2. Pérez Alcázar, G.A., da Silva, E.G.: J. Phys. F. Met. Phys. **17**, 2323 (1987)
3. Pérez Alcázar, G.A., da Silva, E.G., Paduani, C.: Hyp. Int. **66**, 221 (1991)
4. Pérez Alcázar, G.A.: Revista Academica Colombiana de Ciencias **28**, 265 (2004)
5. Pérez Alcázar, G.A., Plascak, J.A., da Silva, E.G.: Phys. Rev. B **38**(4), 2816 (1988)
6. Pérez Alcázar, G.A., da Silva, E.G.: J. Phys. F. Met. Phys. **17**, 2323 (1987)
7. Galanakis, I.: Phys. Rev. B **71**, 012413–1 (2005)
8. Galanakis, I., Dederichs, P.H., Papanikolaou, N.: Phys. Rev. B **66**, 174429 (2002)
9. Galanakis, N.: Phys. Rev. B **71**, 012413 (2005)
10. Buschow, K.H.J., van Engen, P.G.: J. Magn. Magn. Mater. **25**, 90 (1981)
11. Paduani, C., da Silveira, R.G., dos Santos, R.G.C., Pottker, W.E., Ardisson, J.D., Schaf, J., Takeuchi, A.Y., Yoshida, M.I.: J. Alloys Compd. **457**(1–2), 24 (2008)
12. Paduani, C., Migliavacca, A., Krause, J.C., Ardisson, J.D., Yoshida, M.I.: Braz. J. Phys. **37**(4), 1284 (2007)
13. Paduani, C., Migliavacca, A., Pottker, W.E., Schaf, J., Krause, J.C., Ardisson, J.D., Samudio Pérez, C.A., Yoshida, M.I.: Phys. B Condens. Matter **398**(1), 60 (2007)
14. Paduani, C., Pottker, W.E., Ardisson, J.D., Schaf, J., Takeuchi, A.Y., Yoshida, M.I., Soriano, S., Kalisz, M.J.: Mössbauer effect and magnetization studies of the Fe_{2+x} Cr_{1-x} Al system in the $L2_1$ (X_2YZ) structure. Phys. Condens. Matter **19**, 9, 156204 (2007)
15. Paduani, C., Migliavacca, A., Sebben, M.L., Ardisson, J.D., Yoshida, M.I., Soriano, S., Kalisz, M.: Solid State Commun. **141**(3), 145 (2007)
16. Plogmann, S., et al.: Phys. Rev. B **60**, 6428 (1999)

Hyperfine Interact (2010) 195:205–212
DOI 10.1007/s10751-009-0121-7

Mössbauer analysis of Nd-Co M-type strontium hexaferrite powders with different iron content

Carlos Herme · Silvia E. Jacobo · Paula G. Bercoff · Bibiana Arcondo

Published online: 13 October 2009

Abstract The structural analysis of strontium hexaferrites $SrFe_xO_{19}$ (for $x = 12, 11$ and 10) and substituted samples $Sr_{0.7}Nd_{0.3}Fe_{12-y}Co_{0.3}O_{19}$ (for $y = 0.3, 1.3$ and 2.3) prepared through the citrate precursor method is shown. Nd and Co substitution modifies saturation magnetization (M_S) and increases coercivity (H_c) in samples heat-treated at 1100°C for two hours. Mössbauer analyses show different iron occupancy and the influence of the Fe^{3+} content is particularly emphasized. Hematite segregation is observed for some compositions. Samples with low Fe^{3+} content show the best magnetic properties with no secondary phase segregation.

Keywords Strontium hexaferrites · Nd-Co substitution · Non-stoichiometric hexaferrite

PACS 75.50.G · 75.60.E · 61.18.F

1 Introduction

The magnetic properties of substituted hexaferrites are strongly dependent upon the electronic configuration of the substituting cations as well as on the Fe^{3+} ions five different intersticial crystallographic sites: three octahedral sites (12k, 2a and 4f2), one tetrahedral site (4f1) and one bipyramidal site (2b). The magnetic structure is ferrimagnetic with different sublattices: three parallel (12k, 2a and 2b) and two antiparallel (4f1 and 4f2) which are coupled by superexchange interactions

C. Herme · S. E. Jacobo (✉) · B. Arcondo
Facultad de Ingeniería and INTECIN-Conicet, Universidad de Buenos Aires, Paseo Colón 850, CP1063ACV, Buenos Aires, Argentina
e-mail: sjacobo@fi.uba.ar

P. G. Bercoff
Facultad de Matemática, Astronomía y Física, Universidad Nacional de Córdoba and IFFAMAF, Conicet, Argentina

through the O^{2-} ions [1]. It has recently been shown that the magnetic properties of anisotropic M-type $SrFe_{12}O_{19}$ hexagonal ferrites can be improved by doping with La and Co [2–4]. This improvement is largely related to the coercivity and the magneto-crystalline anisotropy [5]. Wang et al. prepared Nd-substituted Sr hexaferrite particles hydrotermally synthesised [6]. They reported that saturation magnetization M_S values are almost the same than that of Sr-M whereas their coercivities are around 11% higher. The solubility of rare earth ions in Sr-hexaferrites is very low and their introduction leads to the formation of secondary phases, which must be avoided in order to obtain permanent magnets with optimal properties. Secondary phases such as $SrFeO_3$ and $NdFeO_3$ were observed for all the Sr-rare earth ions substitution. La-Co substituted hexagonal ferrite prepared by the citrate self-combustion route has been tried by different researches [3, 5]. However, complete single-phase formation was not observed for annealing temperatures below 1200°C.

In this paper we present the novel results for Sr ferrites with Nd-Co substitution. We have recently reported Nd inclusion in strontium hexaferrites [7]. As different iron compounds were reported to be the main secondary phases when rare-earth ions substitution increases [3], we propose a deficient iron formulation. A further improvement of both intrinsic and magnetic properties of the Nd-Co substituted ferrite can be obtained by adjusting the chemical composition. Powders of strontium hexaferrites ($SrFe_xO_{19}$) for $x = 12, 11$ and 10 and substituted samples ($Sr_{0.7}Nd_{0.3}Fe_{12-y}Co_{0.3}O_{19}$ for $y = 0.3, 1.3$ and 2.3) prepared through the citrate precursor method are investigated by Mössbauer spectrometry. The hyperfine parameters deduced from the fittings are discussed in relation with the substitution effects.

2 Experimental

Samples of composition $SrFe_xO_{19}$ (for $x = 12, 11$ and 10) and substituted samples $Sr_{0.7}Nd_{0.3}\ Fe_{12-y}Co_{0.3}O_{19}$ (for $y = 0.3, 1.3$ and 2.3) were prepared by the self-combustion method. The chemical precursors used for these experiments were $Fe(NO_3)_3 \cdot 9H_2O$, $SrCO_3$, Nd_2O_3, $Co(CH_3COO)_2 \cdot 4\ H_2O$, as it was earlier reported [7, 8]. The ratio citric:nitric acid was fixed in 2:1 for each experiment. Aqueous suspensions were stirred and heated for several hours until the sol turned into a dried gel. Then the dried gel was ignited in a corner and a combustion wave spontaneously propagated through the whole gel converting it into loose magnetic powder. Powders were heat-treated in air at 1100°C for two hours. The non-stoichiometric compositions and sample labels for the powders heated at 1100°C are given in Table 1. The oxygen content was calculated assuming the following valences: Sr^{2+}, Nd^{3+}, Co^{2+} and Fe^{3+}.

The structural analysis of the powders was done by x-ray diffraction using a Philips diffractometer and CuK_α radiation. The patterns were taken between $2\theta = 20°$ and 70° with a step of 0.02°. A vibrating sample magnetometer Lakeshore 7300 was used to measure the magnetic properties at room temperature. Hysteresis loops M *vs* H were measured with a maximum applied field of 15 kOe. Since the loops did not saturate at 15 kOe, the value of saturation magnetization M_S for each sample was calculated by extrapolating M *vs* $1/H$ to 0. Fe Mössbauer spectroscopy (MS) was performed at room temperature, in transmission geometry. $^{57}Co(Rh)$ source was

Table 1 Composition, sample labels and magnetic properties of powders heat-treated at 1100°C for 2 hours

Sample	Nominal composition	Fe defect	H_c [Oe]	M_{max} [emu/g]	M_S[emu/g]
F12N0	$SrFe_{12}O_{19}$	0	2100	72	91
F11N0	$SrFe_{11}O_{17.5}$	1.0	4930	77	97
F10N0	$SrFe_{10}O_{16}$	2.0	2845	71	89
F12N3	$Sr_{0.7}Nd_{0.3}Fe_{11.7}Co_{0.3}O_{19}$	0.3	4613	60	75
F11N3	$Sr_{0.7}Nd_{0.3}Fe_{10.7}Co_{0.3}O_{17.5}$	1.3	5120	70	87
F10N3	$Sr_{0.7}Nd_{0.3}Fe_{9.7}Co_{0.3}O_{16}$	2.3	4885	64	77

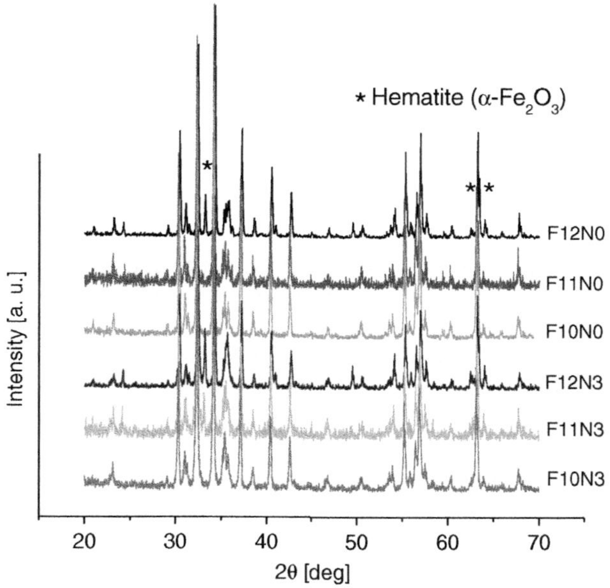

Fig. 1 X-ray diffraction patterns of the studied samples

used with a constant acceleration drive. The spectrometer was calibrated employing an α-Fe foil. Mössbauer spectra were fitted with the NORMOS (SITE) program [9] Isomer shift values are reported relative to $\alpha-$Fe whereas the velocity abscissas of the spectra are relative to the source.

3 Results and discussion

XRD patterns of all the samples are shown in Fig. 1. It is found that they are mainly composed of single-phase magnetoplumbite crystalline structure. Small amounts of hematite are present in samples with higher iron content. In substituted samples neither Nd nor Co compounds were detected indicating that both ions were included in the hexagonal structure. Intermediate phases such as $SrFe_2O_4$ were not observed as it is normally observed in samples synthesized by the ceramic route [10]. This indicates that the solid state reaction is exclusively heterogeneous, and the hexaferrite phase formation proceeds by the formation and growth at the expense of the existing

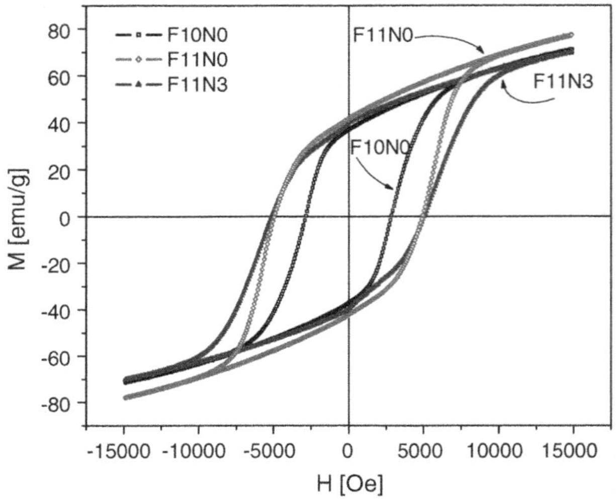

Fig. 2 Hysteresis loops of selected samples

phases—strontium carbonate, hematite and maghemite. This is possible since, in the citrate gel route of synthesis, the mixing of cations takes place at the atomic level, thus reducing the paths of diffusion for various cations, as compared to the ceramic route where the diffusion distances are considerably larger [11]. It may be pointed out that the amount of vacancies at the Fe sites is as high as 16% (for sample F10N0) and yet the hexaferrite structure does not break down nor do any other phases occur indicating that the structure is stable.

Table 1 shows magnetic properties of all samples and Fig. 2 shows the hysteresis loops of some of them. It can be seen that H_c increases with Nd-Co substitution for all Fe composition (y). These magnetic ions modify superexchange interaction in the magnetoplumbite structure. M_S is slightly modified in all samples since its value depends on composition, crystallinity and the presence of non-magnetic phases as hematite.

Mössbauer spectra of $SrFe_xO_{19}$ (see Fig. 3, left) were fitted with five sextets (one for each crystallographic iron site), except for samples were hematite was detected. In these cases, another contribution for hematite was included. The isomer shifts (δ), that may be attributed to Fe^{3+}, follow the sequence 4f2≥12k>2b> 4f1>2a whereas the sequence for hyperfine magnetic fields (B_{hf}) is 4f2≥ 2a>4f1>12k≥2b. Similar results were reported by B. J. Evans et al. [12]. However, isomer shift values corresponding to sites 4f1 and 2a are exchanged. This difference may be attributed either to the different synthesis procedure followed in each case or to the intrinsic difficulties found in fitting these complex spectra due to the poorly resolved 4f2, 2a and 4f1 sites contributions. Hyperfine parameters for these samples are reported in Table 2.

Figure 4 shows the site occupancy O_i estimated from Mössbauer results for non-substituted samples with different iron content as $O_i = \frac{N}{A} A_i$, where N is the theoretical number of Fe atoms in the unit cell, A is the total area of the Mössbauer spectrum and A_i is the area of each site contribution. Corrections for hematite presence were considered. Iron vacancies are not proportionally distributed in all sites. The occupancy of sites 12k, 4f1 and 2b is lower than the theoretical values

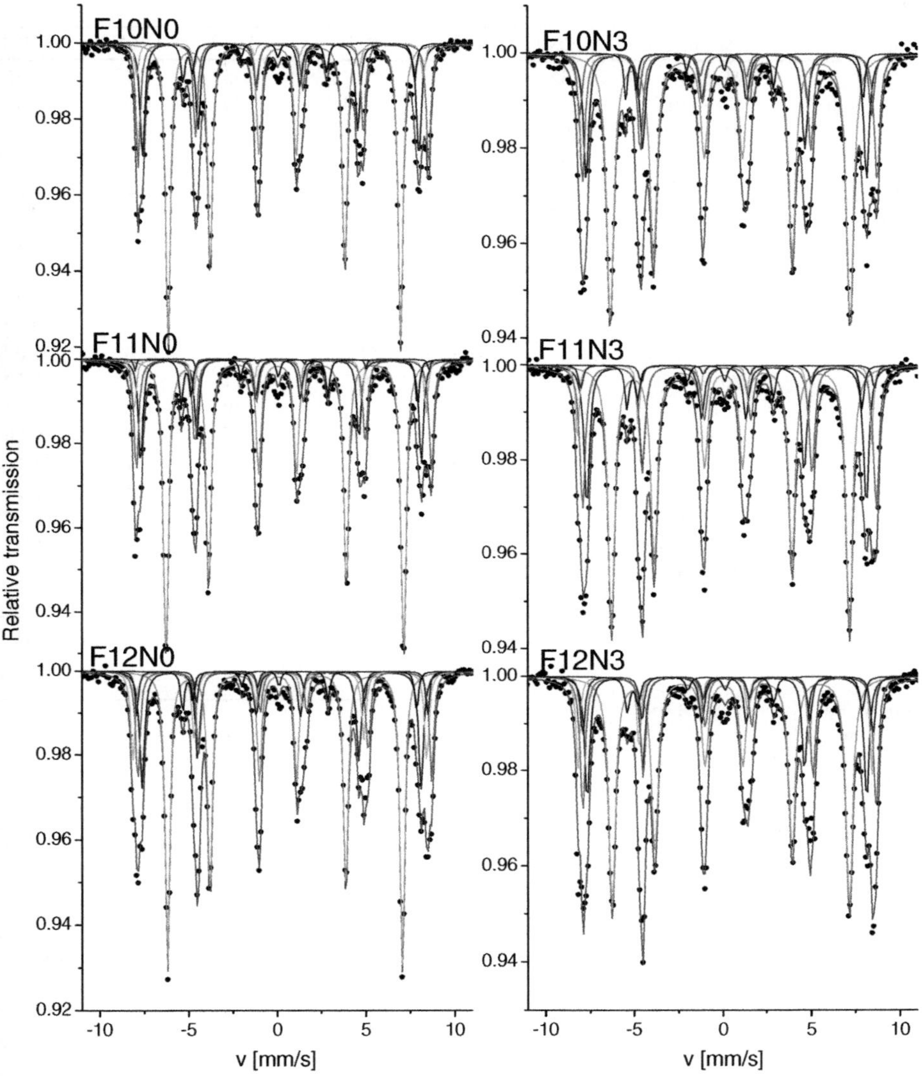

Fig. 3 Room temperature Mössbauer spectra and fittings of $SrFe_xO_{19}$ (left) and of $Sr_{0.7}Nd_{0.3}Fe_{x-0.3}Co_{0.3}O_{19}$ (right) obtained with a $^{57}Co(Rh)$ source

(dashed lines in Fig. 4) and varies with x. From Mössbauer results the effective number of Fe^{3+} spins per cell $n = (12k + 2a + 2b) - (4f1 + 4f2)$ for $x = 12, 11$ and 10 are 7.5, 7.6 and 5.7 respectively, and can be related to M_S (Table 1). The relatively low experimental value of M_S for $x = 12$ (72 emu/g) may be attributed to the presence of hematite (8%) in this sample (F12N0). In our experimental conditions, sample F11N0 shows the best composition for a non-substituted hexaferrite sample as it is monophasic with high M_S and high H_c values (Table 1).

Mössbauer spectra of powders with $x = 12, 11$ and 10 are shown in Fig. 3 (right) and their hyperfine parameters are reported in Table 2. It is observed that B_{hf} and

Table 2 Hyperfine parameters for all samples with different iron content

Site	Parameter	F12N0	F12N3	F11N0	F11N3	F10N0	F10N3
12k ↑	$\delta(\pm 0.01)$	0.35	0.35	0.35	0.35	0.35	0.35
Octahedral	$\Gamma(\pm 0.01)$	0.31	0.45	0.34	0.44	0.32	0.48
	$B_{hf}(\pm 0.1)$	41.0	41.6	41.5	41.7	40.6	42.0
	$\Delta(\pm 0.01)$	0.39	0.39	0.41	0.39	0.39	0.39
4f1 ↓	$\delta(\pm 0.02)$	0.27	0.25	0.26	0.25	0.26	0.27
Tetrahedral	$\Gamma(\pm 0.02)$	0.35	0.35	0.31	0.38	0.32	0.35
	$B_{hf}(\pm 0.2)$	48.7	48.9	49.0	48.9	48.3	49.1
	$\Delta(\pm 0.01)$	0.19	0.17	0.17	0.17	0.15	0.13
4f2 ↓	$\delta(\pm 0.05)$	0.50	0.45	0.40	0.44	0.38	0.41
Octahedral	$\Gamma(\pm 0.02)$	0.30	0.35	0.32	0.38	0.29	0.39
	$B_{hf}(\pm 0.2)$	51.3	51.3	51.6	51.4	50.8	51.6
	$\Delta(\pm 0.05)$	0.07	0.05	0.22	0.11	0.22	0.22
2a ↑	$\delta(\pm 0.03)$	0.24	0.21	0.22	0.19	0.21	0.23
Octahedral	$\Gamma(\pm 0.02)$	0.30	0.35	0.25	0.21	0.27	0.39
	$B_{hf}(\pm 0.3)$	51.3	51.5	51.3	51.5	50.1	51.1
	$\Delta(\pm 0.01)$	0.04	0.03	0.16	0.05	0.18	0.14
2b ↑	$\delta(\pm 0.01)$	0.27	0.27	0.26	0.27	0.25	0.25
Bi-pyramidal	$\Gamma(\pm 0.01)$	0.24	0.37	0.25	0.29	0.27	0.35
	$B_{hf}(\pm 0.1)$	40.9	40.9	41.3	41.1	40.5	41.5
	$\Delta(\pm 0.01)$	2.21	2.21	2.25	2.18	2.22	2.21

Isomer shift δ, quadrupole splitting Δ and line width Γ are in mm/s, hyperfine field B_{hf} is in Tesla

Fig. 4 Fe^{3+} occupancy of the five sites of $SrFe_xO_{19}$ for samples with different iron content (x). Theoretical values for the different sites are indicated by dashed lines

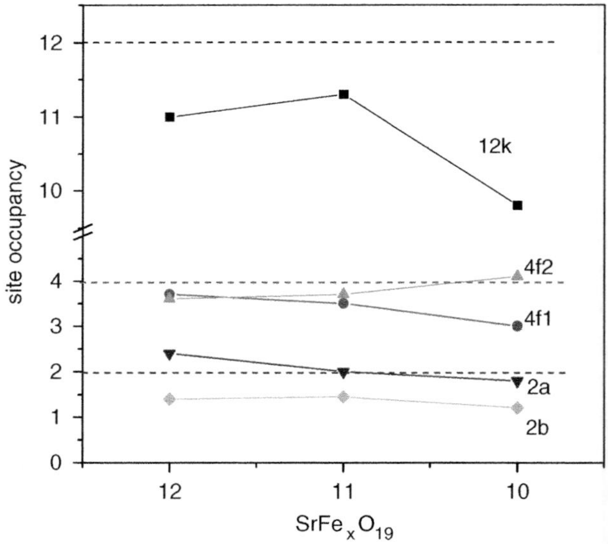

the line width of the 12k contribution increases with substitution, in agreement with an increasing distribution of environments around the 12k site.

The inherent randomness in site occupancy of the different ions and the presence of vacancies (due to non-stoichiometry) appear to be the responsible for the increase in line widths (see Γ in Table 2) whereas the increase in B_{hf} corresponding to 12k sites may be attributed to a perturbation of the superexchange 12k–O^{2}–4f2 and

12k–O^{2}–2a interactions due to the presence of Co^{2+} in the 4f2 and 2a sites. Co substitution in 4f2 sites increases B_{hf} on sites 12k due to its magnetic moment, lower than that of Fe^{3+} and opposite to this one. On the other hand, Co substitution in 2a sites decreases B_{hf} on sites 12k due to the common alignment of both magnetic moments. As the number of 4f2 sites in 12k environment doubles that of 2a sites, the net effect is an increase in B_{hf} corresponding to Fe^{3+} in sites 12k. On 4f2 sites δ decreases with Fe defect. This may be attributed to a decrease of Fe–O distance for 4f2 sites due to vacancy in 2b sites. This behavior is accompanied by an increase of the quadrupole splitting Δ. When substitution takes place, δ decreases for site 4f2 in samples with smaller Fe defect. This decrease is attributed to the presence of Nd^{3+} instead of Sr in the vicinity of the 4f2 site in relation with the electron density in the vicinity of this site. Similar results were reported by Lechevallier et al. for La^{3+} substitution in hexaferrites [13]. On the other hand, in samples with a larger Fe defect δ increases with substitution. This effect may be attributed to the Co^{2+} for Fe substitution in the neighborhood of sites 4f2, with the consequent effect on electron density on the 4f2 site environment.

In sites 2a, a slight tendency towards the diminution of δ with Fe defect is observed. As for 4f2 sites this effect may be attributed to the decrease of the Fe–O distance due to vacancy of Fe in 12k and 4f1 sites. However, the effect of vacancy in both places is not necessary additive. Hyperfine parameters from 4f1 and 2b sites are not affected neither by Fe defect nor by substitution.

4 Conclusions

Nd-Co substituted hexaferrites were prepared by the self combustion method with different iron content. Magnetic properties enhance with substitution in all samples. Samples F11N3 and F11N0 present the best magnetic properties followed by substituted samples F10N3 and F12N3 even when there is a considerable iron vacancy (near 9%). Mössbauer analysis shows that iron occupancy is related to magnetization values and that hyperfine parameters reflect the changes that take place in the hexaferrite structure upon Fe defect and substitution of Fe and Sr by Co and Nd, respectively.

Acknowledgements This work was partially supported by CONICET, SeCyT–UNC and UBA.

References

1. Smith, J., Wijn, H.P.J.: Ferrites. Philips Technical Library (1959)
2. Jonker, H., Wijn, H.P., Braun, P.B.: Phil. Tech. Rev. **18**, 145 (1956)
3. Kools, F., Morel, A., Grössinger, R., Le Breton, J.M., Tenaud, P.: J. Magn. Magn. Mater. **242–245**, 1270–1276 (2002)
4. Lechevallier, L., Le Breton, J.M., Morel, A., Tenaud, P.: J. Magn. Magn. Mater. **316**, e109–e111 (2007)
5. Taguchi, H., Minachi, Y., Masuzawa, K.: In: Proceedings of the Eighth International Conference on Ferrites, p. 405, Kyoto, Japan, 18–21 September 2000
6. Wang, J.F., Ponton, C.B., Harris, I.R.: J. Magn. Magn. Mater. **234**, 233–240 (2001)
7. Jacobo, S.E., Herme, C., Bercoff, P.G.: Influence of the iron content on the formation process of substituted Co-Nd strontium hexaferrite prepared by the citrate precursor method. J. Alloys Compd. (2009, in press)

8. Rezlezcu, N., Doroftei, C., Rezlezcu, E., Popa, P.D.: J. Alloys Compd. (2007). doi:10.1016/j.jallcom.2007.04.102
9. Brand, R.A.: Normos Programs. Laboratorium fuer Angewandte Physik, Universitaet Duisburg (1990)
10. How, H., Zuo, X., Wave, C.V.: IEEE Trans. Magn. **41**, 2349 (2005)
11. Bahadur, D., Fisher, W., Rane, M.V.: Mater. Sci. Eng., A **252**, 109–116 (1998)
12. Evans, B.J., Grandjean, F., Lilot, A.P., Vogel, R.H., Gérard, A.: J. Magn. Magn. Mater. **67**, 123–129 (1987)
13. Lechevallier, L., Le Breton, J.M., Teillet, J., Morel, A., Kools, F., Tenaud, P.: Physica B **327**, 135–139 (2003)

Hyperfine Interact (2010) 195:213–217
DOI 10.1007/s10751-009-0122-6

Quasicrystalline phase formation in the mechanically alloyed $Al_{70}Cu_{20}Fe_{10}$

S. N. de Medeiros · S. Cadore · H. A. Pereira · I. A. Santos · C. C. Colucci · A. Paesano Jr.

Published online: 25 September 2009

Abstract In the present work, the formation of the $Al_{70}Cu_{20}Fe_{10}$ icosahedral phase by mechanical alloying the elemental powders in a high-energy planetary mill was investigated by X-ray diffraction and Mössbauer spectroscopy. It was verified that the sample milled for 80 h produces an icosahedral phase besides Al(Cu, Fe) solid solution (β-phase) and Al_2Cu intermetallic phase. The Mössbauer spectrum for this sample was fitted with a distribution of quadrupole splitting, a doublet and a sextet, revealing the presence of the icosahedral phase, β-phase and α-Fe, respectively. This compound is not a good hydrogen storage. The results of the X-ray diffraction and Mössbauer spectroscopy of the sample milled for 40 h and annealed at 623°C for 16 h shows essentially single i-phase and tetragonal Al_7Cu_2 Fe phase.

Keywords Al–Cu–Fe quasicrystal · X-ray diffraction · Mössbauer spectroscopy and mechanical alloying

1 Introduction

The quasicrystalline state is a metastable state between the amorphous and the crystalline phases first reported in 1984 [1]. Quasicrystalline materials have long-range rotational order but lack of long range translational order and are not truly periodic. Besides being theoretically interesting due to their complex atomic structure, the unique properties of quasicrystalline materials (low electrical and thermal conductivities, unusual optical properties, oxidation resistance, high hardness, etc.) also make them interesting for many applications. Up to now, quasicrystalline

S. N. de Medeiros (✉) · S. Cadore · H. A. Pereira ·
I. A. Santos · C. C. Colucci · A. Paesano Jr.
Departamento de Física, Universidade Estadual de Maringá, Avenida Colombo,
5790-CEP 87020-900, Maringá, Paraná, Brazil
e-mail: snm@dfi.uem.br

phases have been observed in over a 100 different metal alloy systems [2]. Al–Cu–Fe quasicrystalline alloys are more attractive due to their lack of toxicity, easy availability and low costs of their alloying elements [3]. Quasicrystalline alloys are usually prepared by rapid or conventional solidification from the melt. However, in the last few years, it has been reported that quasicrystalline alloys can also be prepared by mechanical alloying (MA) [4]. Studies on the Al–Cu–Fe system prepared by MA have been reported in a number of papers [5–8]. Asahi et al., in the composition $Al_{65}Cu_{20}Fe_{15}$ observed that the formation of the icosahedral phase (i-phase) directly by MA [5], whereas Eckert et al. could produce it only after subsequent annealing [6]. V. Sirinivas et al. have studied the i-phase development in mechanically alloyed $Al_{70}Cu_{20}Fe_{10}$ powder mixture [7]. In a high-energy mill, 40 h milling produces a structure consisting of the icosahedral phase along with the cubic β-phase and a small amount of the Al_2Cu intermetallic phase. Ajay Gupta et al. investigated the effect of high pressure on the short-range order in the stable $Al_{63.5}Cu_{24}Fe_{12.5}$ quasicrystal using Mössbauer spectroscopy (MS) [8]. However, no MS results on $Al_{70}Cu_{20}Fe_{10}$ have been reported so far. In the present work, the formation of icosahedral phase in $Al_{70}Cu_{20}A_{10}$by mechanical alloying of crystalline elemental powders and MA with subsequent annealing were investigated by X-ray diffraction and Mössbauer spectroscopy.

2 Experimental

High purity crystalline powders of Al, Cu and Fe with the nominal composition of $Al_{70}Cu_{20}Fe_{10}$ were mechanically alloyed in a planetary ball mill (Fritsch Pulverisette), with the vial and balls made of hardened steel. The ball to powder weight ratio was maintained at 10:1 and the rotational speed was controlled at 300 rad s^{-1}. Some milled samples were annealed under inert atmosphere (Ar) for different time intervals and temperatures. The X-ray diffraction (XRD) was performed at room temperature, using a Semmens D500 X-ray diffractometer, in the Bragg–Brentano geometry and Cu Kα radiation. The Mössbauer spectroscopy were performed in the transmission geometry, using a conventional Fe^{57} Mössbauer spectrometer operating in a constant acceleration mode. The quasicrystalline sample was submitted to isothermal hydrogenation using the Sieverts technique. The amount of gas absorbed was calculated by the difference in pressure noted at the begin and end of the process.

3 Results and discussion

The Fig. 1 shows the XRD patterns for $Al_{70}Cu_{20}Fe_{10}$ samples after different milling times, which were produced using a high-energy planetary mill. The XRD patterns for milling times of 20, 40, and 60 h show the presence of residual Al, Fe along with an Al(Cu, Fe) solid solution (β-phase) and the Al_2Cu intermetallic phase [7]. In these cases, it is clear from the peak intensities that β-phase is the major phase and that the fraction of Al_2Cu increases with the milling time up to 60 h. Milling up to 80 h results in the evolution for the quasicrystalline phase (i-phase), which is associated with the dissolution of Al and Al_2Cu phases [7], besides the earlier

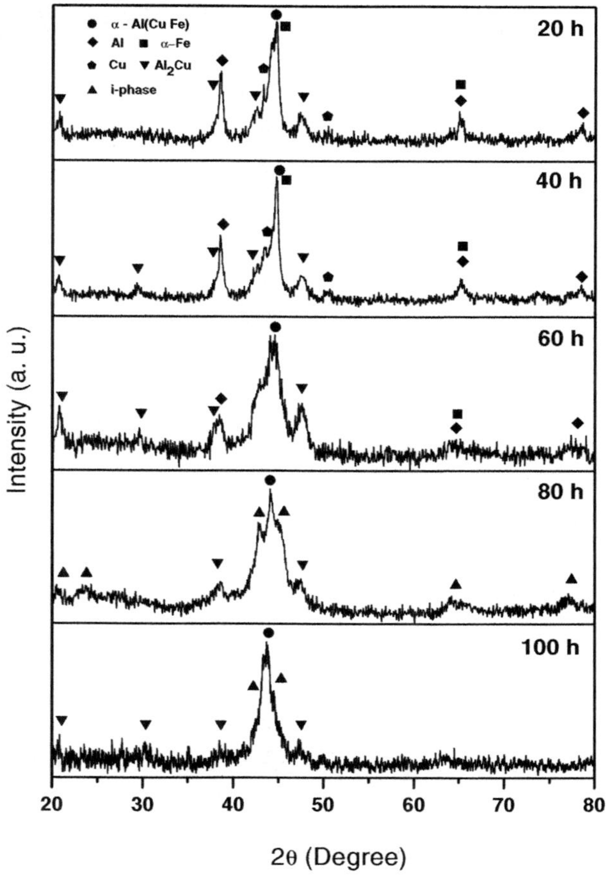

Fig. 1 XRD patterns for the as-milled $Al_{70}Cu_{20}Fe_{10}$ after different milling times

observed phases. The sample milled for 100 h consisted of the β-phase the Al_2Cu compound and i-phase. The Mössbauer spectrum for the sample milled for 80 h can be observed in the Fig. 2. It was fitted with a distribution of quadrupole splitting, a doublet and a sextet, attributed to the i-phase, β-phase and α-phase, respectively. The distribution reveals a variation in the neighborhood of the iron atoms in the i-phase. The parameters for this phase obtained are in good agreement with previously reported results [8]. The X-ray diffraction patterns of the sample milled for 40 h and heat-treated at 623°C for 16 h shows essentially single i-phase and tetragonal Al_7Cu_2Fe phase (Fig. 3).The Mössbauer spectrum for this sample (Fig. 4) was fitted with a distribution of quadrupole splitting and a doublet, revealing the presence of the i-phase and possibility the Al_7Cu_2Fe phase, respectively [9]. The results of the fitting of the Mössbauer spectra for the samples analyzed are showed in the Table 1. This sample was placed in an atmosphere of hydrogen. The kinetics of the reaction at 400°C and pressure of 3 atm was very low to the sample milled for 80 h. After of 24 h, the quantity incorporated of gas is very low. The concentration of the sample were of 0.3 atoms per unit cell. Since a lattice uses to expand under effective hydrogen absorption, and no shift towards left is observed for the peaks of the i-phase (Fig. 5). Therefore, there was not hydride formation from the i-phase. Consequently, the $Al_{70}Cu_{20}Fe_{10}$ compound is not a good hydrogen storage medium.

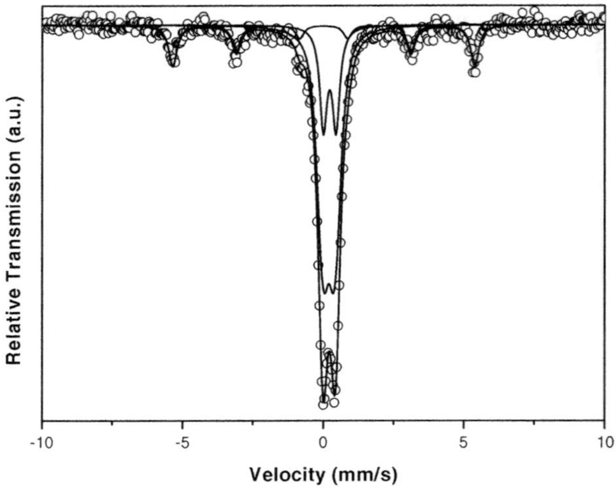

Fig. 2 Mössbauer spectrum for the 80 h milled sample

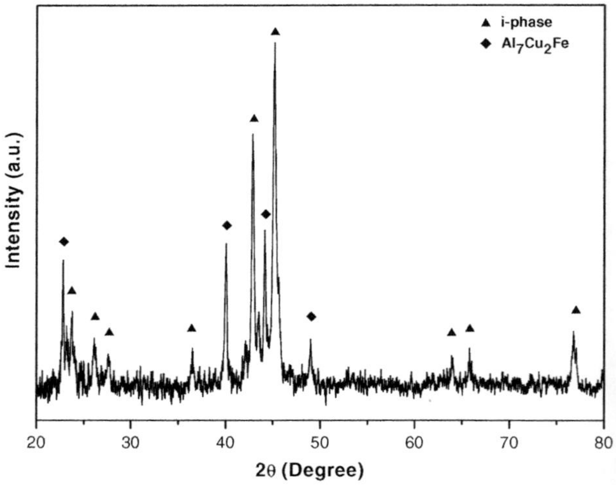

Fig. 3 The X-ray diffraction patterns of the sample milled for 40 h and heat-treated at 623°C for 16 h

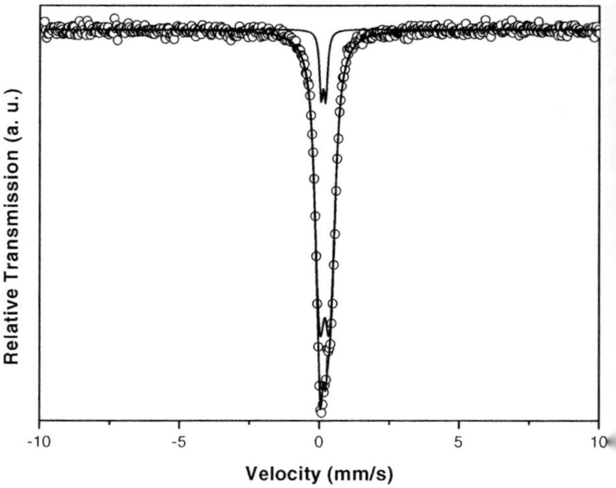

Fig. 4 Mössbauer spectrum for the sample milled for 40 h and heat-treated at 623°C for 16 h

Table 1 Mössbauer hyperfine parameters and relatives areas milled and annealed samples

Milling times		δ(mm/s)	QS (mm/s)	Γ (mm/s)	Bhf (T)	Area (%)
80 h	Sextet (α phase)	0.01	–	0.35	33.0	18.5
	Doublet (β-phase)	0.22	0.44	0.31	–	19.6
	QS DIST. (i-phase)	0.19	0.44	0.35	–	61.9
40 h	Doublet (Al_7Cu_2Fe)	0.14	0.16	0.28	–	8.7
623°C	QS DIST. (i-phase)	0.19	0.36	0.35	–	91.3

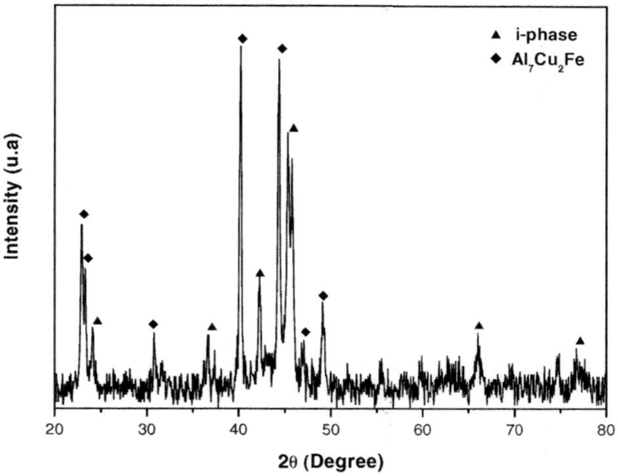

Fig. 5 The X-ray diffraction patterns of the sample after hydrogenation

4 Conclusions

An icosahedral quasicrystalline phase has been synthesized by mechanical alloying of $Al_{70}Cu_{20}Fe_{10}$ samples. Annealing the mechanically alloyed $Al_{70}Cu_{20}Fe_{10}$ powder gives rise to an i-phase and a tetragonal Al_7Cu_2Fe phase. Hyperfine properties of the quasicrystalline mechanically alloyed $Al_{70}Cu_{20}Fe_{10}$ powders were determined by Mössbauer spectroscopy. Studies concerning H_2 incorporation on Al–Cu–Fe quasicrystal phase reveled that this sample is not a good hydrogen storage medium.

References

1. Schechtman, D., Blech, I., Gratias, D., Cahn, J.W.: Phys. Rev. Lett. **53**, 1951 (1984)
2. Huttunen-Saarivirta, E.: J. Alloys Compd. **363**, 150 (2004)
3. Bloom, P.D., Baikerikar, K.G., Otaigbe, J.U., Sheares, V.V.: Mater. Sci. Eng. **294–296**, 156 (2000)
4. Cohen, N.S., Pankhurst, O.A., Barquin, L.F.: J. Phys. Condens. Matter. **11**, 8839 (1999)
5. Asahi, N., Maki, T., Masumoto, S., Sawai, T.: Mater. Sci. Eng. A **181–182**, 841 (1994)
6. Eckert, J., Schultz, L., Urban, K.: Eur. Phys. Lett. **13**, 349 (1991)
7. Srinivas, V., Barua, P., Murty, B.S.: Mater. Sci. Eng. A **294–296**, 65 (2000)
8. Gupta, A., Neelima, P., Vijaykumar, V., Godwal, B.K.: J. Non-Cryst. Solids **334–335**, 343 (2004)
9. Faudot, F., Quivy, A., Calvayrac, Y., Gratias, D., Harmelin, M.: Mater. Sci. Eng. A **133**, 383 (1991)

Hyperfine Interact (2010) 195:219–226
DOI 10.1007/s10751-009-0123-5

Structural and magnetic properties of Fe–Ni mecanosynthesized alloys

J. F. Valderruten · G. A. Pérez Alcázar · J. M. Greneche

Published online: 25 September 2009

Abstract $Fe_{100-x}Ni_x$ samples with $x = 22.5$, 30.0 and 40.0 at.% Ni were prepared by mechanical alloying (MA) with milling times of 10, 24, 48 and 72 h, a ball mass to powder mass (BM/PM) ratio of 20:1 and rotation velocity of 280 rev/min. Then the samples were sintered at 1,000°C and characterized by X-ray diffraction (XRD) and transmission Mössbauer spectrometry (TMS). From the refinement of the X ray patterns we found in this composition range two crystalline phases, one body centered cubic (BCC), one face centered cubic (FCC) and some samples show FeO and Fe_3O_4 phases. The obtained grain size of the samples shows their nanostructured character. Mössbauer spectra were fitted using a model with two hyperfine magnetic field distributions (HMFDs), and a narrow singlet. One hyperfine field distribution corresponds to the ferromagnetic BCC grains, the other to the ferromagnetic FCC grains (Taenite), and the narrow singlet to the paramagnetic FCC grains (antitaenite). Some samples shows a paramagnetic doublet which corresponds to FeO and two sextets corresponding to the ferrimagnetic Fe_3O_4 phase. In this fit model we used a texture correction in order to take into account the interaction between the particles with flake shape and the Mössbauer γ-rays.

Keywords Invar alloys · Magnetic nanostructured powders · Mechanical alloying · Mössbauer spectrometry · X-ray diffraction

PACS 64.70.ND · 76.80._Y · 75.50.TT · 61.05.CP

J. F. Valderruten (✉) · G. A. Pérez Alcázar
Departamento de Física, Universidad del Valle, A. A. 25360, Cali, Colombia
e-mail: jfvalder@yahoo.com

J. M. Greneche
Laboratoire de Physique del'Etat Condensé, UMR CNRS 6087, Université du Maine, 72085 Le Mans, Cedex 9, France

1 Introduction

Fe–Ni system has attracted the attention of researchers because it originates the formation of a great number of alloys with special thermal and magnetic properties [1–14]. The use of high energy mills and other new technological methods of preparation joined, in some cases, by heat treatments, enable to obtain samples with different structures and novel properties. This is the current field of the nanotechnology, which allows preparing nanostructured Fe–Ni alloys and other different materials [15, 16]. Their small grain size gives rise to a high ratio between the number of atoms located in the grain boundary and those in the interior. In addition, these materials are very interesting from the magnetic point of view because their reduced grain size approaches to the size of a magnetic domain and offer the possibility to eliminate the influence of the domain walls. It is very well established that the intermetallic compounds prepared by mechanical alloying (MA) have a high structural disorder and are unstable [17–19]. These factors implied unusual physical properties, which are very different compared with those of bulk microcrystalline materials. For this reason, several studies have been carried out investigating the structural properties of the Fe–Ni alloys prepared by MA [6, 7]. Kaloshkin et al. [12], after annealing at 650°C samples of the $Fe_{1-x}Ni_x$ system prepared by MA, showed that the concentration ranges of the phase existence shift to the low nickel content side of the phase diagram of melted alloys, and additionally showed the formation of a non-ferromagnetic alloys very unusual for the Fe–Ni system. More recently Valderrutem et al. [20] have reported a MS and XRD study of samples from the $Fe_{100-x}Ni_x$ system, with x varying from 22.5 up to 40, prepared by MA. They found that all the samples present the coexistence of some phases: the ferromagnetic Fe–Ni BCC, the ferromagnetic Fe–Ni FCC (taenite) and the paramagnetic Fe–Ni FCC phase (antitaenite). They proposed a new fit model which includes two HMFDs which correspond to the two ferromagnetic phases and a paramagnetic site which corresponds to the antitaenite phase.

The aim of the present work is to compare the phase composition and hyperfine properties of three samples of the $Fe_{1-x}Ni_x$ system, obtained by mechanical alloying and then sintering. According to our knowledge, no work has been reported previously on these alloys prepared by this route.

2 Experimental method

Pure carbonyl Fe powder (99.9%) and electrolytic Ni powder (99.9%) were used as the starting materials. $Fe_{100-x}Ni_x$ samples with $22.5 \leq x \leq 40.0$ were alloyed under vacuum for 10, 24, 48 and 72 h by MA in a planetary ball mill (Fritsch "Pulverisette 5") using hardened chromium steel vials and balls. The ball mass-to-powder mass (BM/PM) ratio was about 20:1. Part of the final powders were used for be sintered at 1,000°C during 1 h. The MA powders and those obtained after sintering were characterized by X-ray diffraction (XRD) using a Rigaku diffractometer with the CuKα radiation and Mössbauer spectrometry (MS) by collecting the spectra at room temperature (RT) with a conventional transmission spectrometer using a ^{57}Co (Rh) source and an α-Fe foil as calibration sample. The XRD patterns were refined by the

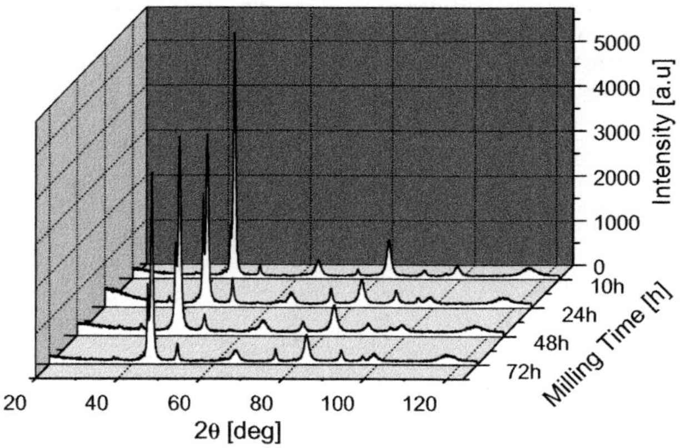

Fig. 1 XRD patterns for $Fe_{70.0}$–$Ni_{30.0}$ samples alloyed during 10, 24, 48 and 72 h

Rietveld Method using the MAUD program [21] and the Mössbauer spectra were fitted by using the MOSFIT program [22].

3 Experimental results and discussion

Figure 1 illustrated the XRD patterns of the 30.0 at.% Ni samples milled during different times and then sintered. It can be noted the coexistence of two phases in all the studied samples, BCC and FCC and no appreciable changes with the milling time can be observed. In some sintered samples small quantities of FeO and Fe_3O_4 oxides appear. Different microstructural models were proposed for the refinement of the patterns and the better description is obtained assuming pseudo-cubic grains for BCC and FCC phases.

Figure 2 shows the behavior of the volumetric percentage of the detected phases as a function of the Ni concentration. The coexistence of the BCC and FCC phases is obtained for all the samples and of the FeO and Fe_3O_4 oxides in some ones.

The comparison between the obtained grain sizes for the BCC and FCC phases of the samples obtained by MA and those sintered is shown in Fig. 3. Clearly it can be noted that the grain sizes of the sintered samples (between 20 and 40 nm) are larger than that obtained for the AM samples (between 8 and 20 nm). This enhancement of the grain size is induced by the thermal treatment. Besides, the grain size systematically decreases between 10 and 48 h of mill and then it increases between 48 and 72 h. The reduction of the grains of the phases BCC and FCC can be related with the presence of oxides, which increase the fragility of the particles of the powder.

Figure 4 shows the behavior of the volumetric percentage of the detected phases as a function of the milling time for the samples with 22.5 and 40.0 at.% Ni MA and sintered. It can be noted that the BCC and FCC phases coexist for the sample with $x = 22.5$, the content of which decreases and increases as the Ni content increases, respectively. For the sample with $x = 40.0$ the BCC phase is not

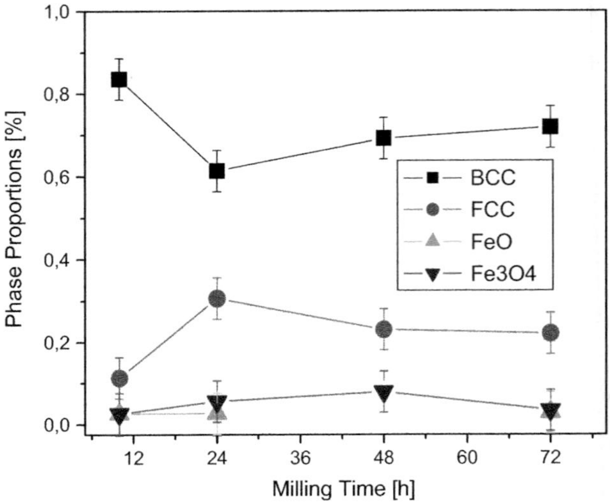

Fig. 2 Phase proportions obtained from the XRD patterns of the $Fe_{70}Ni_{30}$ MA and then sintered samples

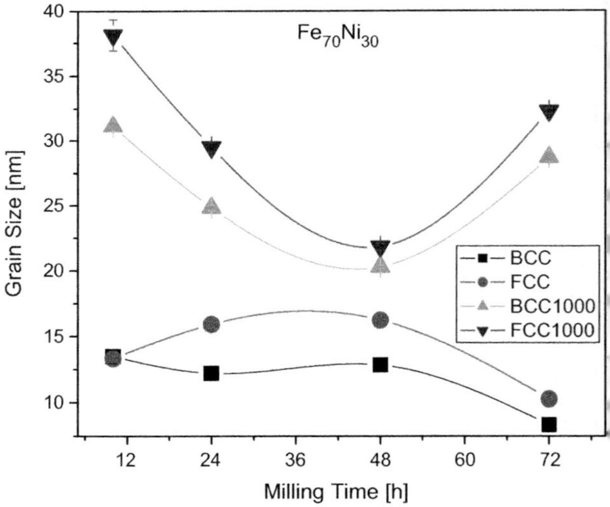

Fig. 3 Grain size vs. milling time of the Fe–Ni (BCC) and Fe–Ni (FCC) phases obtained from the refinement of the XRD patterns of MA, and MA plus sintering $Fe_{70}Ni_{30}$ samples

detected. Some samples show the FeO and Fe_3O_4 oxides, which must be due to fails in the Argon flux during the thermal treatment and/or trapped air inside the powders.

The RT Mössbauer spectra which are presented in Fig. 5 were fitted following a recent method proposed for this type of samples [20], which include two hyperfine magnetic field distributions (HMFDs) and a singlet, in addition to a small preferential orientation of the Fe magnetic moments. Some samples shows the paramagnetic doublet corresponding to FeO and two sextets corresponding to the ferrimagnetic

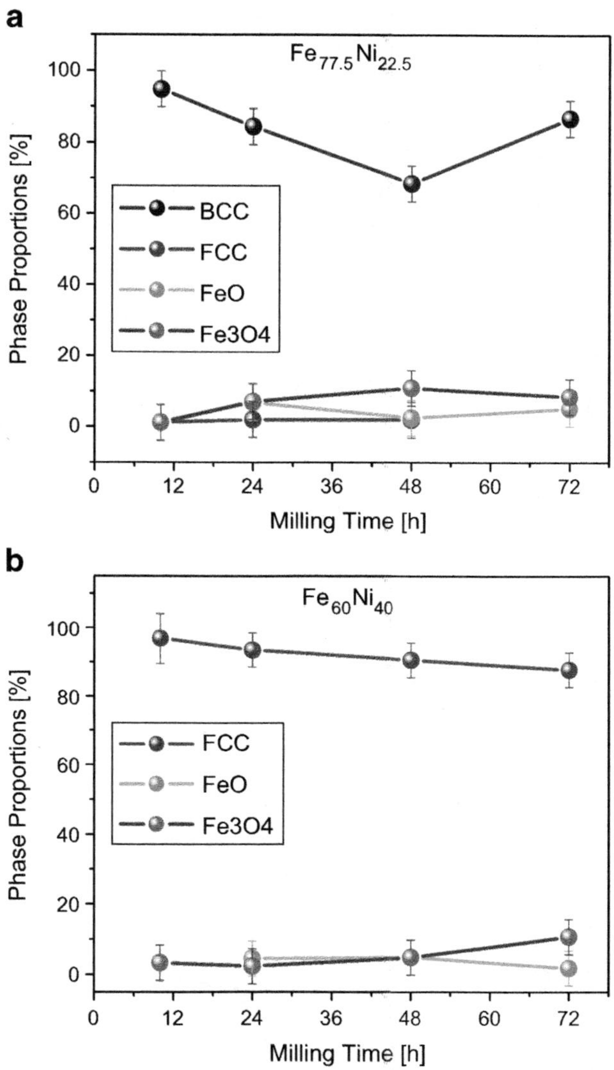

Fig. 4 Phase proportions vs. milling time obtained from XRD patterns for the $Fe_{77.5}Ni_{22.5}$ and $Fe_{60.0}Ni_{40.0}$ sintered samples

Fe_3O_4 phase in according with XRD results. The best agreement between the experimental and the calculated spectra was obtained when a maximum field of about 32T was taking for the FCC HMFD, a small overlap between the HMFDs was permitted and two different isomer shifts values were used. Finally, the single line is due to the FCC Fe–Ni paramagnetic grains with high Fe content [8, 11, 21]. It can be noted also that for the sample with lower Ni content (22.5 at.%) the bigger contribution to the spectral area corresponds to the HMFD of the BCC phase. For the case of the sample with bigger Ni content (40 at.%) the mayor spectral area corresponds to the HMFD of the FCC phase. The MS results obtained for the samples with $x = 22.5$ and 30 are in according with those obtained by XRD. For the $x = 40$ sample the BCC ferromagnetic phase was detected by MS and was not

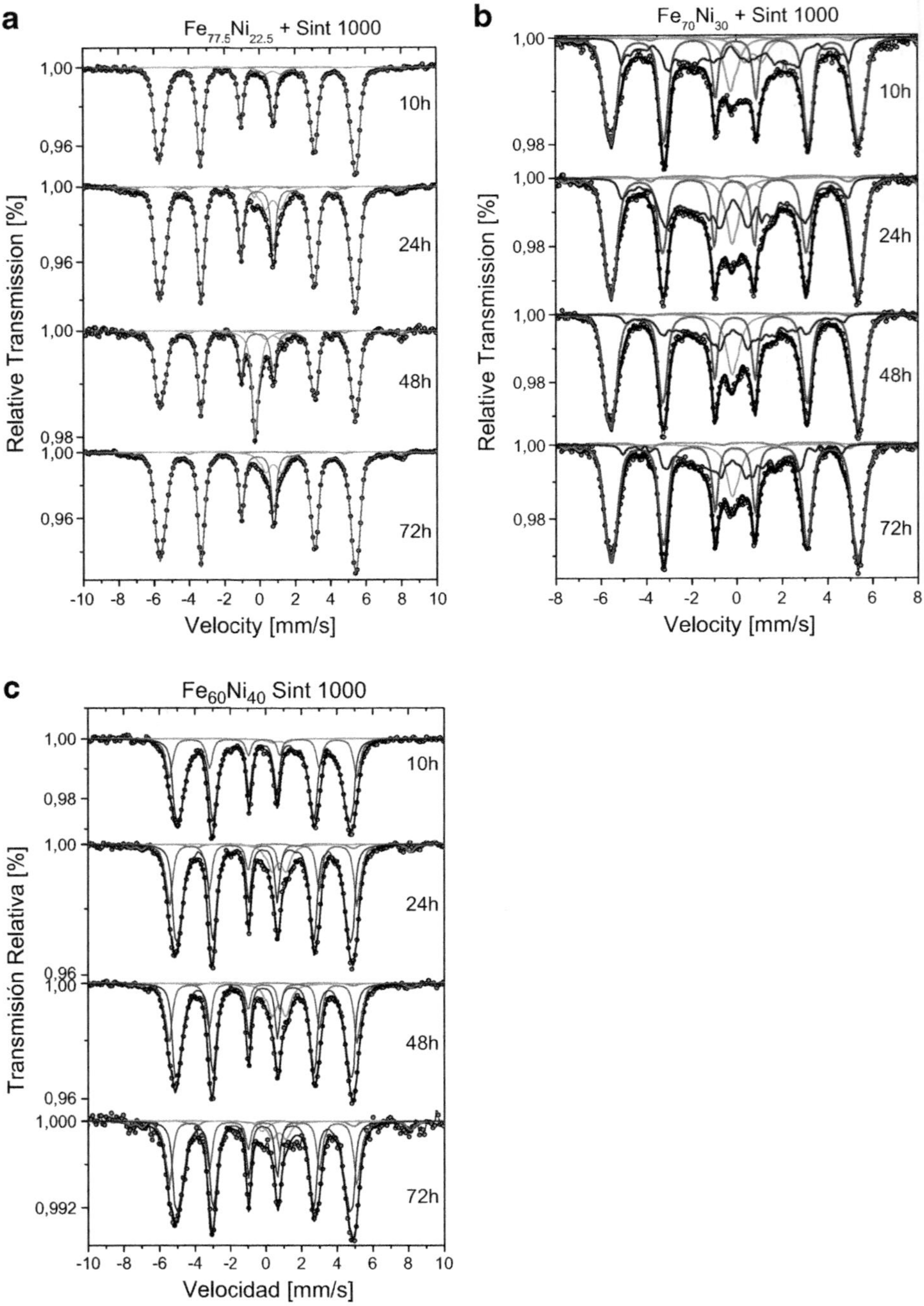

Fig. 5 RT Mössbauer spectra of $Fe_{77.5}Ni_{22.5}$, $Fe_{70.0}Ni_{30.0}$ and $Fe_{60.0}Ni_{40.0}$ sintered samples

detected by XRD. Then this phase appears with a volume fraction lower than 2%, but this phase presents more Fe content than that of the FCC phase explaining in this way its appearing in the MS.

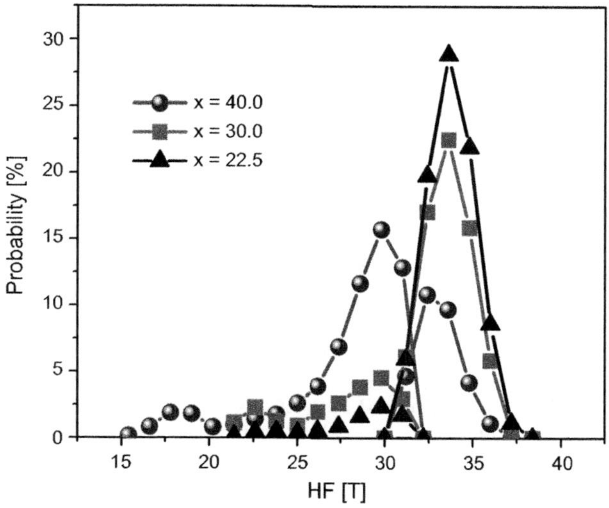

Fig. 6 HMFDs obtained for the $Fe_{100-x}Ni_x$ samples with 22.5, 30.0 and 40.0 milled during 10 h and sintered. The lines are a guide for the eyes

Figure 6 shows the HMFDs estimated from the proposed fit model and after considering the texture effect. It can be noted two principal domains: the high field component is attributed to the BCC Fe–Ni ferromagnetic grains [9], while the other one (lower field values) is ascribed to the FCC Fe–Ni ferromagnetic grains [9].

4 Conclusions

The XRD and MS studies of samples of the $Fe_{100-x}Ni_x$ system with $x = 22.5$, 30.0 and 40.0 prepared by MA, and MA and then sintering show that the structural and magnetic properties of samples prepared by these two routes are very similar. However, two differences can be detected as a consequence of the heat treatment: it stabilize the FCC phase and increase the grain size. These small differences permit to conclude that the MA samples of the Fe–Ni system will preserve their magnetic and structural properties if they will be compacted and sintered in order to manufacture pieces for different applications.

Acknowledgements The authors would like to thank ECOS-NORTE and Colciencias, Colombian Agency, and Universidad del Valle, for financial support given to the Excellence Center for Novel Materials (ECNM), under contract no.043–2005. They are very grateful to A.M. Mercier from UMR CNRS 6010 Université du Maine for performing XRD measurements.

References

1. Johnson, C.E., et al.: Proc. Phys. Soc. Lond. **81**, 1079 (1963)
2. Gonser, Y., et al.: J. Magn. Magn. Mater. **10**, 244 (1979)
3. Window, B.: J.Phys. F **4**, 329 (1974)
4. Billard, L., et al.: Solid State Commun. **17**, 113 (1975)
5. Hesse, J., Müller, J.B.: Solid State Commun. **22**, 637 (1977)
6. Kuhrt, C., Schultz, L.: J. Appl. Phys. **73**, 1975 (1993)
7. Kuhrt, C., Schultz, L.: J. Appl. Phys. **73**, 6588 (1993)

8. Scorzelli, R.B.: Hyperfine Interact. **110**, 143 (1997)
9. Hong, L., Fultz, B.: J. Appl. Phys. **79**, 3946 (1993)
10. Baldokhin, Yu.V., Tcherdyntsev, V.V., Kaloshin, S.D., Kochetov, G.A., Pustov, Yu.A.: J. Magn. Magn. Mater. **203**, 313 (1999)
11. Lapina, T.M., Shabashov, V.A., Sagaradze, V.V., Arbuzov, V.L.: Mater. Sci. Forum **294–296**, 767 (1999)
12. Kaloshin, S.D., Tcherdyntsev, V.V., Tomilin, I.A., Baldokhin, Yu.V., Shelekhov, E.V.: Physica B **299**, 236 (2001)
13. Tcherdyntsev, V.V., Kaloshin, S.D., Tomilin, I.A., Shelekhov, E.V., Baldokhin, Yu.V.: Nanostruct. Mater. **12**, 139 (1999)
14. Luborsky, F.E.: J. Appl. Phys. **32**, 171S (1961)
15. Herzer, G.: J. Magn. Magn. Mater. **112**, 258 (1992)
16. Benjamin, J.S.: Sci. Am. **234**, 40 (1976)
17. Hellstern, E., Schultz, L.: J. Appl. Phys. **63**, 1408 (1988)
18. Suryanarayana, C.: Prog. Mater. Sci. **46**, 1–184 (2001)
19. Hellstern, E., Fecht, H.J., Fu, Z., Johnson, W.L.: J. Appl. Phys. **65**, 305 (1989)
20. Valderutem, J.F., Pérez Alcázar, G.A., Greneche, J.M.: J. Phys. Condens. Matter **20**, 485204 (2008)
21. Lutterotti, L., Matthies, S., Wenk, H.R.: In: Proceedings of the Twelfth International Conference on Texture of Materials (ICOTOM-12), vol. 2, p. 1599 (1999)
22. Varret, F., Teillet, J.: Unpublished Mosfit Program (1976)

Hyperfine Interact (2010) 195:227–233
DOI 10.1007/s10751-009-0130-6

Mössbauer study on $Zn_{1-x}Fe_xO$ semiconductors prepared by high energy ball milling

L. C. Damonte · M. Meyer · L. Baum · L. A. Mendoza-Zélis

Published online: 11 November 2009

Abstract We present the preparation of massive $Zn_{1-x}Fe_xO$ ternary oxides using the mechanical mill. The Fe atom is a particular dopant since it presents two different oxidation states which allow us to vary the starting materials: Fe_2O_3, α-Fe or FeO. Parameters such as initial concentrations, atmosphere and milling times were varied. X-ray diffraction and ^{57}Fe Mössbauer spectrometry (MS) were applied in order to analyze the structure evolution and iron incorporation in the wurtzite crystalline structure with milling time. At final stages, Fe atoms seem to be incorporated in the ZnO structure for those samples milled under Ar atmosphere. In all cases, two paramagnetic components, attributed to Fe atoms in both valence states, were observed by MS.

Keywords Mechanical milling · Mössbauer spectroscopy · Fe-doped ZnO · Diluted magnetic semiconductors

1 Introduction

During the last years, the wide bandgap semiconductor ZnO doped with small quantities of magnetic atoms (Mn, Co, Fe and Ni) has attracted the attention since it may show room temperature ferromagnetism. However, the mechanism responsible of such singular ferromagnetic behaviour is not yet well understood and controversial experimental evidences of the magnetic character were reported [1–6].

L. C. Damonte · M. Meyer · L. Baum · L. A. Mendoza-Zélis
Departamento de Física, Universidad Nacional de La Plata,
C.C.67 (1900), La Plata, Argentina

L. C. Damonte (✉)
IDF, Departamento de Física Aplicada, Universidad Politécnica de Valencia,
Camí de Vera s/n, 46071, Valencia, Spain
e-mail: damonte@fisica.unlp.edu.ar

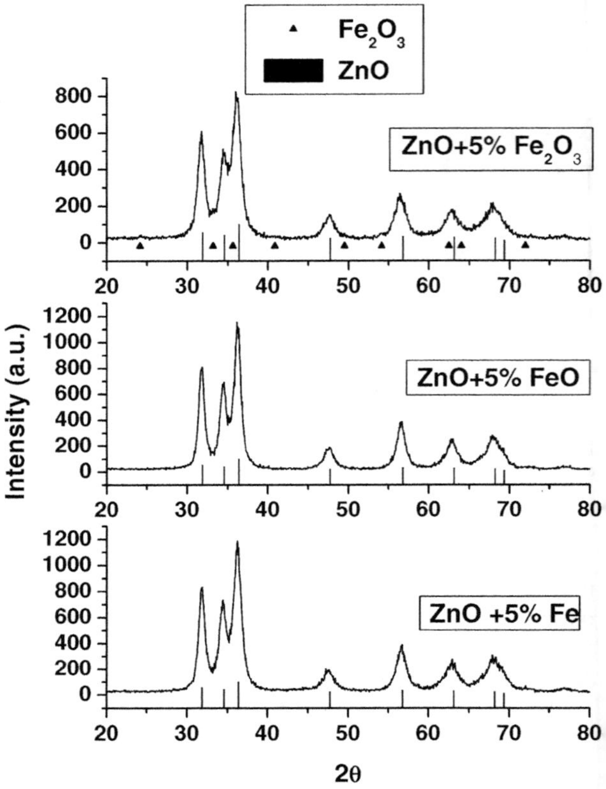

Fig. 1 XRD patterns for ZnO powders with 5 at.% α-Fe, FeO and Fe_2O_3 after 16 h of milling under Ar atmosphere. *Column bar* indicates the diffraction peaks for ZnO hexagonal structure ($P6_3mc$), *filled triangle* corresponds to Fe_2O_3 phase

We initiated our work in these diluted magnetic semiconductors by preparing Co-doped ZnO samples [7] and massive $Zn_{1-x}Fe_xO$ ternary oxides [8] using the mechanical mill.

The Fe atom is a particular dopant since it presents two different oxidation states which allow us to vary the starting materials: Fe_2O_3, α-Fe or FeO. On the other hand, high energy ball milling has proved to be a useful and versatile solid-state powder processing technique [9] which has been previously used with similar aims [10, 11].

Continuing that work and contributing to the comprehension of the magnetic behaviour of these semiconductors, we investigated the preparation of massive ternary oxides $Zn_{1-x}Fe_xO$ using mechanical work in Ar atmosphere and different starting powder mixtures.

Progressive iron incorporation in the wurtzite crystalline structure was studied by X-ray diffraction. ^{57}Fe Mössbauer spectrometry was applied in order to characterize the different ion phases formed and their magnetic properties.

2 Experimental

Samples from mixtures of ZnO (Alfa Aesar, Johnson Matthey Co., 99.99) with 5 and 10 at.% of α-Fe (Merk, 99.5) or FeO (Sigma Aldrich, 99.9, −10 mesh) or Fe_2O_3 (Johnson Matthey Co., 99.99, −15 mesh) powders were prepared. All samples were

Table 1 Lattice parameter (a, b) for doped-ZnO after 16 h of milling obtained from Rietveld refinement (determined from the fit with best agreement factors)

Sample	a(Å)	b(Å)
ZnO + Fe	3.249_1	5.208_1
ZnO + FeO	3.251_1	5.208_1
ZnO + Fe_2O_3	3.271_1	5.206_1
ZnO as received	3.248_1	5.204_1
ZnO m.m. 1 h	3.251_1	5.205_1
ZnO m.m. 5 h	3.252_1	5.204_1

Results for non milled and milled pure ZnO (m.m. stands for mechanical milled) are included for comparison [8]

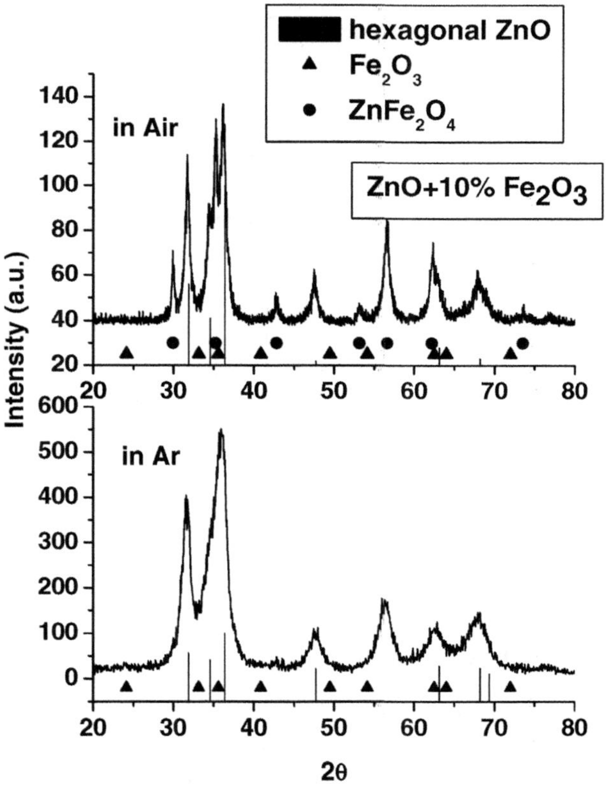

Fig. 2 XRD patterns for ZnO powders with 10 at.% Fe_2O_3 after 16 h of milling under air (*top*) or Ar (*bottom*) atmosphere. *Column bar* indicates the diffraction peaks for ZnO hexagonal structure (P6₃mc), *filled triangle* corresponds to Fe_2O_3 phase and *filled circle* corresponds to spinel $ZnFe_2O_4$ phase

manipulated in a Controlled Atmosphere Chamber (O_2 content less than a few ppm), introduced in a cylindrical steel milling chamber together with one steel ball (ϕ = 12 mm), filled with Ar at 0.2 MPa and sealed with an O ring. The ball to sample mass ratio was 11.5:1 and progressive milling was carried on using a horizontal oscillatory mill Retsch, at a fixed oscillation frequency of 32 Hz. X-ray diffraction patterns were obtained with CuK_α radiation in the $20° \le 2\theta \le 80°$ range at 0.02°/s using a Philips PW1710 diffractometer. ^{57}Fe Mössbauer spectra were measured using a conventional constant acceleration spectrometer at room temperature employing a ^{57}Co*Rh* source, in transmission geometry. All quoted isomer shifts are given relative to α-Fe.

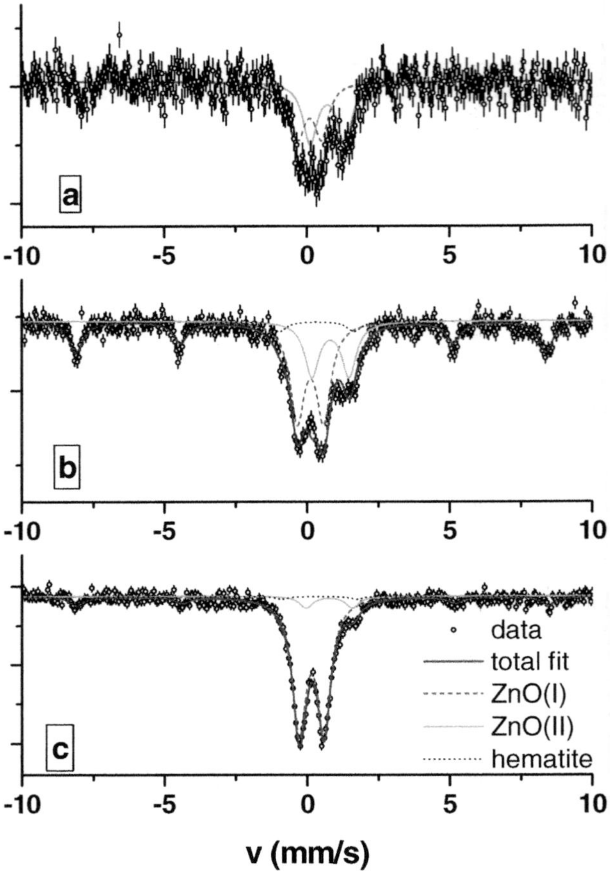

Fig. 3 Mössbauer spectra for: **a** ZnO+5 at.% FeO, **b** ZnO+5 at.% Fe_2O_3, and **c** ZnO+10 at.% Fe_2O_3 milled in Ar atmosphere

3 Results and discussion

3.1 XRD measurements

For all samples, as milling proceeds two main features can be observed: broadening of diffraction peaks due to grain size reduction and progressive disappearance of the initial added Fe-based materials. Figure 1 shows the resulting X-ray patterns for ZnO powders samples with 5 at.% α-Fe, FeO and α-Fe_2O_3 after 16 h of milling. The characteristic diffraction peaks of ZnO hexagonal structure ($P6_3mc$) are observed for all samples. For those samples milled with hematite, a slight contribution of this phase can still be appreciated. Lattice parameters obtained from Rietveld refinements show no significant changes with iron content with the exception of a slight increase in parameter *a* (Table 1) a similar behavior observed previously for milled pure ZnO [8] .

Another sample of ZnO with 10 at.% Fe_2O_3 was also prepared in Ar atmosphere, in order to compare with previous results done under air conditions [9]. While after prolonged mechanical milling in air the spinel structure $ZnFe_2O_4$ forms, instead under Ar, the wurtzite structure is obtained (Fig. 2).

Table 2 Fitted hyperfine parameters for the Mössbauer spectra of 16 h ball-milled samples

Sample	Phase or site	Isomer shift δ (mm/s)	Quadrupole splitting Δ or shift 2ε (mm/s)	Line width Γ (mm/s)	Hyperfine magnetic field (T)	Absorption area proportion (%)
ZnO+5%FeO	ZnO(I)	0.25 ± 0.03	0.82 ± 0.05	0.60 ± 0.06	–	46 ± 17
	ZnO(II)	0.87 ± 0.02	1.20 ± 0.04	0.60 ± 0.06	–	54 ± 17
ZnO+5%Fe_2O_3	ZnO(I)	0.26 ± 0.01	0.92 ± 0.02	0.61 ± 0.02	–	46 ± 6
	ZnO(II)	0.97 ± 0.02	1.29 ± 0.03	0.61 ± 0.02	–	25 ± 7
	Fe_2O_3	0.38 ± 0.02	-0.21 ± 0.04	–	51.3 ± 0.2	29 ± 6
ZnO+10%Fe_2O_3	ZnO(I)	0.26 ± 0.01	0.86 ± 0.01	0.57 ± 0.01	–	82 ± 4
	ZnO(II)	0.89 ± 0.04	1.62 ± 0.09	0.57 ± 0.01	–	7 ± 5
	Fe_2O_3	0.40 ± 0.04	-0.23 ± 0.08	–	51.4 ± 0.2	11 ± 4
ZnO+10%Fe_2O_3	ZnO(I)	0.32 ± 0.02	0.57 ± 0.03	0.58 ± 0.06	–	45 ± 10
milled in air [9]	$ZnFe_2O_4$	0.47 ± 0.02	2.64 ± 0.3	0.51 ± 0.2	–	56 ± 17

3.2 Mössbauer spectrometry

For all samples, a progressive evolution with milling time from the characteristic hyperfine features for FeO or Fe_2O_3 to two similar paramagnetic signals is observed. In Fig. 3, the Mössbauer spectra corresponding to the most prolonged milling time for ZnO mixtures with 5 at.% FeO and Fe_2O_3, are displayed. Both cases can be successfully described with similar quadrupole interactions. For the case of hematite less than a 30% of Fe atoms experienced the characteristic sextet for this oxide, assuming the same values of recoilless f factor. In consequence, there is a good agreement with X-ray diffraction results. The resulting hyperfine parameters obtained from least squared fittings are displayed in Table 2.

The observed two quadrupole interactions can be identified with two ion sites in the wurtzite ZnO structure in Fe^{2+} (ZnO(I)) and Fe^{+3} (ZnO(II)) state. These values are consistent with the proposed fit B in a previous work on mixtures of ZnO with 10% of additional Fe-based phase [9]. The (ZnO(I)) called values are consistent with Fe $^{2+}$ ions in tetrahedral coordination, as expected for substitutional sites in ZnO. Similar results were observed by Ahn et al. [12] in samples prepared by solid state reaction at 1,200°C and Lin et al. [9] for Fe-doped ZnO prepared by mechanical milling at low iron concentrations. As it was discussed by Verdier et al. [5] on the formation on Fe^{2+} during milling hematite with ZnO to obtain the spinel $ZnFe_2O_4$, we must analyzed the presence of both valence state of Fe in our results. Initially Fe is on the valence state Fe^{+2} in FeO and Fe^{+3} in Fe_2O_3 and in state 0 if coming from the milling tools. At the final step of milling both valence states are found for both experiments, and when the milling was done under air as well. It was argued [5] that ZnO presence and tools contamination may be responsible of such redox reduction reaction, but since these facts occur independently of atmosphere and initial precursors, we state that the mechanical work itself, by the introduction of different kinds of defects, does originates the variety of valence states in Fe ions. However some additional experiments and *ab initio* calculations will be done in order to elucidate this point.

Under air milling conditions favour the formation of spinel structure $ZnFe_2O_4$, in agreement with results of previous section. Figure 4 exhibits the Mössbauer spectrum

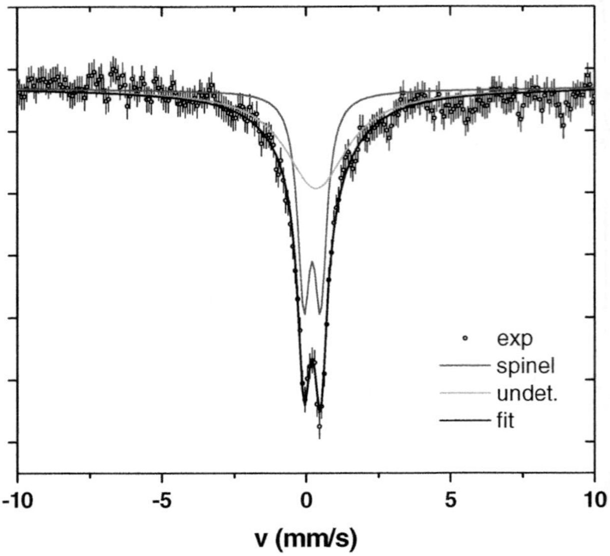

Fig. 4 Mössbauer spectra for $ZnO+10$ at.%Fe_2O_3 milled in air [8]

of $ZnO+10$ at.%Fe_2O_3 after 16 h of air-milling: the obtained parameters agree with those reported in literature for this phase [13].

4 Conclusions

Fe-doped ZnO oxides were successfully synthesized by high energy ball milling under Ar atmosphere irrespective of concentration and kind of starting materials (FeO or α-Fe_2O_3). Among the milling conditions, the atmosphere is the more important parameter for the final product phases. Fe atoms with different ionization states, substitute Zn atoms in tetragonal sites leading to two quadrupole splitting sites: Zn(I) Fe^{+2} and ZnO(II) Fe^{+3}. In all cases, no final magnetic hyperfine structures were observed in the studied composition range.

References

1. Coey, J.M.D., Venkatesan, M., Fitzgerald, C.B.: Nat. Matters **4**, 172 (2005)
2. Norton, D.P., Pearton, S.J., Hebard, A.F., Theeodoroponlow, N., Boatner, L.A., Wilson, R.G.: Appl. Phys. Lett. **82**, 239 (2003)
3. Lin, Y., Jiang, D., Lin, F., Shi, W., Ma, X.: J. Alloys Compd. **436**, 30–33 (2007)
4. Potzger, K., Zhou, S., Reuther, H., Mücklich, A., Eichhom, F., Schell, N., Skorupa, W., Helm, M., Fassbender, J., Hermannsdörfer, T., Papageorgiou, T.P.: Appl. Phys. Lett. **88**, 052508 (2006)
5. Weyer, G., Gunnlaugsson, H.P., Mantovan, R., Fanciulli, M., Naidoo, D., Bharuth-Ramand, K., Agne, T.: J. Appl. Phys. **102**, 113915 (2007)
6. Ahn, G.Y., Park, S.-I., Kim, C.S.: J. Magn. Magn. Mater. **303**, e329–e331 (2006)
7. Damonte, L.C., Hernández-Fenollosa, M.A., Meyer, M., Mendoza-Zélis, L., Marí, B.: Phys. B Condens. Matter **398**, 380–384 (2007)
8. Damonte, L.C., Mendoza-Zélis, L., Marí, B., Hernández-Fenollosa, M.A.: Powder Technol. **148**, 15–19 (2004)

9. Baum, L., Meyer, M., Richard, D., Damonte, L.C., Mendoza-Zélis, L.A.: Hyperfine Interact. **176**, 87–92 (2007)
10. Suryanarayana, C.: Prog. Mater. Sci. **46**, 1–184 (2001)
11. Verdier, T., Nachbaur, V., Jean, M.: J. Solid State Chem. **178**, 3243–3250 (2005)
12. Ahn, G.Y., Park, S.I., Kim, S.J., Kim, C.S.: Magn, J., Magn. Mater. **304**, e498–e500 (2006)
13. Chinnasamy, C.N., Narayanasamy, A., Ponpandian, N., Chattopadhyay, K., Guérault, H., Greneche, J.-M.: Phys, J.: Condens. Matter **12**, 7795–7805 (2000)

Hyperfine Interact (2010) 195:235–240
DOI 10.1007/s10751-009-0124-4

Hyperfine, structural and electrical transport properties of the high-energy milled (1 − X).ZnO–X.FeO system

E. C. Reisdoerfer · F. F. Ivashita · J. V. Bellini · A. Paesano Jr. · A. C. S. da Costa · S. A. Pianaro · B. Hallouche

Published online: 8 October 2009

Abstract The system (1 − X).ZnO+X.FeO was high-energy ball-milled for 24 h and characterized by X-ray diffraction and Mössbauer spectroscopy. Sintered pellets were also prepared for $X \leq 0.07$ by further thermal annealing pressed milled samples. These samples were also characterized regarding their I–V behavior. The results revealed the monophase formation of a $(Zn_{1-X}Fe_X)O$ solid solution isomorphous to the zincite and a spinel-like phase of the $Zn_{1-Y}Fe_{2+Y}O_4$ type. The sintered samples showed I–V curves typical of poor varistor systems and the resistivity increasing with X.

Keywords ZnO · Ball-milling · Diluted magnetic semiconductor · Mössbauer spectroscopy

1 Introduction

Extended solid solution can be prepared by high-energy ball-milling metal or ceramic precursors, eventually increasing the mutual solubility of the milled materials far beyond the equilibrium [1].

E. C. Reisdoerfer · F. F. Ivashita (✉) · J. V. Bellini · A. Paesano Jr.
Departamento de Física, UEM, Maringá, Brazil
e-mail: Fivashita@yahoo.com.br

A. C. S. da Costa
Departamento de Agronomia, UEM, Maringá, Brazil

S. A. Pianaro
Departamento de Engenharia de Materiais, UEPG, Ponta Grossa, Brazil

B. Hallouche
Departamento de Química e Física, UNISC, Santa Cruz do Sul, Brazil

The metastable combination mechanically induced between transition metals oxides, for instance, has frequently been investigated due to the importance of the mineralogical or technological aspects of these solid solutions.

The mechanosynthesis of metastable solid solutions made from monoxides—i.e., (M, M′)O—with different structures, for instance, is particularly interesting since new materials, not always achievable by chemical methods, may eventually result.

Thus, we investigated the properties of the high-energy ball-milled $(1-X).ZnO$–$X.FeO$ binary system. Wustite (FeO) and zincite (ZnO) have cubic and hexagonal structures, respectively, and the as-milled powders were characterized regarding structural and hyperfine properties. We report in this paper the results obtained specifically for the $0 < X \leq 0.12$ concentration range.

Interestingly, the system $(Zn_{1-\delta}Fe_{\delta})O$ has been pointed out as a diluted magnetic semiconductor (DMS) [2–4]. These materials have been increasingly investigated because of their potential technological applications such as in spintronic devices [5]. In addition, the $(Zn_{1-\delta}Fe_{\delta})O$ system, in particular, has been studied also as a promising Fe-doped ZnO based varistor [6].

Thus, motivated by these additional reasons, we also prepared some sintered samples from the milled powders and characterized them respectively for their electric resistance. The electric transport results are explained on the basis of a varistor model.

2 Experimental

The samples were prepared by high-energy ball-milling a $(ZnO)_{1-X}(FeO)_X$ powder mixture in free atmosphere, with starting nominal compositions within the $0 < X \leq 0.12$ range. The milling was conducted in a planetary mill, with a rotation speed of 300 rpm, using hardened steel vial and balls. The ball-to-powder mass ratio was ~30:1.

The $X = 0.03$ and 0.07 milled powders were uniaxially compacted in the form of pellets, with a pressure of 60 MPa. Further, the pellets were sintered in air at 1,000°C/1 h, using a resistive furnace, with a heating rate of 10°C/min and slow cooling inside the furnace.

All the samples were characterized by X-ray diffraction (XRD) and Mössbauer spectroscopy (MS). The sintered samples were electrically characterized after electric contacts were provided through the Au-deposition by sputtering on both faces of the pellets.

The XRD analysis was carried out in a conventional diffractometer, with a Cu-K_{α} beam. The scanning step and time of pulse accumulation were 0.02° (2θ) and 0.6 s, respectively.

The MS characterizations were performed at room temperature (RT), in transmission geometry, using a conventional Mössbauer spectrometer, in a constant acceleration mode. The γ-rays were provided by a $^{57}Co(Rh)$ source. The Mössbauer spectra were analyzed with a non-linear least-square routine, with Lorentzian line shapes. All isomer shift (IS) data given are relative to α-Fe throughout this paper.

The $I \times V$ measurements (i.e., current x voltage) were performed by using a high voltage source with a GPIB interface for computer. The $J \times E$ (i.e., current

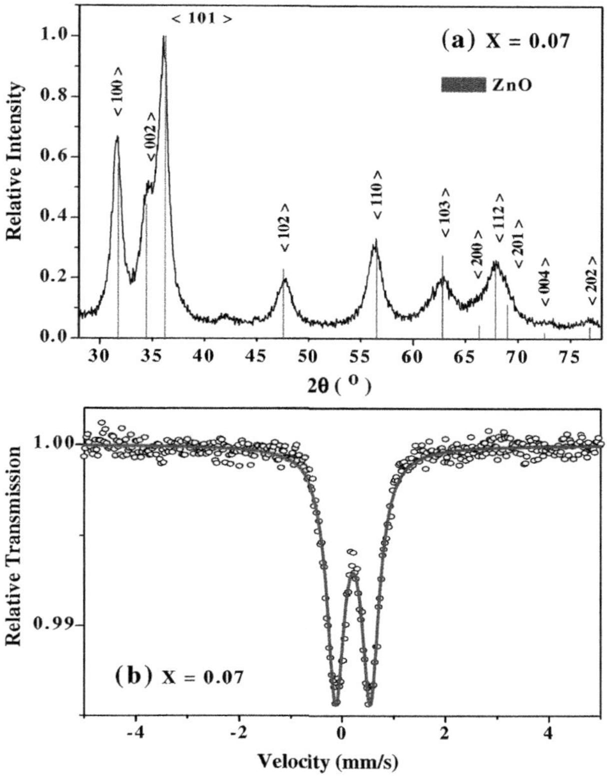

Fig. 1 X-ray diffraction pattern (**a**) and Mössbauer spectrum (**b**) for the X = 0.07 as-milled sample

density × electric field) curves were traced after carefully measuring the geometrical dimensions of the sintered samples.

3 Results and discussions

Figure 1 shows a representative diffractogram and the respective RT Mössbauer spectrum obtained for the samples prepared in the $X \leq 0.07$. The colored vertical bars in the X-ray pattern are positioned according to the peak positions for the zincite phase (JCPDS—File no. 80-0075). The reflection peaks are broadened, as expected for milled samples that consist of very fine and disordered crystallites. No other structure is clearly identifiable in the diffractogram, besides the zincite hexagonal one, although the small peak at ~43° could be attributed to a spinel structure (JCPDS—File no. 22-1012). The Mössbauer spectra for the samples prepared in this concentration range were fitted with only one doublet, which is attributed to a solid solution of the $(Zn_{1-X}Fe_X)O$ type [2, 7]. Table 1 presents the hyperfine parameters for these and all other samples.

Typical diffractometric and Mössbauer results obtained for the samples prepared in the $0.09 \leq X \leq 0.12$ concentration range are shown in Fig. 2. The diffractogram reveals that a spinel-like phase of the $Zn_{1-Y}Fe_{2+Y}O_4$ type formed together with the

Table 1 Mössbauer hyperfine parameters and subspectral areas for the as-milled and sintered samples

Sample	X	Phase	IS (mm/s)	QS (mm/s)	Γ (mm/s)
As-milled	0.01	$(Zn_{1-X}Fe_X)O$	0.22	0.49	0.37
	0.03	"	0.22	0.66	0.44
	0.05	"	0.22	0.62	0.39
	0.07	"	0.22	0.67	0.41
	0.09	$(Zn_{1-X'}Fe_{X'})O$	0.22	0.71	0.38
		$Zn_{1-Y}Fe_{2+Y}O_4$	0.25	0.37	0.27
	0.12	$(Zn_{1-X'}Fe_{X'})O$	0.22	0.63	0.43
		$Zn_{1-Y}Fe_{2+Y}O_4$	0.27	0.35	0.25
Sintered	0.03	$ZnFe_2O_4$	0.40	0.48	0.39
	0.07	"	0.41	0.50	0.38

IS = isomer shift, *QS* = quadrupole splitting, *Γ* = linewidth;

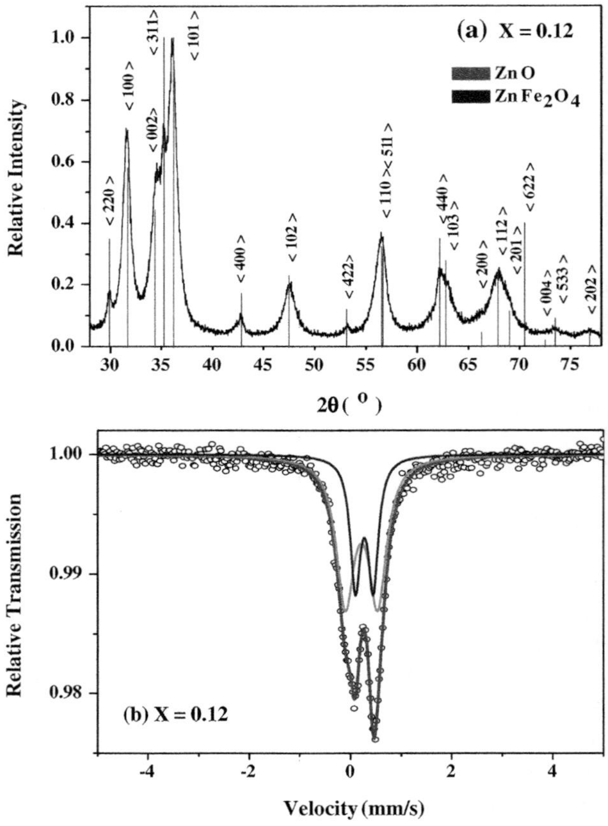

Fig. 2 X-ray diffraction pattern (**a**) and Mössbauer spectrum (**b**) for the X = 0.12 as-milled sample

$(Zn_{1-X'}Fe_{X'})O$ solid solution (with $X' < X$). For these samples, fitting the Mössbauer spectra required two doublets, one related to the iron doped zincite phase and the other to the spinel phase, which showed hyperfine parameters—specially the quadrupole splittings—similar to those of milled franklinite ($ZnFe_2O_4$) [8]. According to the

Fig. 3 X-ray diffraction pattern (**a**) and Mössbauer spectrum (**b**) for the X = 0.07 sintered sample

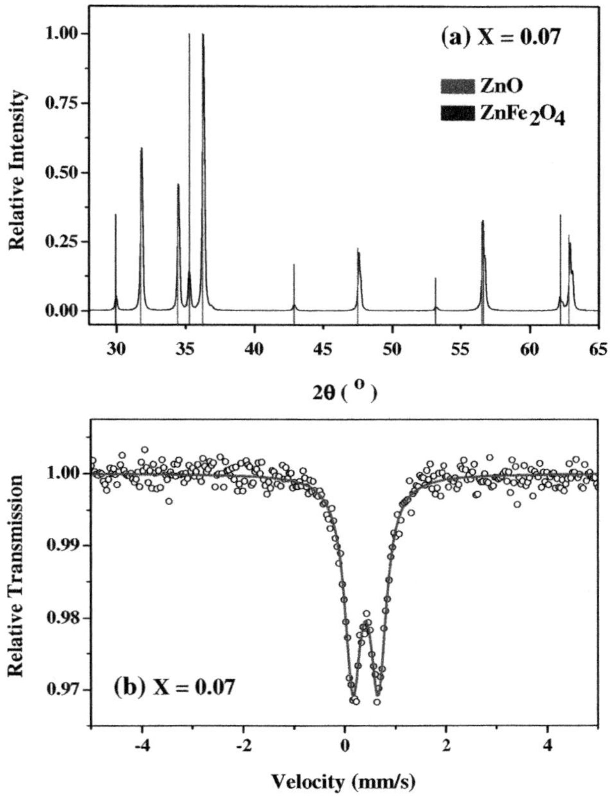

subspectral areas, and supposing similar *f* factors for both phases, one third of the iron atoms present in these samples belongs to the spinel phase.

Figure 3 presents the XRD and MS results for the X = 0.07 sintered sample. The peaks are now much better defined as a consequence of the re-crystallization induced by the thermal annealing. The phases identified by XRD for the two annealed samples were ZnO and $ZnFe_2O_4$. The Mössbauer spectra showed again only one doublet, which, according to the X-ray results, belongs to franklinite. Small discrepancies regarding the hyperfine parameters reported earlier [9, 10] may be attributed to the differences in the synthesis routes. Therefore, sintering gave rise to a phase separation in which all, or almost all, the iron dissolved in the ZnO matrix was consumed in the $ZnFe_2O_4$ formation. Earlier results revealed solubility of 1% for iron in ZnO samples heat-treated in air at 1,000°C [9]. It is also worthy of note that the isomer shifts for franklinite in the sintered samples presented a positive variation, respectively to the same phase in the as-milled samples, which can be attributed to oxidation (i.e., a decrease in the effective iron valence).

The $J \times E$ curves for the sintered samples are shown in Fig. 4. It can be observed that when the iron content increased from X = 0.03 to X = 0.07, the resistivity of the samples also increased. From these curves, and considering the nonlinear equation $J = K \times E^{\alpha}$, where K is a constant related with the material microstructure, the α values were found to be (from the angular coefficients of the straight lines plotted

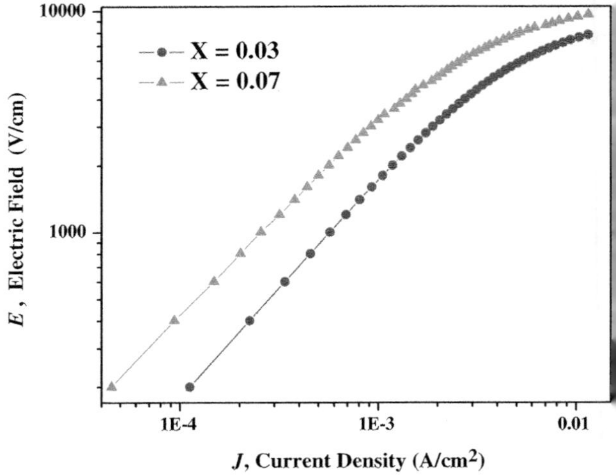

Fig. 4 *J–E* curves for the sintered samples

on logarithmic scale) 2.8 and 4.0 for the X = 0.3 and 0.7 samples, respectively. This indicates a poor varistor behavior [11].

The breakdown electric field, E_r, obtained at J = 1 mA/cm^2, also varied with X, increasing from 1742 V/cm, for X = 0.03, to 3220 V/cm, for X = 0.07. This is probably related with the lower grain size resultant in the microstructure of the material with higher X, due to the pinning induced by the $ZnFe_2O_4$ precipitates.

4 Conclusions

The properties of the high-energy as-milled (1 − X).ZnO–X.FeO powders depend on the concentration range. For $0 < X \leq 0.07$, a single $(Zn_{1-X}Fe_X)O$ phase was mechanosynthesized, whereas for $0.09 \leq X \leq 0.12$, a $Zn_{1-Y}Fe_{2+Y}O_4$ spinel phase was formed in addition to the $(Zn_{1-X'}Fe_{X'})O$ solid solution. Sintering the $0 < X \leq 0.07$ milled samples induced the segregation of $ZnFe_2O_4$ precipitates in the ZnO matrix, in a ceramic body with poor varistor behavior.

References

1. Suryanarayana, C.: Prog. Mater. Sci. **46**, 1 (2001)
2. Lin, Y., Jiang, D., Lin, F., Shi, W., Ma, X.: J. Alloys Compd. **436**, 30 (2007)
3. Zhang, H.-W., Wei, Z.-R., Li, Z.-Q., Dong, G.-Y.: Mater. Lett. **61**, 3605 (2007)
4. Wang, Y.Q., Yuan, S.L., Liu, L., Li, P., Lan, X.X., Tian, Z.M., He, J.H., Yin, S.Y.: J. Magn. Magn. Mater. **320**, 1423 (2008)
5. Wolf, S.A., Awschalom, D.D., Buhrman, R.A., Daughton, J.M., Molnár, S., Roukes, M.L., Chtchelkanova, A.Y., Treger, D.M.: Science **294**, 1488 (2001)
6. Sedky, A., Abu-Abdeen, M., Almulhem, A.A.: Phys. B Condens. Matter **388**, 266 (2007)
7. Ahn, G.Y., Park, S.-I., Shim, I.-B., Kim, C.S.: J. Magn. Magn. Mater. **282**, 166 (2004)
8. Ehrhardt, H., Campbell, S.J., Hofmann, M.: J. Alloys Compd. **339**, 255 (2002)
9. Goya, G.F., Stewart, S.J., Mercader, R.C.: Sol. State Comm. **96**, 485 (1995)
10. Jiang, J.Z., Wynn, P., Mørup, S., Okada, T., Berry, F.J.: Nanostruct. Mater. **12**, 737 (1999)
11. Pianaro, S.A., Bueno, P.R., Olivi, P., Longo, E., Varela, J.A.: J. Mater. Sci. **9**, 159 (1998)

Hyperfine Interact (2010) 195:241–247
DOI 10.1007/s10751-009-0125-3

Effect of Ni on the lattice parameter and the magnetic hyperfine field in $(Fe_{70}Al_{30})_{100-x}Ni_x$ alloys

D. C. Palacio · J. F. Vallderruten · L. E. Zamora · G. A. Pérez Alcázar · J. A. Tabares

Published online: 6 October 2009

Abstract Samples of the $(Fe_{70}Al_{30})_{100-x}Ni_x$ system, with $x = 5$, 10, 15, and 20, were prepared by mechanical alloying using milling times of 12, 24, 36, 48, and 72 h and characterized by X-ray diffraction (XRD) and Mössbauer spectrometry (MS). XRD of all the samples allowed us to identify the BCC structure as the main component. A decreasing lattice parameter, as the milling time and Ni content increase, was obtained. The MS experiments were carried out at room temperature. The spectra were fitted with a hyperfine magnetic field distribution (HMFD). The Mean Hyperfine Fields (MHF) are ranged between 27 and 29 T with a small dependence with the milling time and was not strongly influenced by Ni content. An additional paramagnetic single line in the spectra of samples with $x = 20$ and that of sample with $x = 15$ and milled during 72 h must be included. The Mössbauer spectral area of this phase increases when the milling time increases and it was associated to an FCC phase.

Keywords Mössbauer spectroscopy · X-ray diffraction · Mechanical alloying

1 Introduction

Mechanical alloying (MA) is an useful and interesting technique that allows to produce a large variety of new powdered materials with technological applications mainly in the nanomaterial field. It is well known that reactions occur far from equilibrium conditions and a high amorphization in the products can be achieved and some special physical properties have been found [1, 2]. Among the alloys produced by MA it can be mentioned here the Fe–Al, Fe–Ni and Ni–Al ones. The Fe–Al alloys are cheap magnetic materials used in transformer core; Fe–Al had shown a

D. C. Palacio · J. F. Vallderruten · L. E. Zamora ·
G. A. Pérez Alcázar · J. A. Tabares (✉)
Departamento de Física, Universidad del Valle, AA 25360, Cali, Colombia
e-mail: jatabare@calima.univalle.edu.co

disordered ferromagnetic character at concentrations less than 18 at.% Al [2] and an order-disorder transition, detected by XRD, was found in $Fe_{70}Al_{30}$ alloys when milling time increases, transition that improves their ferromagnetic behavior [3]. Otherwise, Fe–Ni alloys are soft magnetic materials that exhibit low coercive field, high saturation magnetization, low permeability but high electrical resistance [4]. A simultaneous occurrence of FCC and BCC phases, between 22.5 and 40 at.% Ni, was reported for milled Fe–Ni alloys, where the BCC phase shows a ferromagnetic character in contrast to the FCC phase which exhibits both the paramagnetic and ferromagnetic behavior [5]. The Ni–Al alloys show an ordered phase, with low density, high melting point, excellent oxidation and corrosion resistance, good electric conductivity but a brittle behavior at room temperature [6, 7]. In order to improve the different properties of the mentioned alloys, some studies of the ternary Fe–Al–Ni alloys had been reported [6, 8–12]. In 2002, Principi et al. [6] alloyed mechanically powders of amperite ($Ni_{69}Al_{31}$) with 10, 20, and 30 wt.% Fe, by using ceramic and stainless steel jars. Their XRD results showed a BCC phase, whose lattice parameter decreases when the milling time increase and the values are higher as the Fe content increases. In these experiments the samples were also thermally treated. The Mössbauer spectra corresponding to the untreated ones were fitted with a sextet, associated to metallic iron, an HMFD, corresponding to different Fe environments, and a poorly resolved doublet, maybe corresponding to Fe atoms in antisites of the NiAl alloy. Although, the samples alloyed in ceramic jars and those annealed at 973 K in Ar atmosphere, during 1 h, showed a singlet with an IS $= -0.009$ mm/s. This singlet is associated to an FCC-Fe phase, induced by milling time and temperature. This result is in accord with the small FCC phase detected by XRD in the 20 wt.% Fe using ceramic jars. Z. G. Liu et al. [12] reported that in $Ni_{50}Al_{50-x}Fe_x$ (for $x = 20$, 25 and 30) alloys, after annealing at 973 K, the coexistence of BCC and FCC phases is obtained, and the FCC phase increases as the Fe content increases. The samples without annealing show the Fe–Al–Ni BCC structure. It is clear, according to the previous discussion, which a lot of work has to be done concerning the properties of the ternary Fe–Al–Ni alloys.

The present work is focused to analyze the behavior of the ternary alloy $(Fe_{70}Al_{30})_{100-x}Ni_x$, synthesized by MA at different Ni concentrations and different milling times. The results obtained by using XRD and MS are presented and the changes in the structural and magnetic properties, according to the milling time and the Ni concentration, are discussed.

2 Experimental procedure

The sample series $(Fe_{70}Al_{30})_{100-x}Ni_x$ ($x = 5$, 10, 15 and 20) were prepared by MA with high purity Fe, Al and Ni powders, in a Fritsch–Pulverisette 5 planetary ball mill (vacuum of 4.5×10^{-2} Torr and frequency of 280 rpm). Milling times of 12, 24, 36, 48, and 72 h and ball mass to powder mass ratio of 15:1 were considered. The Mössbauer measurements were performed using a conventional constant acceleration Mössbauer spectrometer at RT with a 10 mCi γ-source of $^{57}Co(Rh)$. The spectra were fitted using the MOSFIT program while the IS values are referred to the α-Fe at RT. The XRD patterns of the samples were performed at RT using the Cu/K_α

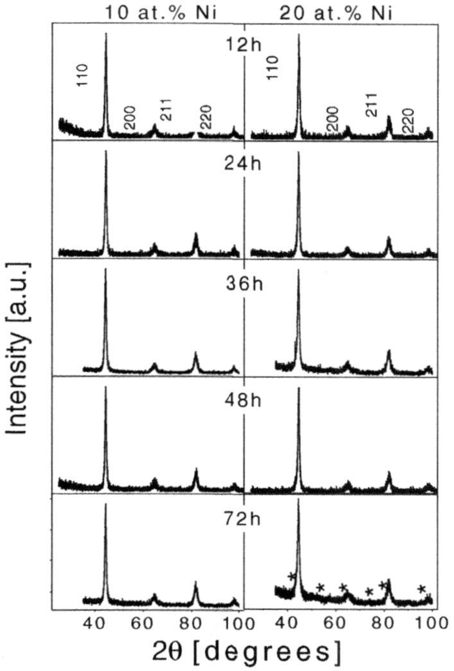

Fig. 1 Diffraction patterns of samples with 10 and 20 at.% Ni and different milling times showing peaks of a BCC phase. Sample with 20 at.% Ni and 72 h of milling exhibits additional peaks of a FCC phase (*asterisks*)

radiation and were refined by using the MAUD and GSAS [13] programs, allowing the structural phases, the lattice parameters, and the grain sizes to be determined.

3 Experimental results and discussion

3.1 XRD

Figure 1 shows the XRD patterns (normalized intensity) for samples with $x = 10$ and 20 and milling times of 12, 24, 36, 48, and 72 h. It is clear that they correspond to the BCC structure. Same results (no shown here) were found with 5 and 15 at.% Ni and 12, 24 and 48 h. These results were expected, mainly due to the high Fe content, and are in accordance with the results reported by Principi et al. [6]. In Fig. 2a the behavior of the lattice parameter with milling time for $x = 10$ and 20 is showed. The lattice parameters decrease when milling time increases, due to the high energy transferred in the collisions permitting shortening the distance between particles in the solid. This means that the Ni diffuses into the FeAl phase. In Fig. 2b is presented the behavior of samples with $x = 5, 10, 15$ and 20 and 12, 24, and 48 h of milling. As Ni concentration increases the lattice parameter decreases, due the substitution of Al, with higher atomic radius, by Ni atoms.

A close observation of the XRD pattern of the sample alloyed during 72 h and with $x = 20$ allows to identify a second phase, associated to an FCC phase with a calculated phase percent near to 5%. The presence of this FCC phase is justified as a result of the Fe substitution in the Al and Ni cells (both FCC) that is evident as

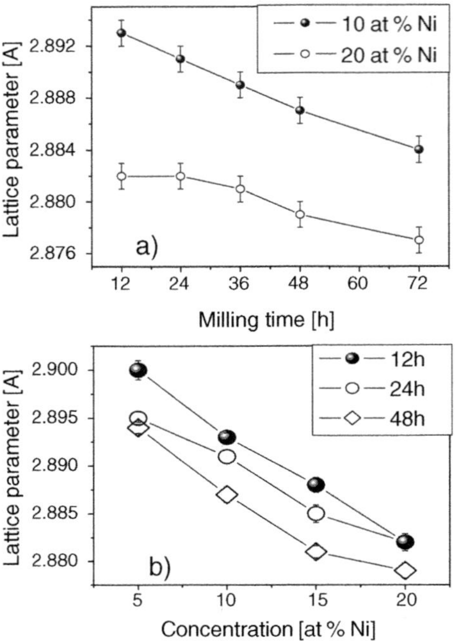

Fig. 2 Behavior of the lattice parameter of the BCC phase versus: **a** milling time and **b** Ni concentration

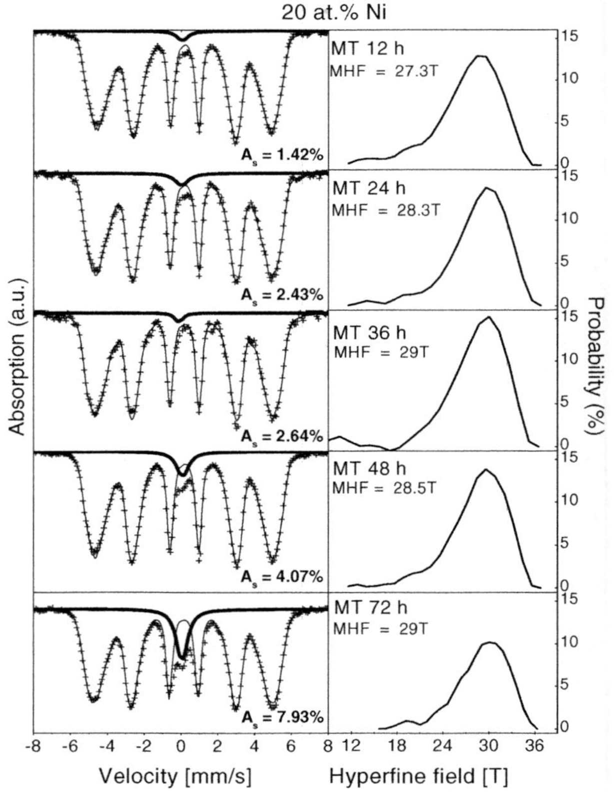

Fig. 3 Mössbauer spectra and their corresponding HFD, of the sample with $x = 20$ and milling times (MT) of 12, 24, 36, 48, and 72 h of milling time (MT). A_s is the spectral area of the single line in the Mössbauer spectra

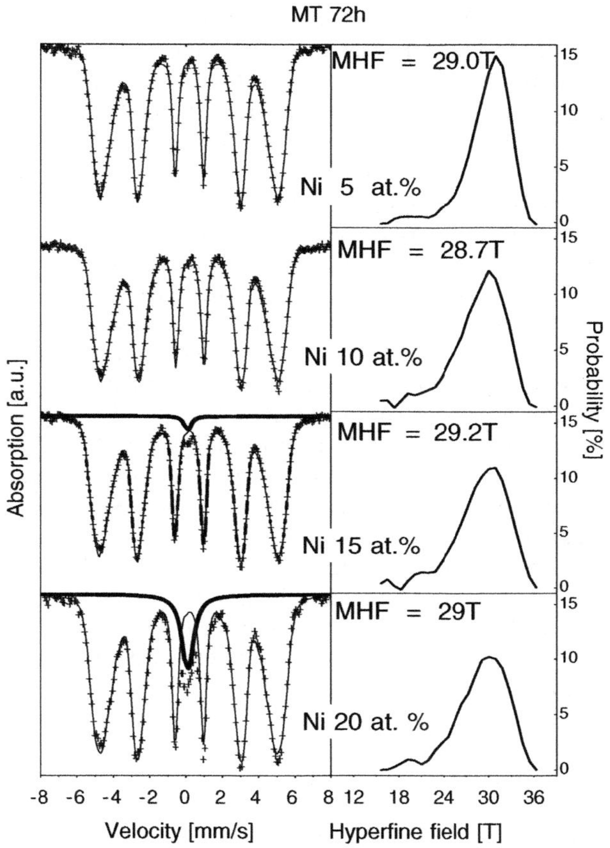

Fig. 4 Mössbauer spectra for the samples milled during 72 h and their corresponding hyperfine field distributions

milling time increases according to the energy delivered in the milling process. This FCC phase for the highest Ni concentration was reported by G. Principi et al. [6] and Z.G. Liu et al. [12]. Despite to the low phase fraction (less than 5%), a Rietveld refinement allowed us to get a lattice parameter of 2.947 Å for this FCC structure, higher than the lattice parameter for the BCC phase (2.877 Å). This behavior is in according to the increase of a paramagnetic single line Mössbauer spectral area (discussed below).

The mean grain size of the alloys in the parallel and perpendicular directions to the incident X-rays was calculated using the GSAS program. The obtained values indicate that the shape of the grains is that of an oblate spheroid with the major axis length between 14 and 40 nm and their aspect ratio decreasing with Ni content tending to a disk shape.

3.2 Mössbauer

Figure 3 shows the spectra for $x = 20$ and milled during 12, 24, 36, 48 and 72 h. They were fitted with an HMFD whose IS tends to decrease with milling time to a value of 0.06 mm/s for all Ni concentrations and QS values close to zero and negatives. Additionally, a paramagnetic single line whose IS = −0.004 mm/s, very similar to that reported by G. Principi et al. [6], must be included, associated to an FCC phase. The spectral area of this phase increases as the milling time increases. Figure 4 shows some typical Mössbauer spectra of alloys for 72 h of milling time, and their corresponding HMFDs. The Mean Hyperfine Field (MHF) values are ranged between 27 and 29 T, with a small dependence with the milling time and not strongly influenced by Ni content. These values are very similar to the value of 29 T reported by Jartych et al. [14] for the $Fe_{70}Al_{30}$ alloy prepared by mechanosynthesis and attributed to the high moment state as the most probable state in their alloys.

It is interesting to note that the HMFD for 5 at.% Ni is highly symmetric, indicating a high grade of disorder, but as the Ni increases a small peak close to 19 T was observed, which can be associated to texture effects of particles with preferential orientation maybe associated to the grain flattening, as was observed by XRD.

The paramagnetic single line (singlet), also detected for the sample with $x = 15$ and milled during 72 h, is attributed to the FCC-FeAlNi phase detected by XRD in the 20 at.% Ni samples and discussed above. Two types of interpretation can be postulated to explain this behavior. The first one is attributed to the paramagnetic behavior of Fe atoms in the FCC phase (γ-Fe) at RT which at low temperatures behaves as antiferromagnetic [15]. The second one, recalling that FCC-FeNi alloys with compositions near the Invar one, shows a paramagnetic site as a consequence of the low spin of Fe atoms in this region which interact antiferromagnetically [16, 17]. Low temperature experiments are now in progress in order to elucidate the appropriate interpretation.

4 Conclusions

XRD results showed a BCC structure for all samples and a small signal corresponding to an FCC structure in the sample with 20 at.% Ni and milled during 72 h. As Ni concentration increases the lattice parameter decreases, due to the substitution of high atomic radius (Al atoms) by Ni atoms. The grain size is of nanometric order (≅28 nm), at high Ni concentration. The fitting of the Mössbauer spectra suggested that alloys are ferromagnetic and disordered with an MHF ranging from 27 to 29 T. For the 20 at.% Ni samples a paramagnetic phase, associated to the FCC phase obtained by XRD, which increases as milling time increases, must be included and justified by the paramagnetic behavior of the Fe atoms inside an FCC structure or to the decrease of its magnetic moment in this composition region.

Acknowledgements The authors would like to thank Universidad del Valle, the Centro de Excelencia de Nuevos Materiales (CENM) and Colciencias for the financial support of this work.

References

1. Edelstein, A.S., Cammarata, R.C.: Nanomaterials: synthesis, properties and applications, 1st edn. Institute of Physics Publications, Bristol (1996)
2. Lozano Oyola, D., Martínez Rojas, Y., Bustos y, H., Pérez Alcázar, G.A.: Rev. Col. de Física **34** 385 (2002)
3. Apiñaniz, E., Plazaola, F., Garitaonandia, J.S., Martín, D., Jiménez, J.A.: J. Appl. Phys. **93**, 7649 (2003)

4. Liu, Y., Zhang, J., Yu, L., Jia, G., Jing, C., Cao, S.: J. Magn. Magn. Mater. **285**, 138 (2005)
5. Valderruten, J.F., Pérez Alcázar, G.A., Greneche, J.M.: J. Phys., Condens. Matter. **20**, 485204 (2008)
6. Principi, G., Spataru, T., Maddalen, A., Gialanella, S.: Hyperfine Interact **139**, 315 (2002)
7. D'Angelo, L., González, G., Ochoa, J.: J. Alloys Compd. **434–435**, 348 (2007)
8. Vincze, I.: Phys. Rev. B **7**, 54 (1973)
9. Fu, C.L., Zouf, J.: Acta Mater. **44**, 1471 (1996)
10. Bozzolo, G.H., Khalil, J., Noebe, R.D.: Comput. Mater. Sci. **24**, 457 (2002)
11. Golberg, D., Shevakin, A.: Intermetallics **3**, 293 (1995)
12. Liu, Z.G., Guo, J.Y., Zhou, L.Z., Hu, Z.Q., Unemoto, M.: J. Mater. Sci. **32**, 4857 (1997)
13. Larson, A.C., Von Dreele, R.B.: GSAS, General Structure Analysis System, pp. 86–748. Los Alamos National Laboratory report LAUR (2004)
14. Jartych, E., Zuravicz, J.K., Oleszak, D., Pekala, M., Sarsynski, J., Budzynski, M.: J. Magn. Magn. Mater. **186**, 299 (1998)
15. Gonser, U., Wagner, H.G.: Hyperfine Interact. **26**, 769 (1985)
16. Rancourt, D.G., Scorzelli, R.B.: J. Magn. Magn. Mater. **174**, 324 (1997)
17. Kaloshkin, S.D., Tomilin, I.A., Andrianov, G.A., Baldokhin, U.V., Shelekhov, E.V.: Mater. Sci. Forum. **235–238**, 565 (1997)

Hyperfine Interact (2010) 195:249–255
DOI 10.1007/s10751-009-0126-2

Characterization of scraps produced by the industrial laser cutting of steels

R. C. Mercader · S. G. Marchetti · J. F. Bengoa · G. Punte · E. D. Cabanillas

Published online: 20 October 2009

Abstract We have studied some properties relevant for technological applications of the debris produced by industrial laser cutting of steels. The investigated material is made up of spheroidal particles, hollow and solid, which get oxidized over the cutting process, and that we have reduced afterwards in a H_2 atmosphere. The samples, before and after the reduction, were characterized by X-ray diffraction, scanning electron microscopy, specific surface area and Mössbauer spectroscopy. We have found that, after the reduction treatment, the shape remains unchanged but that the chemical composition and the physical properties of the external and internal surface structures are modified. In particular, the specific surface area of the material increased by one order of magnitude.

Keywords Laser cutting · Microstructured particles · Magnetite

1 Introduction

During the process of laser cutting and shaping of steels, a significant amount of scrap material is produced and usually discarded. Along the line of our previous investigations on the subject [1, 2] we have paid attention to the waste itself and

R. C. Mercader · G. Punte
Departamento de Física, IFLP-CCT-La Plata, Facultad de Ciencias Exactas, Universidad Nacional de La Plata, 115 y 49, 1900 La Plata, Argentina

S. G. Marchetti (✉) · J. F. Bengoa
CINDECA. Fac. Cs. Exactas, UNLP-CICPBA-CCT-CONICET, 47 N° 257, La Plata 1900, Argentina
e-mail: march@quimica.unlp.edu.ar

E. D. Cabanillas
CONICET and Departamento de Combustibles Nucleares, Comisión Nacional de Energía Atómica, Av. del Libertador 8250, 1429 Buenos Aires, Argentina

Fig. 1 External surface of L270

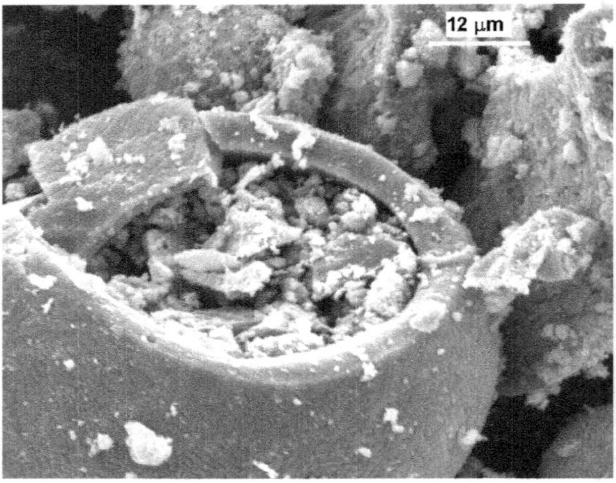

Fig. 2 Inner of L270

have found that it is made of spherical iron oxide, nano and microparticles, solid and hollow. This material has potential industrial applications as carrier for different substances, like pesticides, fertilizers or medical drugs and contrast enhancers in NMR imaging. Besides, if it is reduced, it could be used as a Fischer–Tropsch catalyst for hydrocarbon synthesis. To assess the feasible application of the material to these prospective uses, we are currently exploring the incorporation of different substances into them. In particular, we have started a project to produce a pesticide-carrying painting that slowly releases the drug over several months. To reach higher substance loadings we have reduced and passivated the particles to increase their specific surface area. In this work we report the characterization of the shape, size and nature of the material, before and after the reduction, by scanning electron microscopy, specific surface area (BET), X-ray diffraction and ^{57}Fe Mössbauer spectroscopy.

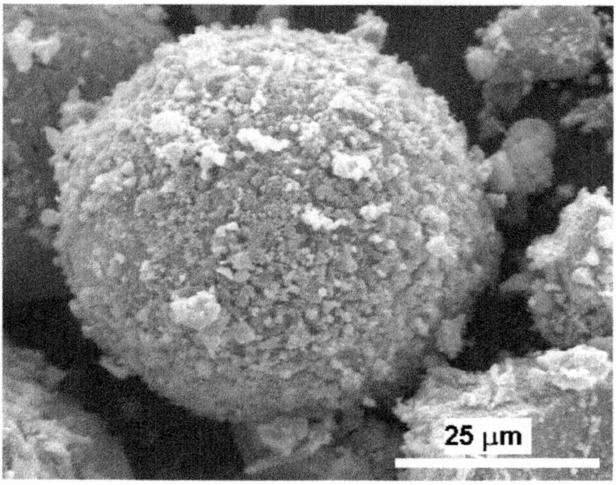

Fig. 3 External surface of L270r

Fig. 4 Inner of L270r

2 Experimental

In an industrial environment, an IR gaseous CO_2 laser of 1.65 kW was tuned to a wavelength of 10.6 μm. As the contributing gas for cutting, 99.9% pure oxygen at 69 kPa was used. The gas flow was 6.67×10^{-4} m^3/s. The laser cutting was carried out in open air on a 6.35 mm thick SAE 1010 hot rolled steel plate. The cutting velocity was 31.75 mm/s. The ejected particles were collected, sieved and classified by size. The particles we presented here are those of sizes lying between 53 and 75 μm (mesh 270). This sample was called L270. A portion of L270 was reduced in ultra pure H_2 (60 cm^3/min), up to 698 K, with a heating rate of 5 K/min, remaining for 26 h

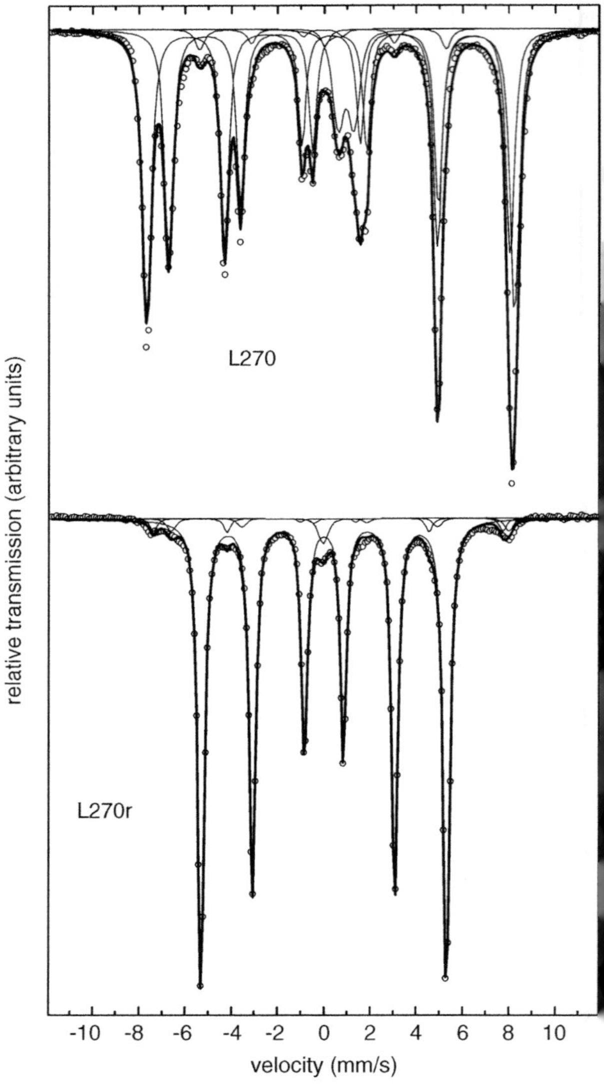

Fig. 5 Mössbauer spectra of L270 and L270r

at the highest temperature. The hydrogen stream was previously filtered in black Pd and a molecular sieve 4 A, to eliminate O_2 and H_2O traces. After reduction, the solid was passivated using a N_2:O_2 mixture, with growing O_2 quantities. We called L270r the sample of reduced and passivated solid.

L270 and L270r were analyzed by X-ray diffraction and Mössbauer spectroscopy. XRD data were obtained in the range $10 \leq \theta \leq 95°$ with a PANalytical's X'Pert PRO X-ray diffractometer system with a diffracted beam monochromator, using Cu-Kα radiation. The Mössbauer spectra at 298 K were taken in transmission geometry with a constant acceleration spectrometer. A source of ^{57}Co in Rh matrix of nominally 50 mCi was used. The velocity calibration was performed against a metallic 12 μm-thick α-Fe foil. All isomer shifts, δ, are referred to this standard. The Mössbauer spectra

Table 1 Mössbauer hyperfine parameters of L270 and L270r at 298 K

Species	Parameters	L270	L270r
Fe_3O_4–Fe in site A	H (T)	49.4 ± 0.6	46.9^a
	δ (mm/s)	0.30 ± 0.01	0.18^a
	2ε (mm/s)	-0.1 ± 0.01	0^a
	%	47.7 ± 0.2	2.7 ± 0.2
Fe_3O_4–Fe in site B	H (T)	45.8 ± 0.5	45.3 ± 0.1
	δ (mm/s)	0.67 ± 0.01	0.69^a
	2ε (mm/s)	0^a	0^a
	%	39.6 ± 0.2	2.7 ± 0.2
α–Fe	H (T)	33.1 ± 0.1	32.94 ± 0.01
	δ (mm/s)	-0.02 ± 0.04	0.001 ± 0.001
	2ε (mm/s)	0^a	0.001 ± 0.001
	%	3.0 ± 0.2	93.1 ± 0.2
FeO	Δ (mm/s)	0.68 ± 0.29	–
	δ (mm/s)	0.94 ± 0.20	–
	%	9.0 ± 0.9	–
Fe–Cr impurity	δ (mm/s)	-0.04 ± 0.04	-0.03 ± 0.01
	%	0.7 ± 0.1	1.5 ± 0.1

H hyperfine magnetic field in Tesla, δ isomer shift (all the isomer shifts are referred to α-Fe at 298 K), 2ε quadrupole shift, Δ quadrupole splitting

[a]Parameter held fixed while fitting

were fitted using the Recoil computer code [3]. Lorentzian lines with equal linewidths were considered for each spectrum component. Both solids were characterized by scanning electron microscopy (SEM). To observe their inner parts, the particles were embedded in an epoxy resin, polished by emery paper and diamond powder and finally etched with Nital (ethanol 96% and 4% vol. nitric acid) The specific surface area BET was determined using a Micromeritics® ASAP 2020 analyzer.

3 Results and discussion

The SEM micrographs show that L270 consists of spherical particles (Fig. 1) with a nearly smooth surface, some of them are solid while others are hollow (Fig. 2) with creviced shells of different thicknesses. Figure 3 demonstrates that the shape has been maintained in L270r after the reduction treatment; however, the external surface shows a rougher appearance. In addition, the inner parts of the particles exhibit a dendritic-like structure without the evidence of ferritic grains, as could be expected after the Nital etching (Fig. 4).

The Mössbauer spectrum of L270 displays magnetic split signals and a central doublet (Fig. 5). This spectrum was fitted with three sextets, one doublet and one singlet, and their hyperfine parameters are shown in Table 1. The less intense sextuplet was assigned to α-Fe and the other two sextets were assigned to magnetite [4]. According to the ratio of the areas belonging to sites A and B ($S(B)/S(A) = 0.78$) this compound would be a non-stoichiometric magnetite or magnetite with γ-Fe_2O_3 as impurity [5]. The central doublet can be assigned to wüstite [6]. The remaining signal of small relative area has hyperfine parameters that can be attributed to an Fe–Cr impurity [7]. From the analysis of the percentages of the different species we

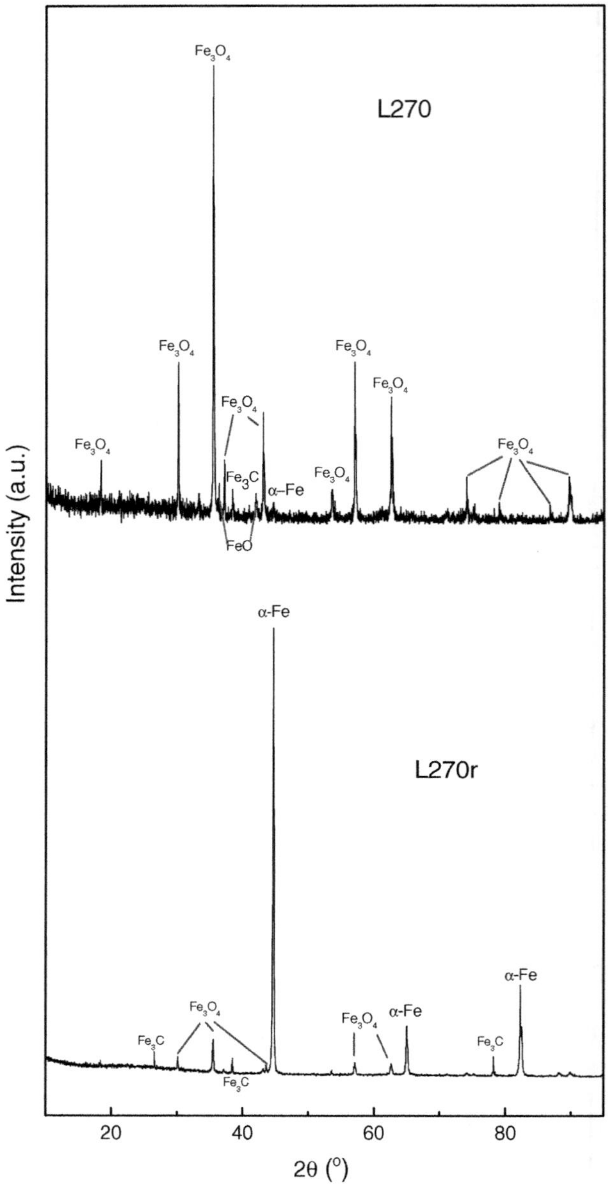

Fig. 6 XRD of L270 and L270r

can conclude that the particles are constituted mainly by magnetite (87.3 ± 0.4%). When the sample is reduced and passivated, the Mössbauer spectrum displays a very intense sextuplet, two less intense magnetic split components, and a very small central singlet (Fig. 5). It was fitted with three sextets and one singlet (Table 1). The more intense sextet (93.1 ± 0.2%) belongs to α-Fe and the other two ones to magnetite likely produced over the passivation treatment (5.4 ± 0.4%). The central singlet belongs to the Fe–Cr impurity mentioned above.

The different iron species assigned by Mössbauer spectroscopy in L270 and L270r, and a small quantity of cementite, were observed in the XRD results (Fig. 6). The sample heterogeneity, could explain why cementite was not detected by Mössbauer spectroscopy.

The values of the specific surface area determined by N_2 adsorption using the BET method were 0.6 m^2/g for L270 and 5.1 m^2/g for L270r. Therefore, a very simple treatment like the reduction in H_2 stream followed by a passivation with N_2/air, increases ten times the available surface, producing channels that allow the access of different substrates inside the spheres.

4 Conclusions

This work demonstrates that after subjecting the scraps produced by the industrial laser cutting of steels to a simple reduction-passivation treatment, the original spherical shape is maintained but the porous structure is significantly increased yielding a specific surface area about one order of magnitude higher than before the treatment. The process renders the reduced material as an almost ideal candidate to be used as a bulk catalyst for hydrocarbons synthesis through the catalytic hydrogenation of carbon monoxide (Fischer–Tropsch reaction) or to be used as a carrier for different substances, like pesticides, fertilizers or medical drugs. Our results indicate that the passivated particles—that have a thin magnetite layer on their surface—are stable for handling, which favors their potential technological applications.

Acknowledgements We thank the Materials Department of CNEA for facilitating the equipment, Lic. R. Kempf for specific area measurements and the partial economic support of Consejo Nacional de Investigaciones Científicas y Técnicas, Comisión de Investigaciones Científicas de la Provincia de Buenos Aires, Universidad Nacional de La Plata and Agencia Nacional de Promoción Científica y Tecnológica (PICT no. 14-11267).

References

1. Cabanillas, E.D., Creus, M.F., Mercader, R.C.: J. Mater. Sci. Lett. **40**, 519 (2005)
2. Cabanillas, E.D.: J. Mater. Sci. Lett. **39**, 3821 (2004)
3. Lagarec, K., Rancourt, D.G.: Mössbauer Spectral Analysis Software, Version 1.0. Department of Physics, University of Ottawa (1998)
4. Vandenberghe, R.E., De Grave, E., Landuydt, C., Bowen, L.H.: Hyperfine Interact. **53**, 175 (1990)
5. Da Costa, G.M., De Grave, E., De Bakker, L.P.M.A., Vandenberghe, R.E.: Clays Clay Miner. **43**(6), 656 (1995)
6. McCammon, C.A., Price, D.C.: Phys. Chem. Miner. **11**, 250 (1985)
7. Nammas, S., Al-Omari, I.A., Mahmood, S.H.: J. Alloy. Compd. **353**, 53 (2003)

Hyperfine Interact (2010) 195:257–263
DOI 10.1007/s10751-009-0127-1

Mössbauer spectroscopy analysis on a tempered martensitic 9% Cr steel

C. P. Ramos · A. Sztrajman · R. Bianchi · C. A. Danón · C. Saragovi

Published online: 1 October 2009

Abstract We report an analysis by means of Mössbauer spectroscopy, on a 9% Cr steel submitted to a tempering process at 780°C. Spectra of several samples with different tempering times obtained at room temperature are fitted, and a study on existent phases is made. From the former analysis, we infer the existence of different neighbors of the Fe atom governed by the concentration of alloying elements. In particular, we analyze the behavior of Cr, as the main substitutional atom. An additional measurement of samples at low velocities is made, aiming to reveal the existence of precipitated carbides. Finally a comparison between samples is performed, in order to obtain a detailed study of the effect of tempering time on hyperfine parameters.

Keywords Martensitic steels · Tempering · Mössbauer spectroscopy

C. P. Ramos · C. Saragovi
Departamento de Física, Centro Atómico Constituyentes,
CNEA Av. Gral. Paz 1499, 1650, Buenos Aires, Argentina

A. Sztrajman · R. Bianchi
Facultad de Ciencias Exactas y Naturales,
Universidad de Buenos Aires, Intendente Güiraldes, 2160,
Ciudad Universitaria, 1428, Cap. Fed., Argentina

C. A. Danón
Departamento de Materiales, Centro Atómico Constituyentes,
CNEA Av. Gral. Paz 1499, 1650, Buenos Aires, Argentina

C. P. Ramos (✉)
CONICET, Buenos Aires, Argentina
e-mail: ciramos@cnea.gov.ar

Table 1 Composition in wt% of the T91 steel sample studied

C	Si	Mn	P	S	Cr	Mo	Ni	Al	Nb	V	N
0.11	0.28	0.48	0.018	0.002	8.28	0.93	0.11	0.015	0.07	0.21	0.045

1 Introduction

Ferritic–martensitic steels of the 9%Cr1%Mo type have been extensively used in power plant components, heat exchangers, piping and tubing, etc., due to an excellent combination of properties such as creep resistance, toughness and resistance to oxidation at high temperatures. The continuous improvement of the properties of 9%Cr materials in the last decades has allowed a substantial increment of their benefits: increase of the service temperatures—with the consequent increase in efficiency—and increasingly important values of resistance to rupture. From the environmental point of view, the increase of the efficiency also implies a reduction of CO_2 emissions.

The basic idea of improving the strength and toughness of martensitic steels at high temperatures by addition of strong carbide or nitride formers was proposed in the early 80's [1]. The application of this idea to industrial components triggered a broad corpus of research that derived in the introduction and registration of the ASTM A213 grade T91 steel (9%Cr1%Mo0.2%V0.08%Nb0.04%N).

The manufacturing steps of this steel include normalizing at 1050–1060°C and tempering at 750–760°C. The resulting microstructure is a lath martensite and the major precipitated phase is constituted by the $M_{23}C_6$ carbides.

In this work we report an analysis by means of Mössbauer spectroscopy, on a 9% Cr steel submitted to a tempering process at 780°C. From the analysis, we infer the existence of different neighborhoods of the Fe atom governed by the concentration of alloying elements. In particular, we analyze the behavior of Cr, as the main substitutional atom.

2 Experimental

Samples consisted of a T91 steel normalized at 1060°C and then submitted to a tempering process at 780°C during different time intervals (M1: 40 min, M2: 1 h, M3: 4 h, M4: 6 h and M5: 7 h). The wt% composition of the steel is given in Table 1. Measurements were performed at RT using a transmission geometry on a conventional Mössbauer Spectrometer set in constant acceleration mode with a ^{57}Co source on a Rh matrix.

Spectra were recorded at 11 mm/s and also at 2 mm/s to obtain a detailed insight of particular spectral regions concerning the precipitation products and then fitted by using the Normos program developed by Brand [2]. Isomer shift (IS) values are given relative to that of α-Fe at room temperature.

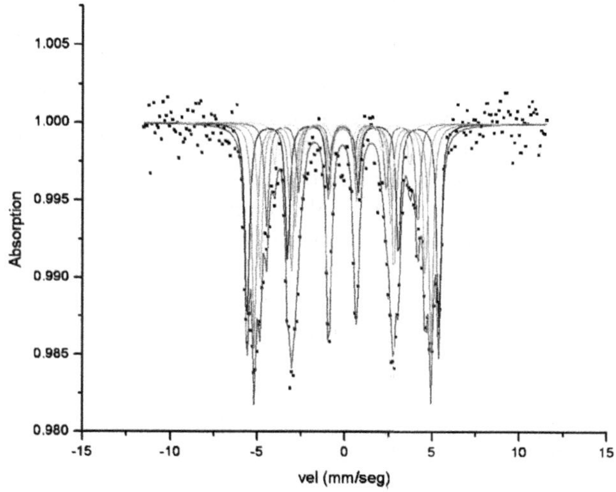

Fig. 1 Fitted Mössbauer spectrum for M1

Table 2 Hyperfine parameters. Bhf: hyperfine field [T], IS: isomer shift [mm/s]

Subspectra	Sample (tempering time)	M1 (40′)	M2 (1 h)	M3 (4 h)	M4 (6 h)	M5 (7 h)
S0	B_{hf}	34.1	33.5	33.9	33.8	33.7
	IS	0.00	0.00	0.00	0.00	0.00
S1	B_{hf}	31.4	30.6	31.4	30.9	30.9
	IS	0.01	0.01	0.01	0.01	0.01
S2	B_{hf}	29.4	28.5	29.6	29.5	28.8
	IS	0.02	0.02	0.02	0.02	0.02
S3	B_{hf}	26.8	26.6	27.8	27.2	26.5
	IS	0.03	0.03	0.03	0.03	0.03
S4	B_{hf}	24.1	24.5	25.9	24.8	24.3
	IS	0.04	0.04	0.04	0.04	0.04

3 Results and discussion

Figure 1 displays the fitted Mössbauer spectrum at RT and at 11 mm/s obtained for M1. This sample corresponds to the commercially standard material tempered during 40 min (as-received material). The other samples, M2, M3, M4 and M5, correspond to additional laboratory tempering in order to achieve 1 h, 4 h, 6 h and 7 h of total tempering time. Their corresponding spectra are similar to that displayed in Fig. 1 for M1.

Table 2 shows the hyperfine parameters obtained after the fitting routine for the different subspectra (S0, S1, S2, S3, S4 and S5) in each one of the samples.

Spectra were fitted to a superposition of five sextets with different hyperfine parameters, depending on the number of iron neighbors of the ^{57}Fe atom. These ferromagnetic components represent the martensitic matrix.

From now on some approximations should be considered. Mössbauer spectroscopy by itself cannot distinguish between the single phase ferrite and martensite from the ferrite in tempered martensite [3]. Martensite crystal structure is tetragonal bct; however, under normal circumstances it is difficult to detect any tetragonality in martensite in steels with less than 0.2 wt% C. In our case, the T91 steel studied has a low carbon and nitrogen content. During tempering of the as-quenched martensite, the tetragonal structure relaxes due to the interstitial diffusion of carbon atoms bounded to lath boundaries and dislocations; carbon diffusion also accounts for carbide precipitation. So, as our first approximation the matrix will be analyzed as bcc-Fe as a whole.

We also consider that the hyperfine structure is only caused by Cr, the principal alloying element. For this reason we are going to refer to a work of Wertheim et al. [4] in which binary iron-reach alloys of the type FeX (X = Cr, Mn, V, Si, Al, Ti, Co, Ru, Ga or Sn) were studied by means of Mössbauer Spectroscopy. From an analysis of the observed structure of the magnetic splittings as a function of the solute concentration, Wertheim concluded that the hyperfine field at a given Fe site is reduced by an amount proportional to the number of impurity near neighbors (nn) and next near neighbors (nnn) alone, the proportionality constant being different for nn and nnn occupancy but independent of concentration c. The computations assume that the effects of any number of impurity neighbors in the first two shells are additive and that the relative location of one neighbor with respect to another is of no import. Wertheim proposed that the hyperfine field for an Fe atom with n impurity nn and m impurity nnn in an alloy with a fractional impurity concentration c is expressed for low concentrations as,

$$H(m, n) = H_{Fe}(1 + a.n + b.m)(1 + kc), \tag{1}$$

where a and b give the fractional change in hyperfine field per nn and nnn impurity atom, respectively. The concentration-dependent factor may contain unresolved effects of more distant neighbors. H_{Fe} is the bcc-Fe field.

For bcc structure, each atom has eight nn and six nnn sites. Assuming that the relative abundancies of the five sextets are in agreement with a Cr random distribution in the first and second neighborhoods of a given Fe atom in the bcc lattice and that the difference of the distance to the first and second neighbor is small, the probability of each configuration $P(c, n)$ would be expressed as,

$$P(c, n) = \binom{14}{n}.c^n.(1-c)^{14-n}, \tag{2}$$

where n is the number of Cr nn and nnn of an iron atom and c is the fractional Cr concentration. The proportions of each one of the ferromagnetic components experimentally obtained by fitting the Mössbauer spectra were about 30, 31, 20, 12 and 7% in all the samples for S0, S1, S2, S3 and S4 respectively. Comparing these data with the $P(c, n)$ calculated values using Eq. 2 and considering c = 0.088 (Table 3), we can infer that the model agrees with a random Cr distribution. This way each n value was assigned to each one of the values obtained for the hyperfine field (Table 2).

Table 3 Calculated probabilities from Eq. 2

n	0	1	2	3	4
P(c, n)	28	37	23	9	3

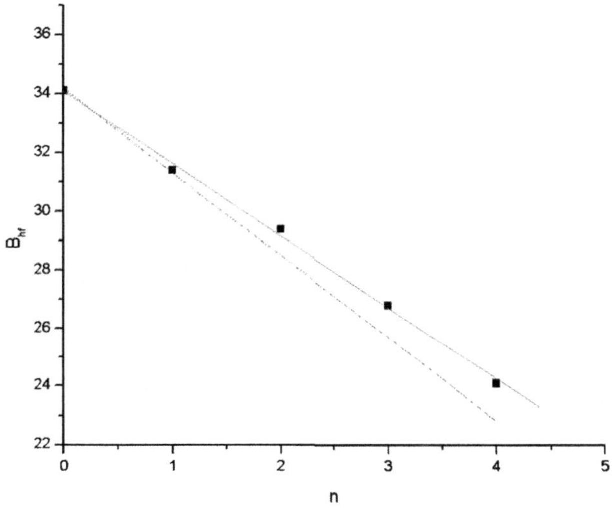

Fig. 2 Hyperfine field (B_{hf}) as a function of the Cr number of neighbors n for M1 (*solid line*). Wertheim et al. [3] results calculated from Eq. 1 (*dash dots*)

Therefore, the five sextets S_0–S_4 are the subspectra corresponding to Fe atoms having a different number of substitutional atoms as near neighbors in the α phase. Then we can interpret the five components as coming from Fe atoms with no Cr atoms in the two first coordination shells (S_0) and from those having 1, 2, 3 and 4 Cr atoms (S_1, S_2, S_3 and S_4), respectively. Note that S_0 and S_1 prevail.

Figure 2 displays, as an example, the hyperfine field as a function of the Cr number of neighbors (n) in the matrix for sample M1. It is clear that it behaves linearly ($R \approx 0.999$, solid line in Fig. 2), the magnetic field decreases with the increase of the number of Cr neighbor atoms. This conclusion is also valid for the other samples and is consistent with Wertheim et al. results calculated from Eq. 1 (dash dots in Fig. 2). Differences between both curves probably arising from another impurity species, rather than Cr, not considered in our approximations and from alloying elements in the third and further coordination spheres.

From Table 2 we also derive that IS values decrease with the number of Cr impurities, in a similar way to Fe–Cr and Fe–V alloys [5, 6]. Every impurity atom decreases the IS value in about 0.01 mm/s. This fact means that the Cr presence in the Fe neighborhood increases the effective charge density of the s-type electrons in the Fe site, resulting in a difference between Fe and Cr electronegativities.

In Fig. 3 a comparison of the relative intensities of the spectra for different tempering times is depicted where we can appreciate a difference which could be related to the Fe-bearing precipitates presence.

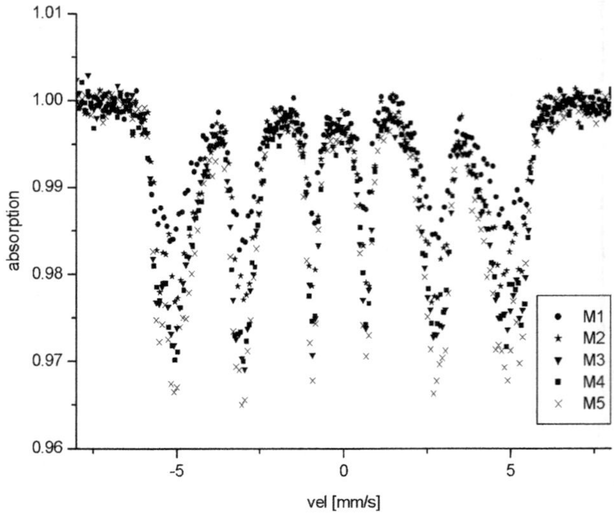

Fig. 3 Comparison of the relative intensities of the spectra from M1 to M5

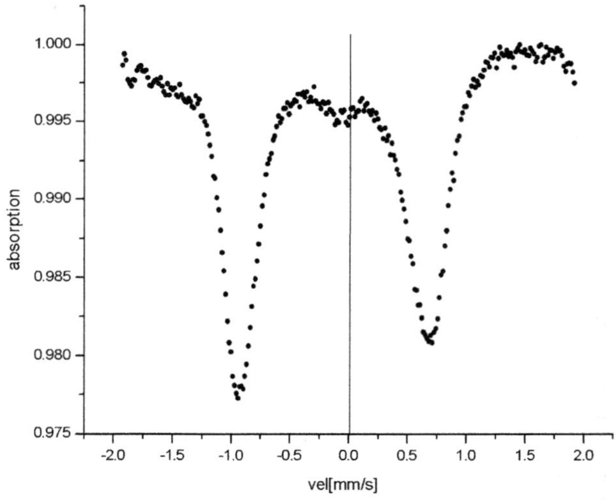

Fig. 4 Low velocity spectrum (2 mm/s) for M4

The carbide precipitation process is rather complex and depends to a greater extent on the chemical composition of the steel. A typical carbide transformation sequence in 9Cr steel during tempering being of the type

$$(Fe, Cr)_3\,C \rightarrow M_7C_3 \rightarrow M_{23}C_6,$$

where M symbolizes a mixture of transition metals like Cr, Fe, Mo. In this work Cr is the main component on the largely major precipitated phase $M_{23}C_6$, so let us simplify using M as $(Cr_{1-x}Fe_x)$.

It is visible the difference (about 2% of the effect) between the spectrum corresponding to M1 and that to M5 (Fig. 3), denoting more Fe in solution for M5, although the atoms distribution probability is the same. Cr replaces Fe with the same probabilities for the different neighborhoods in all of the samples.

To solve more accurately the central zone of the spectrum a finer velocity scale was used (2 mm/s) and the presence of precipitates was confirmed (see for example Fig. 4).

The type of precipitates described above are characterized by paramagnetic hyperfine spectra [7, 8]. Due to the low content of precipitates present in our samples the paramagnetic signal is weak facing those corresponding to the ferromagnetic components of the matrix; even taking into account the low velocity spectra and having an excellent count-rate statistics. This fact inhibits the possibility of obtaining accurate hyperfine parameters from the fit. More work is in progress to isolate the precipitates for a better Mössbauer analysis. However, considering M4 (Fig. 4) which showed the best available resolution in the recorded spectra among all the studied samples, we observe at first sight a little broad doublet (peak half width about 0.4 mm/s) with QS and IS values characteristic of $M_{23}C_6$-type precipitates (about 0.1 and 0.06 mm/s respectively) [7]. Other even lower iron-bearing precipitates contributions are not discarded. A detailed study of the precipitates will be the subject of our future work.

4 Conclusions

- After tempering at 780°C Mössbauer spectra indicate the existence of different neighborhoods of the Fe atom.
- The magnetic Fe surroundings in the matrix are controlled by the alloy elements concentration. In fact, the hyperfine magnetic field linearly decreases with the number of Cr neighbor atoms whereas the isomer shift values decrease.
- The qualitative behavior for the matrix does not show a dependence on tempering time from 40 min to 7 h. The percentage of each different Fe neighborhood is almost constant as a function of tempering time.
- During tempering Fe-bearing precipitates of the $M_{23}C_6$-type are obtained.

References

1. Cerjak, H., Hofer, P., Schaffernak, B.: ISIJ Int. **39**, 874 (1999)
2. Brand, R.A.: Normos program. Internat. Rep. Angewandte Physic, Univ. Duisburg. (1987)
3. Jirascova, O.W.F.Y., Svoboda, J.: Appl. Surf. Sci. **239**, 132 (2004)
4. Wertheim, J.G.K., Jaccarino, V., Buchanan, D.N.E.: Phys. Rev. Lett. **12**, 24 (1964)
5. Dubiel, S.M., Zukrowski, J.: J. Magn. Magn. Mater. **23**, 214 (1981)
6. Dubiel, S.M., Zinn, W.: J. Magn. Magn. Mater. **37**, 237 (1983)
7. Principi, G., Frattini, R., Magrini, M.: Gazz. Chim. Ital. **113**, 281 (1983)
8. Schaaf, P., Krämer, A., Wiesen, S., Gonser, U.: Acta Metall. Mater. **42**(N° 9), 3077 (1994)

Hyperfine Interact (2010) 195:265–274
DOI 10.1007/s10751-009-0128-0

Synthesis and characterization of silica-coated nanoparticles of magnetite

R. V. Ferreira · I. L. S. Pereira · L. C. D. Cavalcante · L. F. Gamarra · S. M. Carneiro · E. Amaro Jr. · J. D. Fabris · R. Z. Domingues · A. L. Andrade

Published online: 26 September 2009

Abstract Magnetic nanoparticles coated with silica have been subjected of extensive, and, in many aspects, also intensive investigations because of their potential application in different technological fields, particularly in biomedicine. This work was conceived and is being carried out in two main parts: (1) synthesis of the ferrimagnetic nanoparticles, specifically magnetite, and (2) coating these particles with tetraethyl orthosilicate (TEOS). The nanosized magnetite sample was prepared by the reduction–precipitation and the nanomagnetite particles were coated by the sol-gel method, based on the hydrolysis of tetraethyl orthosilicate (TEOS). The so obtained materials were characterized with powder X-ray diffraction (XRD), FTIR spectroscopy, saturation magnetization measurements, and ^{57}Fe Mössbauer spectroscopy at room temperature.

Keywords Nanomagnetite · Silica · Coating

R. V. Ferreira (✉) · I. L. S. Pereira · L. C. D. Cavalcante · J. D. Fabris · R. Z. Domingues
Departamento de Química, Universidade Federal de Minas Gerais, Campus-Pampulha, 31270-901 Belo Horizonte, Minas Gerais, Brazil
e-mail: robertavia@gmail.com

L. F. Gamarra · E. Amaro Jr.
Instituto Israelita de Ensino e Pesquisa Albert Einstein, IIEPAE, São Paulo, Brazil

S. M. Carneiro
Instituto Butantan, São Paulo 05503-900, Brazil

A. L. Andrade
Departamento de Química/s10, Universidade Federal de Minas Gerais, Campus-Pampulha, CEP 31270-901, Minas Gerais, Brazil
e-mail: angelala01@hotmail.com

1 Introduction

Magnetic nanoparticles have been subjected of extensive, and, in many aspects, also intensive investigations, stimulated by their potential application in different technological fields, particularly in biomedicine and bioengineering, as in magnetohyperthermia treatments, magnetically assisted drug delivery, cell isolation, molecular recognition, biomacromolecule purification, biosensors and quality enhancement of nuclear magnetic resonance images [1–3]. Nanomagnetite is particularly interesting for those purposes because it behaves superparmagnetically, and presents a relatively high magnetic saturation and high magnetic susceptibility, biocompatibility, and is less sensitive to oxidation than are magnetic transition metals, as cobalt, iron and nickel [2, 3]. However, nanomagnetite tends to aggregate into large clusters due anisotropic dipolar attraction, making them to lose specific properties associated to single-domains of its magnetic nanostructures. The reduced dimensions increase the reactivity for these particles: when they are exposed to biological environments they may undergo rapid biodegradation [4].

For in-vivo applications, it is important embedding magnetic nanoparticles into a nonmagnetic matrix to prevent aggregation and sedimentation, allowing them to functionalize for specific applications. They may also enable efficient excretion, which helps protecting the human body from toxicity. The most interesting inorganic nonmagnetic particle coat for this purpose, taking into account its relatively low cost and wide availability, is silica. On the surface of magnetic nanoparticles, silica, which itself is inert, tends to favor dispersing magnetic nanoparticles in the liquid media, and to more effectively keep the surface of silica-coated magnetic nanoparticles chemically functional [5].

There are several methods currently used to prepare silica-coated magnetic nanoparticles. One of these is the sol–gel method, which has been adopted more widely to prepare silica-coated magnetic nanoparticles, as being relatively mildness and for allowing a surfactant-free reaction, at relatively low cost. In such a preparation, the silica phase is formed on colloidal magnetic nanoparticles in a basic alcohol/water mixture.

This work was conceived in two main parts: (1) synthesis of the ferrimagnetic nanoparticles, specifically of magnetite, by the modified reduction-precipitation method, and (2) coating these particles *via* the method described by Stöber et al. [6] and Deng et al. [5], with some modifications. The so obtained materials were characterized by powder X-ray diffraction (XRD), FTIR spectroscopy, saturation magnetization measurements and ^{57}Fe Mössbauer spectroscopy at room temperature.

2 Experimental section

2.1 Synthesis of magnetite nanoparticles

The magnetite sample (labeled Mag) was obtained by reduction - precipitation of ferric chloride through reaction with Na_2SO_3, after the method reported by Qu et al. [7], with some modifications. The experiment consisted on the mixture of

$FeCl_3 \cdot 6H_2O$, Na_2SO_3 with diluted ammonium hydroxide solution, under a nitrogen atmosphere. The black precipitate so formed was washed with water and centrifuged, to separate the solid phase. The obtained product was finally dried.

2.2 Synthesis of silica coated magnetite

The sample (labeled MagSi) of coated magnetite with silica was prepared according to methods by Stöber et al. [6] and Deng et al. [5], with minor modifications. A mass of 0.04 g of the magnetite powder was added to ethyl alcohol and the dispersion was homogenized in an ultrasonic water bath device. After 20 min dispersion under those conditions, it was added water, aqueous ammonia and 1 mL of TEOS, which was mechanically stirred for 24 h. The obtained product was placed under humidified atmosphere overnight. Then it was placed under dry atmosphere for 24 h.

2.3 Synthesis of silica glass

A silica glass control sample was used as a comparative sample. TEOS was added to an acidified (HCl—pH 1.7) aqueous solution, and then to ethanol, in a molar ratio of $H_2O/TEOS/CH_3CH_2OH$ of 4:1:4. The solution was magnetically stirred until the solution became apparently viscous. The material was then placed under humidified atmosphere overnight, followed by dry atmosphere for 24 h.

2.4 Characterizations

All samples were characterized by powder XRD in Rigaku model Geigerflex equipment using Co(Kα) radiation, scanning from 20 to 80°(2θ) at scan rate of 4°min^{-1}, with silicon as an external standard. FTIR spectra were used to compare differences in patterns for the samples of magnetite and of the prepared silica-coated magnetite nanoparticles. The samples were also analyzed in transmission mode in a Perkin-Elmer Spectrum GX spectrophotometer. Mössbauer spectra were collected in constant acceleration transmission mode with ~15 mCi Co^{57}/Rh gamma-ray source. A spectrometer equipped with a transducer (CMTE model MA250) controlled by a linear function driving unit (CMTE model MR351) was used to obtain the spectra at 298 K. Mössbauer isomer shifts are quoted relatively to an α-Fe foil at room temperature. The experimental reflections were fitted to Lorentzian functions by least-square fitting with software NORMOS^{TM}-90. Magnetization measurements were performed in a portable magnetometer with a fixed magnetic field of 0.3 T [8] calibrated with Ni metal. Zeta Potential (ξ) was determined in a NANO SIZE ZS apparatus. For that, the ground material was suspended in water and homogenized under ultrasound for 15 min. After this time, another suspension was made by adding drops of the previous suspension to an aqueous solution of KCL 10^{-3} mol L^{-1}. The pH of this solution was measured and adjusted with either aqueous NaOH 10^{-3} mol L^{-1} or HNO_3 10^{-3} mol L^{-1} ξ was measured as a pH function. To study the morphology of the nanoparticles samples powder were dispersed over copper grids and then they were examined in a Leo 906E (Zeiss) TEM at 80 kV.

Table 1 Saturation magnetization of the samples magnetite and silica-coated magnetite

Sample	σ/J T^{-1} kg^{-1}
Mag	67.00 ± 0.01
MagSi	57.51 ± 0.05

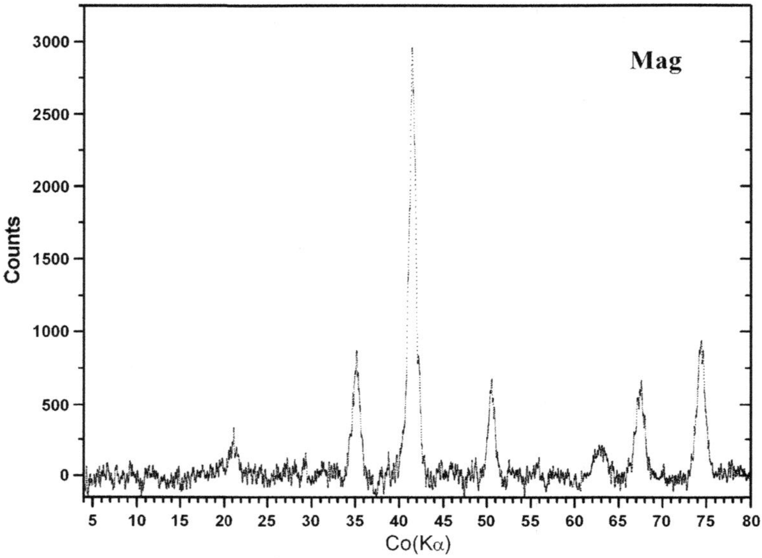

Fig. 1 X-ray diffraction pattern for the sample Mag

3 Results and discussion

The saturation magnetization of the synthesized magnetite nanoparticles is smaller than the theoretical value for Fe_3O_4 (92 J T^{-1} kg^{-1}) and it is reduced in the magnetite coated (Table 1). This behavior may be due to the oxidization of the nanoparticles or the thickness of the silica coating. Since no γ-Fe_2O_3 was detected by XDR or FTIR, probably the silica coating layer is responsible for this reduction in the saturation magnetization.

X-ray diffractograms for Mag (Figs. 1 and 2) showed the occurrence of essentially a single crystallographic phase with a cubic lattice parameters $a = 0.83885$ nm. This value is in good agreement with the reference file JCPDS card # 19–629 ($a =$ 0.8396 nm) [9]. Similar diffractogram was obtained for the silica coated magnetite ($a = 0.83904$); this is expected since the silica layer formed magnetite is amorphous.

Infrared spectroscopy was used to characterize the magnetite nanoparticles and the silica coated magnetite. The magnetite present the inverse spinel structure $Fe^{III}(Fe^{II}Fe^{III})O_4$ and the group analysis show the existence of four IR active modes in the vibrational spectra for this structure, but in most cases two of them are very weak e only observed at lower wavenumbers [10]. The FTIR spectra (Fig. 3) for sample Mag present two main absorption bands at 560 and 395 cm^{-1}, characteristic

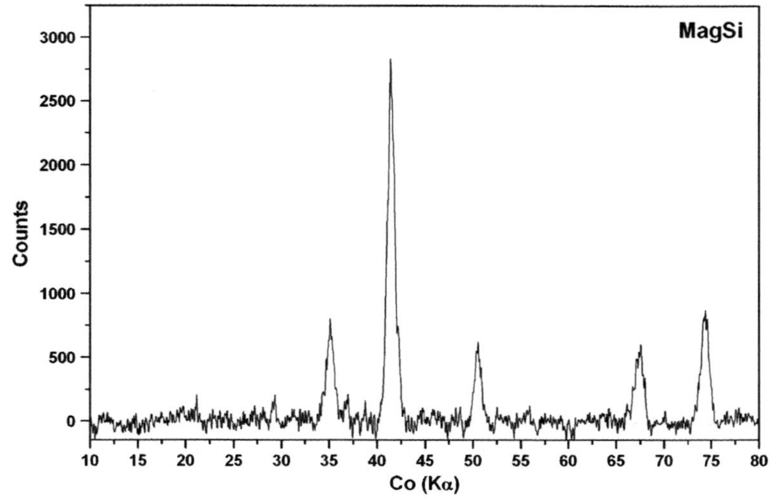

Fig. 2 X-ray diffraction patterns for the sample MagSi

Fig. 3 FTIR spectra for: **a** silica-glass, **b** Mag, and **c** MagSi

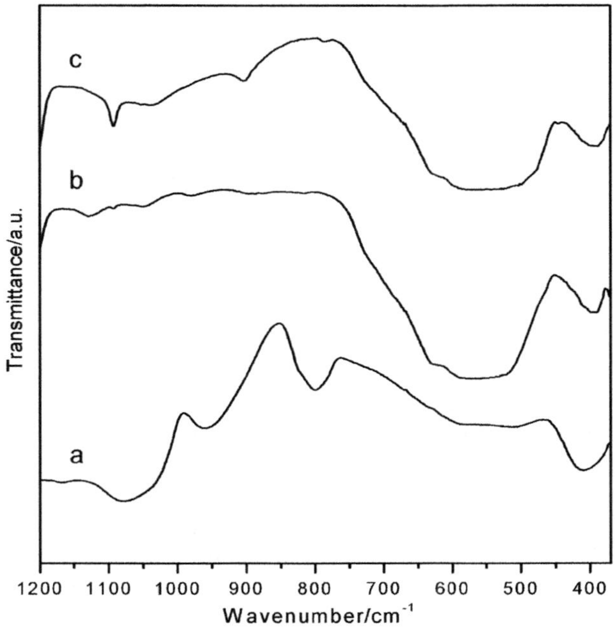

vibrations to the Fe–O stretching modes of the magnetite lattice, in good agreement with other reported data [10, 11].

The FTIR spectrum of the silica glass showed the characteristic absorption bands for the Si–O–Si bonds at 460, 800 and 1,080 cm^{-1} and a band at 950 cm^{-1} due to the Si–OH vibrations. MagSi presented bands related to silica glass at 1,088 and 900 cm^{-1}

Table 2 298 K-Mössbauer parameters deduced from model-independent hyperfine-field distributions

Sample	Assignment	$\bar{\delta}/\alpha Fe$	B_{hf}^{max}	*RA*/%
Mag	$Fe^{3+/2+}$	0.77(4)	13.1(2)	5.43(1)
		0.70(4)	33.2(1)	19.96(2)
		0.67(4)	42.4(1)	15.22(2)
	Fe^{3+}	0.16(2)	30.7(3)	7.39(2)
		0.23(2)	40.9(3)	25.8(1)
		0.27(2)	45.5(4)	26.2(1)
MagSi	$Fe^{3+/2+}$	0.62(4)	14.1(3)	7.00(3)
		0.62(4)	33.4(3)	22.70(4)
		0.62(4)	42.6(4)	21.19(3)
	Fe^{3+}	0.15(4)	30.4(1)	5.07(1)
		0.24(4)	41.0(2)	22.75(4)
		0.28(4)	45.6(3)	21.29(4)

$Fe^{3+/2+}$ = mixed valence iron in octahedral sites; Fe^{3+} iron in tetrahedral sites; $\bar{\delta}/\alpha Fe$ = averaged isomer shift relative to αFe; B_{hf}^{max} = maximum-probability hyperfine field and *RA* = relative subspectral area

besides the band at 560 and 395 cm^{-1} associated with magnetite [12]. Comparing the FTIR spectra of these two samples, no difference can be readily observed.

Magnetite is a ferrimagnetic iron oxide with iron atoms occupying two different coordination sites: one with tetrahedral (A-sites) and other with octahedral (B-sites) symmetry. These two sites are characterized by well distinguishable hyperfine fields. At room temperature, Fe^{2+} and Fe^{3+} ions in the octahedral sites exchange electrons at a hopping rate that the corresponding Mössbauer lines appear as for a single state, in such a way that the resulting ^{57}Fe Mössbauer spectrum typically shows then sextets, one of them corresponding to the contribution of the magnetic hyperfine interaction of Fe^{3+} in A-sites and other to $[Fe^{3+} \leftrightarrow Fe^{2+}]$ coupling in B-sites of the spinel structure (AB_2O_4).

The experimental spectrum for each presently studied sample was least-squared fitted with two blocks of model-independent hyperfine field distribution: one of them for the A-sites and other for the B-sites in magnetite. The resonance lines in all cases are rather asymmetrically broadened, if compared with the typical symmetrical Lorentzian-shape of a bulk reference magnetite sample: in the present case, values of hyperfine field at maximum probability, B_{hf}^{max}, isomer shift relative to αFe, $\bar{\delta}/Fe$, and relative areas of subspectra for octahedral sites and tetrahedral sites are presented in Table 2. The hyperfine parameters for tetrahedral iron $B_{hf} = 45.5$ T and isomer shift $\bar{\delta}/Fe = 0.26\,mm\,s^{-1}$, and for octahedral iron with $B_{hf} = 42.4$ T, and $\delta = 0.67\,mm\,s^{-1}$ though lower than characteristic values, together with XRD and magnetization (Table 1) data, indicate the presence of magnetite in sample Mag.

The differences between spectrum obtained from this magnetite sample and a typical spectrum for a bulk material can be due to several effects. As reportedly discussed in the literature [11, 13] the small particle size distribution may lead to an inhomogeneous hyperfine coupling and asymmetrically line broadening. Similarly above the blocking temperature T_B for the superparamagnetic relaxation effects and collective magnetic excitation cause significant line broadening and asymmetric line shapes [13].

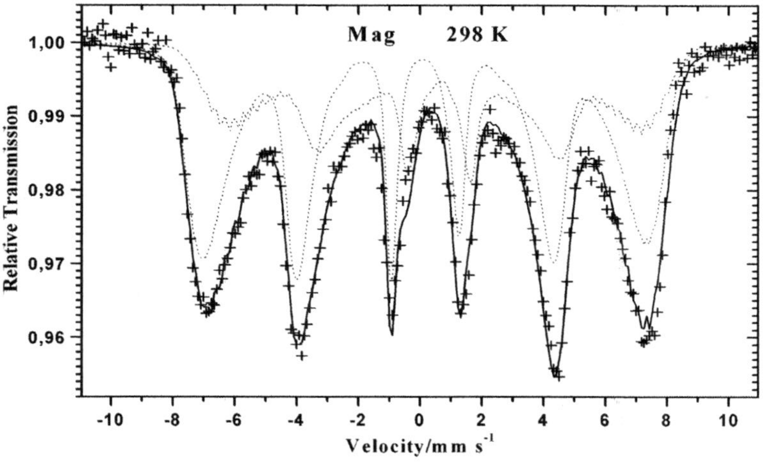

Fig. 4 Mössbauer spectrum at 298 K for the sample Mag

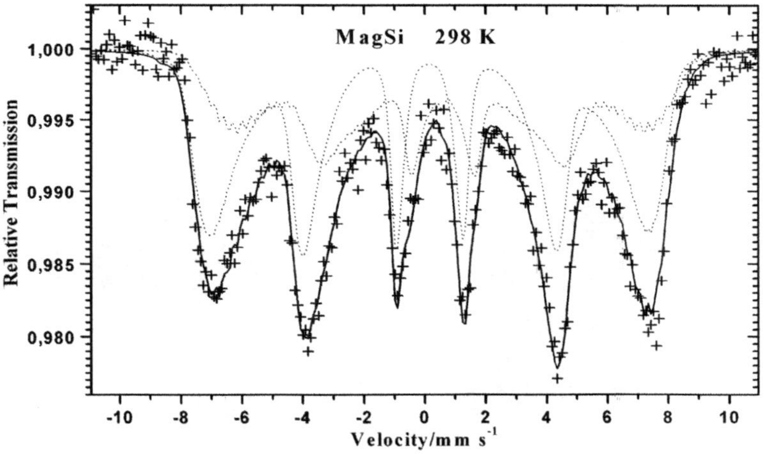

Fig. 5 Mössbauer spectrum at 298 K for the sample MagSi

Spectrum for the Mag sample at 298 K (Fig. 4; corresponding fitted parameters, in Table 2) was very similar to that obtained for the MagSi sample (Fig. 5 and Table 2). No substantial differences were observed in Mössbauer parameters after silica coating layer, indicating that the oxidation states for iron ions were preserved during the coating process. This behavior is in good agreement with previously reported data from other authors [14].

The zeta potential measurement was used to verify the coating efficiency. The silica-coated and non-coated particles were measured as a function of the solutions pH by zeta potential (Fig. 6). The pH corresponding to the isoelectric point (pH_{IEP}) to the Mag sample was found to be about 5,1 and the coated particle showed a pH dependency similar to that of pure silica. The pure silica pH_{IEP} was 2,1, while silica coated magnetite showed pH_{IEP} equal to 2.3. This differences could show the

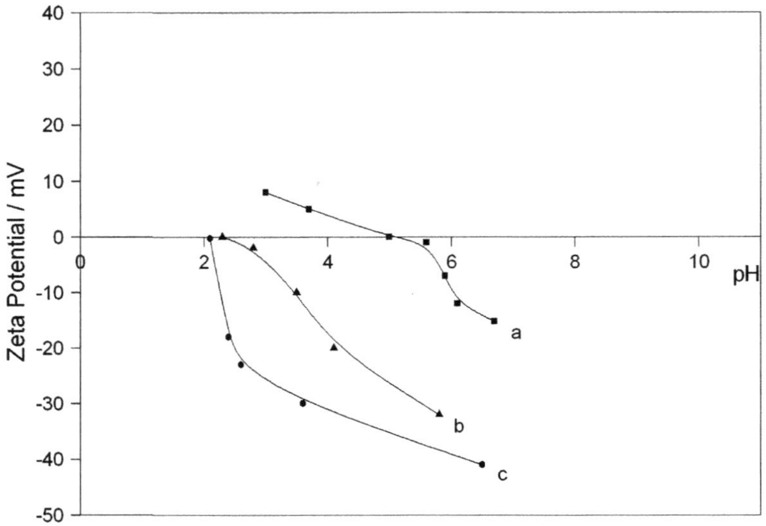

Fig. 6 Zeta Potential as a function of pH for **a** Mag, **b** silica-glass and **c** MagSi

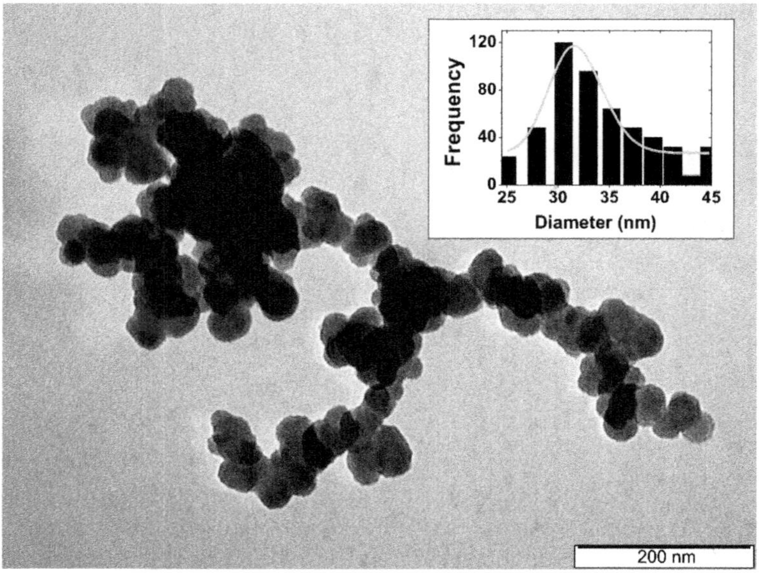

Fig. 7 Micrograph of the sample Mag. *Inset* Histogram for the distribution of nanoparticles sizes

effectiveness of the coating process, because the surface charge properties of the MagSi are close to those of pure silica.

The size poly-dispersity of the nanoparticles was analyzed from the TEM digitized micrographs using an image analysis software Java version of Image J v 1.33u [15].

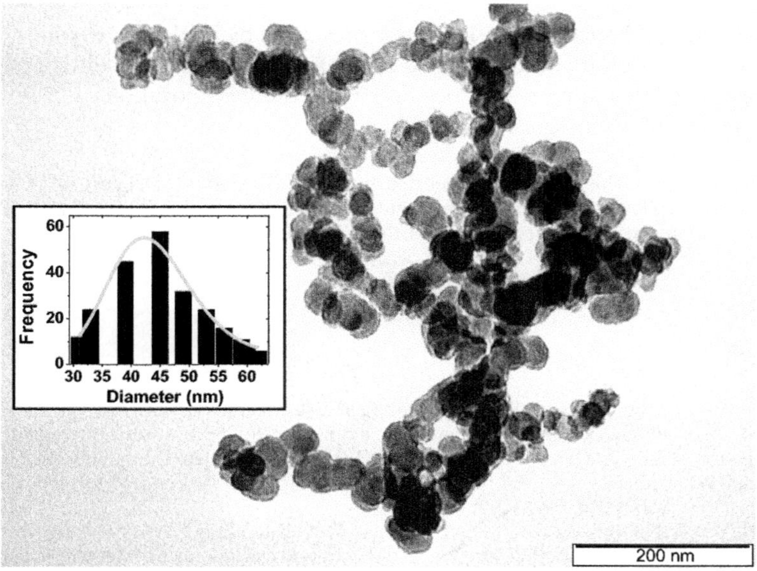

Fig. 8 Micrograph of the sample MagSi. *Inset* Histogram for the distribution of nanoparticles sizes

Mean diameters were obtained by fitting the experimental data with a lognormal distribution function, as suggested by O'Grady and Bradbury [16]

$$f(D_P) = \frac{1}{\sqrt{2\pi}\, D_P \omega_P} \exp\left(-\frac{\left(\text{In}\, D_P - \text{In}\, D_P^0\right)^2}{2\omega_P^2}\right), \tag{1}$$

with mean diameter $<D_P>= D_P^0 \exp\left(\omega_P^2/2\right)$ and ω_p as the standard deviation around $\text{In}\, D_P^0$. The standard deviation of the mean diameter is $\sigma_P = D_P^0\left[\exp\left(2\omega_P^2\right) - \exp\left(2\omega_P^2\right)\right]^{1/2}$.

The size poly-dispersity analysis for the sample Mag and their distribution indicated that the average diameter is $\langle D_P \rangle$31.5 nm and standard deviation $\sigma_P =$ 0.3 nm. For the sample MagSi were obtained average diameter $<D_p> = 43.3$ nm and standard deviation $\sigma_P = 0.6$ nm. The analysis were based in the histogram (inset Figs. 7 and 8) where was used more the 500 particles for sample Mag and more the 230 particles for silica coated magnetite.

4 Conclusions

In this work, a magnetite sample was synthesized by a reduction-precipitation method. Magnetization measurements, XRD diffraction, and Mössbauer analysis confirmed that the magnetic phase formed is actually magnetite. The particles size obtained with TEM images was 31.5 nm for the magnetite and 43.3 nm for the silica coated magnetite.

The silica-coated magnetite was synthesized by a simple and effective method, in room temperature. The effectiveness of the coating process was show for the Zeta

Potential and TEM microscopy. The dates presented by XDR, FTIR and Mössbauer spectroscopy showed that the coating do not affect the oxidation state of magnetite. These are very interesting results because allow the incorporation of drugs into the coating material for biological applications.

Acknowledgements This work was supported by FAPEMIG and Capes (Brazil). CNPq/Prosul (grant # 490132/2006-5) also supported the international mission by JDF to attend the Latin American Conference on the Applications of the Mössbauer Effect, LACAME'2008, in La Plata, Argentina.

References

1. Berry, C.C., Curtis, A.S.G.: J. Phys. D. Appl. Phys. **36**, 198 (2003)
2. Harris, A.L.: Polymer stabilized magnetite nanoparticles and poly(propylene oxide) modified styrene-dimethacrylate networks. Ph.D. thesis, Virginia Tech, Blacksburg (2002)
3. Souza, D.M., Andrade, A.L., Fabris, J.D., Valério, P., Góes, A.M., Leite, M.F., Domingues, R.Z.: J. Non-Cryst. Solids **354**, 4894 (2008)
4. Lu, Y., Yin, Y.D., Mayers, B.T., Xia, Y.N.: Nano Lett. **2**, 183 (2002)
5. Deng, Y.H., Wang, C.C., Hu, J.H., Yang, W.L., Fu, S.K.: Colloids Surf. A Physicochem. Eng. Asp. **262**, 87 (2005)
6. Stöber, W., Fink, A., Bohn, E.: J. Colloid Interface Sci. **26**, 62 (1968)
7. Qu, S., Yang, H., Ren, D., Kan, S., Zou, G., Li, D., Li, M.: J. Colloid Interface Sci. **215**, 190 (1999)
8. Fabris, J.D., Coey, J.M.D.: Espectroscopia Mössbauer do ^{57}Fe e Medidas Magnéticas na Análise de Geomateriais. Tópicos em Ciências do Solo. Viçosa, Sociedade Brasileira de Ciência do Solo, vol. 2, pp. 47–102 (2002)
9. JCPDS—Joint Committee on Power Diffraction Standards. Mineral Powder Diffraction Files Data Book, Swarthmore, Pennsylvania (1980)
10. Guillot, B., Jemmali, F., Rousset, A.: J. Solid State Chem. **50**, 138 (1983)
11. Roca, A.G., Marco, J.F., Morales, M.P., Serna, C.J.: J. Phys. Chem. C **111**, 18577 (2007)
12. Bruni, S., Cariat, F., Casu, M., Lai, A., Musinu, A., Piccalug, G., Solinas, S.: Nanostruct. Mater. **11**, 573 (1999)
13. Morup, S., Topsoe, H., Lipka, J.: J. Phys. (Paris) Colloq. **12**, 287 (1976)
14. Pârvulescu, V.I., Pârvulescu, V., Endruschat, U., Filoti, G., Wagner, F.G., Kubel, C., Richards, R.: Chem. Eur. J. **12**, 2343 (2006)
15. Rasvand, W.: Image processing and analysis in Java. http://rsb.info.nig.gov/ij. Accessed 6 Oct 2004
16. O'Grady, K., Bradbury, A.: J. Magn. Magn. Mater. **39**, 91 (1983)

Hyperfine Interact (2010) 195:275–280
DOI 10.1007/s10751-009-0129-z

Synthesis and characterization of superparamagnetic iron oxide nanoparticles for biomedical applications

**L. A. Cano · M. V. Cagnoli · S. J. Stewart ·
E. D. Cabanillas · E. L. Romero ·
S. G. Marchetti**

Published online: 25 September 2009

Abstract Monodisperse iron oxide nanoparticles (NPs) of 4 nm were obtained through high-temperature solution phase reaction of iron (III) acetylacetonate with 1, 2-hexadecanediol in the presence of oleic acid and oleylamine. The as-synthesized iron oxide nanoparticles have been characterized by X-ray diffraction, transmission electron microscopy, Mössbauer spectroscopy and magnetic measurements. The species obtained were Fe_3O_4 and/or γ-Fe_2O_3. These NPs are superparamagnetic at room temperature and even though the reduced particle size they show a high saturation magnetization (MS $\approx$ 90 emu/g).

Keywords Magnetic nanoparticles · Drug delivery systems · Mössbauer spectroscopy · Medical applications · Magnetite

1 Introduction

The use of magnetic NPs in drug delivery offer exciting new opportunities, since it is feasible to produce, and specifically tailor their functional properties for these

L. A. Cano · M. V. Cagnoli · S. G. Marchetti (✉)
CINDECA, Facultad de Ciencias Exactas, UNLP, CCT, CONICET,
Calle 47 N° 257, 1900 La Plata, Argentina
e-mail: march@quimica.unlp.edu.ar

S. J. Stewart
IFLP, CAC–La Plata, CONICET, Departamento de Física,
Facultad de Ciencias Exactas, Universidad Nacional de La Plata,
49 y 115, 1900 La Plata, Argentina

E. D. Cabanillas
Departamento de Materiales, CNEA, Av. del Libertador 8250,
1429 Buenos Aires, Argentina

E. L. Romero
Departamento de Ciencia y Tecnología, Universidad Nacional de Quilmes,
Sáenz Peña 180, Bernal 1876, Buenos Aires, Argentina

applications. Thus, drug targeting to tumours would be desirable since anticancer agents demonstrate non-specific toxicities that significantly limit their therapeutic potentials [1]. In these applications, a drug is bound to a magnetic compound, introduced in the body, and then concentrated in the target area by means of an external magnetic field gradient. Then, the particles must release the drug by simple diffusion or through mechanisms requiring enzymatic activity or changes in physiological conditions such as pH, osmolality, or temperature; drug release can also be magnetically activated using a thermosensitive grafted polymeric system (drug–polymer–magnetic NPs composite) which can be triggered to release the loaded drug with an increase in temperature, induced by an external alternating current magnetic field.

The key parameters in the behavior of magnetic NPs designed for such purposes are related to: (1) surface chemistry, (2) size: magnetic core, hydrodynamic volume size distribution, (3) magnetic properties: superparamagnetism and high saturation magnetization. Sufficiently reduced size of magnetic NPs is adequate to achieve superparamagnetic regime, which is in turn needed to avoid magnetic agglomeration once the magnetic field is removed. But small particle size implies a magnetic response of reduced strength (i.e., a saturation magnetization diminished), making it difficult to direct particles and keep them in the proximity of the target while withstanding the drag of blood flow. On the other hand, negligible remanence and coercivity inhibit agglomeration and the possible embolization of capillary vessels is avoided.

A great number of NPs of different compounds are currently investigated, but the magnetic iron oxide NPs are very attractive for biomedical applications due to their biocompatibility. For this reason the aim of the present work is to verify whether iron oxide magnetic NPs—synthesized following the recipe of Sun et al. [2] with certain modifications to obtain a particle size lower than these authors—are potentially suitable to be used as effective drug delivery system. The as-prepared magnetic NPs are characterized by means of X-ray diffraction (XRD), transmission electron microscopy (TEM), Mössbauer spectroscopy (MS) and magnetic measurements (MM).

2 Experimental

The iron oxide NPs were obtained by mixing iron (III) acetylacetonate (6 mmol), 1, 2-hexadecanediol (30 mmol), oleic acid (18 mmol), oleylamine (18 mmol), and phenyl ether (60 ml). The mixture was magnetically stirred under a flow of nitrogen. Then, it was heated to 473 K for 30 min and after, under a blanket of nitrogen, heated to reflux (518 K) for another 30 min.

It is worth mentioning that to obtain a lowest final particle size; the reflux temperature is 20 K lower than that proposed by Sun et al. [2]. The black–brown mixture was cooled to room temperature by removing the heat source. Under ambient conditions, ethanol (40 ml) was added to the mixture, and a black material was precipitated and separated via centrifugation. The black product was dissolved in hexane (30 ml) in the presence of oleic acid (~0.15 ml) and oleylamine (~0.15 ml). Centrifugation (6,000 rpm, 10 min) was applied to remove any undispersed residue. The product was then precipitated with ethanol, centrifuged (6,000 rpm, 10 min) to remove the solvent, and redispersed into hexane.

XRD patterns were obtained using a standard Philips PW-1710 diffractometer with a scintillation counter and an exit beam graphite monochromator. Cu Kα ($\lambda = 1.5406$ Å) radiation was used to obtain data in the $10° \leq 2\theta \leq 100°$ range.

TEM micrographs were taken using a Philips CM200 microscope working at 160 keV.

Mössbauer spectra were obtained in transmission geometry with a 512-channel constant acceleration spectrometer. A source of ^{57}Co in Rh matrix of nominally 50 mCi was used. Velocity calibration was performed against a 6 μm-thick α-Fe foil. All isomer shifts (δ) are referred to this standard. The temperature between 25 and 298 K was varied using a Displex DE-202 Closed Cycle Cryogenic System. The Mössbauer spectra were fitted using a commercial program named Recoil [3]. Lorentzian lines with equal linewidths were considered for each spectrum component.

The magnetic measurements were carried out using a Multipurpose Physical Magnetic System superconducting quantum interference device from Quantum Design. The magnetization *versus* magnetic field (M–H) loops were recorded at 2, 5 and 300 K up to a maximum magnetic field of 50 kOe.

3 Results and discussion

The XRD data (not shown here) display the pattern corresponding to a cubic spinel structure belonging to maghemite and/or magnetite. No evidence of hematite peaks was detected.

After subtracting the instrumental broadening we applied the integral breadths method [4] to separate size and strain broadening assuming that the first has a Cauchy profile and the second a Gaussian profile. Using a parabolic relationship between the convolution and both profiles proposed by Halder and Wagner [5] and a graphical method, we obtained an average particle size of 3 nm.

TEM micrographs showed a narrow distribution of spherical NPs. Figure 1 shows the corresponding histogram from where an average particle size of 4.0 ± 0.5 nm was estimated. Analysis of the ring diffraction patterns showed that the NPs present a cubic spinel structure, with lattice parameters corresponding to Fe_3O_4/γ-Fe_2O_3.

Strikingly, the room temperature Mössbauer spectrum does not show any absorption signal (Fig. 2). Coincidentally, below the fusion temperature of oleic acid, i.e. $T_f = 281°$K, a central broad signal starts to appear. Thus, the absence of resonance above T_f can be due to the quasi-free motion of particles of small size coated by the surfactant that increases the nucleus recoil energy.

We observe that a magnetic signal is partially resolved at 25 K. Notwithstanding, the presence of a central signal and a curved background indicate that the blocking of particle moments is not complete, i.e., a fraction remains in a superparamagnetic regime. About 45% is in this regime, therefore, blocking temperature is very close to 25 K.

We fitted this spectrum assuming two relaxation states following the Blume and Tjon model [6]. We have considered six components, five of them to simulate the partially magnetically ordered magnetite sites [7] and the other one to account for the central signal. The hyperfine parameters are shown in Table 1. Comparing these parameters with previous reported data [7, 8] for magnetite bulk we assign the B_i

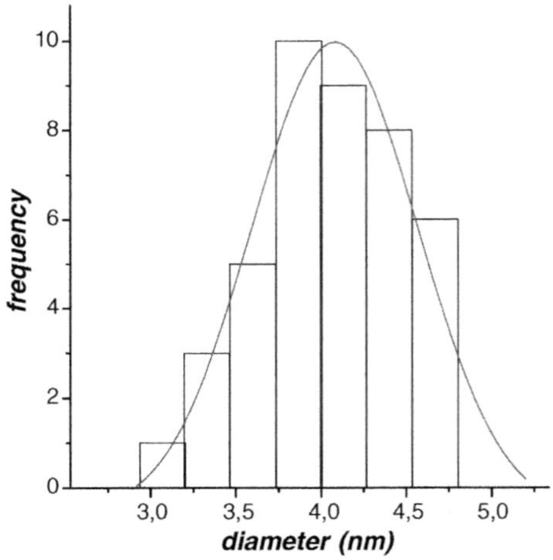

Fig. 1 Particle size histogram from TEM observations

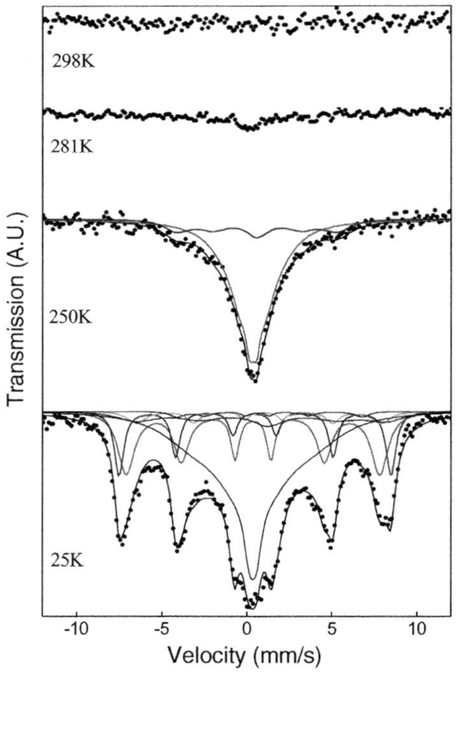

Fig. 2 Mössbauer spectra of magnetic NPs at different temperatures

components to Fe_3O_4. Note that hyperfine magnetic field of all sites are diminished in comparison with the bulk values. This effect could be attributed to the collective magnetic excitations produced by the small particle sizes [9]. The A component can not be assigned to any determined species but considering that their isomer shift is typical of Fe^{3+} it could belong to Fe_3O_4 and/or γ-Fe_2O_3.

The *M–H* loops are shown in Fig. 3. We observe that the curves are reversible down to T = 5 K, where we observe a very low coercive field of $H_C = 2$ Oe (see inset Fig. 1). Decreasing the temperature to 2 K, a hysteretic behavior is observed

Table 1 Hyperfine parameters at 25 K

Species	Sites	H (kOe)	δ (mm/s)	2ε (mm/s)	%
Fe_3O_4 and γ-Fe_2O_3	A	470 ± 4	0.36 ± 0.03	0.001 ± 0.005	25 ± 2
Fe_3O_4	B_1	508 ± 2	0.46 ± 0.02	0.03 ± 0.04	14 ± 2
	B_2	473 ± 14	0.83^a	-0.27^a	5 ± 2
	B_3	466^a	1.03^a	-0.41^a	9 ± 4
	B_4	343 ± 24	0.96^a	0.7 ± 0.5	2 ± 1
Relaxing Fe^{3+}	–	300^a	0.35 ± 0.04	0^a	45 ± 2

H hyperfine magnetic field, δ isomer shift referred to α-Fe at 298 K, 2ε quadrupole shift

[a] Parameter held fixed while fitting

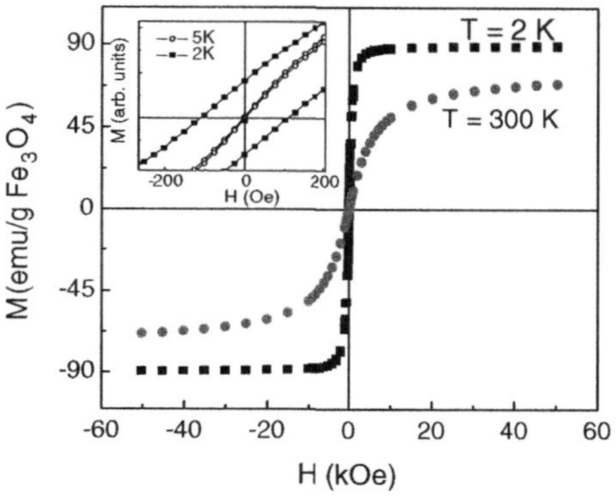

Fig. 3 M–H curves taken at 2 and 300 K. *Inset*: low field region of M vs. H loops taken at 2 and 5 K

with $H_C = 105$ Oe. The M_R/M_S ratio is ≈0.14, which is far from the $M_R/M_S = 0.5$ value predicted for independent NPs with uniaxial anisotropy [10]. Considering the area of the 2K-hysteresis loop we estimated an effective anisotropy constant $K_{\text{eff}} \approx 1.5 \times 10^5$ erg/cm^3, which is within the order of magnitude of the bulk value [11]. Therefore, the surface anisotropy constant could be considered negligible. A saturation magnetization $M_S = 89$ emu/g at T = 2 K was estimated assuming an *M vs. H* dependence of type $M = M_S(1\text{-}\alpha/H)$ at high magnetic fields, where α is a fitting parameter. This M_S, which is similar to that reported for bulk Fe_3O_4 (92 emu/g) [11] reflects the spin disorder absence at NPs surface even though the small particle sizes.

Generally, ferrimagnetic NPs present reduced *Ms* values when compared with their bulk counterparts due to the spin disorder at the surface layer. For instance, nanosized Fe_3O_4 can show a reduction of 80–90% on its M_S value [12, 13]. On the other hand, when these Fe_3O_4 NPs are coated with surfactants, the chemical environment of the coating may also influence on the magnetic properties [11, 14]. The present results are in concordance with that obtained by Guardia et al. [11], confirming that the O^{2-} of the oleic acid molecules bonded covalently to the nanoparticles are able to reduce the surface spin disorder and thus the surface contribution to the anisotropy [11] Considering that these NPs are superparamagnetic at room temperature with a high saturation magnetization, they would have potential applications in biomedicine.

4 Conclusions

We were able to synthesized monodisperse magnetic iron oxide NPs through the reaction of metal acetylacetonate and 1,2-hexadecanediol. Through different characterization techniques we determined a particle diameter of about 4 nm. The NPs composition is a mixture of both ferrimagnetic iron oxides: Fe_3O_4 and γ-Fe_2O_3, determined by Mössbauer spectroscopy. The magnetic properties of these NPs are very similar to that of the bulk system, perhaps due to the oleic coating effect. The hydrophobic NPs could be transformed into hydrophilic ones by mixing with bipolar surfactants, allowing preparation of aqueous NPs dispersions. These iron oxide NPs and their aqueous dispersions could have a great potential in biomedical applications.

Acknowledgements The authors acknowledge support of this work by Consejo Nacional de Investigaciones Científicas y Técnicas (PIP 6524), ANPCyT (PICT 38337) and Universidad Nacional de La Plata (X438), Argentina.

References

1. Jain, T.K., Morales, M.A., Sahoo, S.K., Leslie-Pelecky, D.L., Labhasetwar, V.: Mol. Pharm. **2**(3), 194 (2005)
2. Sun, S., Zeng, H., Robinson, D.B., Raoux, S., Rice, P.M., Wang, S.X., Li, G.: J. Am. Chem. Soc. **126**, 273 (2004)
3. Lagarec, K., Rancourt, D.G.: Mossbauer Spectral Analysis Software, Version 1.0. Department of Physics, University of Ottawa (1998)
4. Klug, H.P., Alexander, L.E.: X-ray Diffraction Procedures. Wiley, New York (1974)
5. Halder, N.C., Wagner, C.N.J.: Acta Crystallogr. **20**, 312 (1966)
6. Blume, M., Tjon, J.A.: Phys. Rev. **165**, 446 (1968)
7. Berry, F.J., Skinner, S., Thomas, M.F.: J. Phys.: Condens. Matter **10**, 215 (1998)
8. Vandenberghe, R.E., De Grave, E.: In: Long, G.J., Grandjean, F. (eds.) Mössbauer Spectroscopy Applied to Inorganic Chemistry, vol. 3, pp. 59. Plenum, New York (1989)
9. Mørup, S., Topsøe, H.: Appl. Phys. **11**, 63 (1976)
10. Stoner, E.C., Wohlfarth, E.P.: Phil. Tran. Roy. Soc. Lond., A **240**, 599 (1948)
11. Guardia, P., Batlle-Brugal, B., Roca, A.G., Iglesis, O., Morales, M.P., Serna, C.J., Labarta, A., Batle, X.: J. Magn. Magn. Mater. **316**, e756–e759 (2007)
12. Fellenz, N.A., Marchetti, S.G., Bengoa, J.F., Mercader, R.C., Stewart, S.J.: J. Magn. Magn. Mat. **306**, 30 (2006)
13. Kumar, R.V., Koltypin, Y., Cohen, Y.S., Cohen, Y., Aurbach, D., Palchik, O., Felner, I., Gedanken, A.: J. Mater. Chem. **10**, 1125 (2000)
14. Hou, Y., Yua, J., Gao, S.: J. Mater. Chem. **13**, 1983 (2003)

Author Index (2010)

GPSR Compliance

The European Union's (EU) General Product Safety Regulation (GPSR) is a set of rules that requires consumer products to be safe and our obligations to ensure this.

If you have any concerns about our products, you can contact us on ProductSafety@springernature.com

In case Publisher is established outside the EU, the EU authorized representative is:

Springer Nature Customer Service Center GmbH
Europaplatz 3
69115 Heidelberg, Germany

Batch number: 10755292

Printed by Printforce, the Netherlands